AF572698

IMPROVING PRODUCTION WITH COOLANTS AND LUBRICANTS

Dennis Zintak
Editor

Published by:
Society of Manufacturing Engineers
Marketing Services Department
One SME Drive
P.O. Box 930
Dearborn, Michigan 48128

IMPROVING PRODUCTION WITH
COOLANTS AND LUBRICANTS

First Edition

Library of Congress Catalog Card Number 82-80849

International Standard Book Number: 0-87263-081-1

Manufactured in the United States of America

SME wishes to express its acknowledgement and appreciation to the following publications for supplying the various articles reprinted within the contents of this book.

Cover illustration courtesy of Van Straaten Chemical Company

Cutting Tool Engineering
Technifax Publications Inc.
120 North Hale
Box 937
Wheaton, Illinois 60187

Industrial Diamond Review
Charters, Sunninghill, Ascot, Berkshire SL5 9PX
ENGLAND

Iron Age
Chilton Company Inc.
Chilton Way
Radnor, Pennsylvania 19089

Journal of Engineering for Industry
American Society of Mechanical Engineers
345 East 47th Street
New York, New York 10017

Manufacturing Engineering
Society of Manufacturing Engineers
One SME Drive
P.O. Box 930
Dearborn, MI 48128

Materials Performance
National Association of Corrosion Engineers
1440 South Creek
Houston, Texas 77084

Modern Machine Shop
Gardner Publications Inc.
600 Main Street
Cincinnati, OH 45202

Precision Metal
Penton/IPC Inc.
614 Superior Avenue West
Cleveland, Ohio 44113

Production
Bramson Publishing Co.
Box 101
Bloomfield Hills, Michigan 48013

Tooling & Production
Huebner Publications, Inc.
5821 Harper Road
Solon, OH 44139

Grateful acknowledgement is also extended to:

American Society of Lubrication Engineers
838 Busse Highway
Park Ridge, Illinois 60068

IIT Research Institute
10 West 35th Street
Chicago, Illinois 60616

MacMillan Accounts & Administration Ltd.
Houndmills Basingstoke
Hampshire RH21 2XS
ENGLAND

Tower Oil & Technology Company
300 West Washington Street
Chicago, Illinois 60606

PREFACE

One of the outstanding opportunities for decreasing costs and increasing productivity rests in the proper selection of coolants and lubricants. However, selection and application of the proper fluid is probably one of the most overlooked opportunities to improve production. It is an industrial oddity because frequently proper fluid selection requires little additional capital investment.

The metalworking chemicals being used in industry today have undergone a tremendous revolutionary process, and the likelihood for continued change exists in the face of both regulatory pressures and potential scarcities or shortages. Starting with simple mineral oil-based fluids to which fatty constituents such as lard oil and fish oil were added, the technology improved to incorporate sulfur and/or chlorine as extreme pressure additives. Subsequently, the technology of emulsion chemistry improved the characteristics, offering products with both the cooling characteristics of water and the lubrication characteristics of oil. This represented the state of the art since World War II and through the 1960's. Modifications were made to enhance filterability, operator acceptance, and to extend the useful life of the fluids.

It wasn't until the potential for scarcities in petroleum stock materials and the impact of environmental concerns for waste disposal of these materials placed new significance on the types of products being used that a new generation of synthetics and microemulsions became prominent in metalworking applications. This evolution is continuing at an increasing rate.

The selection of the material in this book was intended to describe principles and technology for the formulation and selection of metalworking chemicals, followed by a detailed treatment of techniques for the control and maintenance of these fluid or chemical systems. Attention was also given to the selection process for some of the evolving new technological applications requiring new and different formulated products.

This mix of papers represents meaningful treatment of current and evolving technology regarding metalworking chemicals. Several of the papers also give very useful suggestions and guidelines for a reader interested in optimizing the product selection process and the useful life of all their metalworking requirements.

Chapter One "How Fluids Function", briefly discusses two points of fluid functions—the role of carbontetrachloride as a boundary lubricant and the effect of cutting tool temperatures on coolants.

Chapter Two "Classification Of Fluid Types", reviews many types of fluids and discusses how they are applied today. Applications include milling, pressworking, rollforming, cold heading, die casting, turning and drilling.

The book's third chapter, "Selection Criteria", covers a variety of topics useful to the manufacturing engineer evaluating lubricants or coolants to best meet a manufacturing need. A cutting/grinding fluid analysis chart is presented. Recommendations for machine design to improve metalworking fluid performance and for coolant filtration are discussed.

Chapter Four "Extending Useful Fluid Life", explains how the quality of water can affect the quality of product and discusses maintenance costs. The chapter also explores the effect of water soluble cutting fluids on operating conditions in machining and the conservation of hydraulic fluids.

The book's final chapter, "Disposal Of Spent Fluid Systems", presents a general overview of recycling, filtration, disposal and the safety and environmental control factors involved.

I wish to express my gratitude to each of the authors whose work appears in these pages. (Company affiliations of each author are those the author held at the time the paper was written.) My thanks also to the publications who were very generous in supplying material for this volume. They are: *Cutting Tool Engineering, Industrial Diamond Review, Iron Age, Journal of Engineering for Industry, Manufacturing Engineering, Materials Performance, Modern Machine Shop, Precision Metal, Production and Tooling & Production.* My sincere thanks is also extended to the *American Society of Lubrication Engineers, IIT Research Institute, MacMillan Accounts and Administration Limited, Tower Oil and Technology Company* and the *University of Birmingham.*

Finally, my thanks to the staff of the SME Marketing Services Department for their assistance in preparing this book.

Dennis C. Zintak
Senior Product Manager
Van Straaten Chemical Company
Editor

ABOUT THE EDITOR

DENNIS C. ZINTAK

Dennis C. Zintak is the Senior Product Manager of the Cutting And Grinding Fluids Division at Van Straaten Chemical Company in Chicago.

In his 17 years at Van Straaten, Mr. Zintak served six years in Product Engineering, coordinating the field test activities of new product development of cutting and grinding fluids. He also held the position of Eastern Regional Sales Manager. Prior experience includes over three years as a metallurgist with U.S. Steel.

Mr. Zintak holds an M.B.A. from Northwestern University and a Bachelor of Science degree in Metallurgical Engineering from Illinois Institute of Technology. He is a member of the American Society of Lubrication Engineers and the Society of Manufacturing Engineers.

SME

The informative volumes of the Manufacturing Update Series are part of the Society of Manufacturing Engineers' effort to keep its Members better informed on the latest trends and developments in engineering.

With 60,000 members, SME provides a common ground for engineers and managers to share ideas, information and accomplishments.

An overwhelming mass of available information requires engineers to be concerned about keeping up-to-date, in other words, continuing education. SME Members can take advantage of numerous opportunities, in addition to the books of the Manufacturing Update Series, to fulfill their continuing educational goals. These opportunities include:

- Chapter programs through the over 200 chapters which provide SME Members with a foundation for involvement and participation.
- Educational programs including seminars, clinics, programmed learning courses and videotapes.
- Conferences and expositions which enable engineers to see, compare, and consider the newest manufacturing equipment and technology.
- Publications including Manufacturing Engineering, the SME Newsletter, Technical Digest and a wide variety of books including the Tool and Manufacturing Engineers Handbook.
- SME's Manufacturing Engineering Certification Institute formally recognizes manufacturing engineers and technologists for their technical expertise and knowledge acquired through years of experience.

In addition, the Society works continuously with the American National Standards Institute, the International Standards Organization and other organizations to establish the highest possible standards in the field.

SME Members have discovered that their membership broadens their knowledge throughout their career.

In a very real sense, it makes SME the leader in disseminating and publishing technical information for the manufacturing engineer.

MANUFACTURING UPDATE SERIES

Published by the Society of Manufacturing Engineers, the Manufacturing Update Series provides significant, up-to-date information on a variety of topics relating to the manufacturing industry. This series is intended for engineers working in the field, technical and research libraries, and also as reference material for educational institutions.

The information contained in this volume doesn't stop at merely providing the basic data to solve practical shop problems. It also can provide the fundamental concepts for engineers who are reviewing a subject for the first time to discover the state-of-the-art before undertaking new research or application. Each volume of this series is a gathering of journal articles, techical papers and reports that have been reprinted with expressed permission from the various authors, publishers or companies identified within the book.

SME technical committees, which are made up of educators, engineers and managers working within industry, are responsible for the selection of material in this series.

We sincerely hope that the information collected in this publication will be of value to you and your company. If you feel there is a shortage of technical information on a specific manufacturing area, please let us know. Send your thoughts to the Manager of Educational Resources, Marketing Services Department at SME. Your request will be considered for possible publication by SME—the leader in disseminating and publishing technical information for the engineer.

TABLE OF CONTENTS

CHAPTERS

1 HOW FLUIDS FUNCTION

2 CLASSIFICATION OF FLUID TYPES

3 SELECTION CRITERIA

4 EXTENDING USEFUL FLUID LIFE

5 DISPOSAL OF SPENT FLUID SYSTEMS

CHAPTER 1

HOW FLUIDS FUNCTION

Reprinted from the American Society of Mechanical Engineers' Journal of Engineering for Industry, May 1978

T. Shirakashi
Visiting Research Mechanical Engineer,
Asst. Professor, Tokyo Inst. of Technology

R. Komanduri
Asst. Professor Mechanical Engineering

M. C. Shaw[1]
University Professor
Carnegie Mellon University,
Pittsburgh, Pa.

The Conflicting Roles of Carbontetrachloride as a Boundary Lubricant

Under certain conditions, carbontetrachloride is found to be a negative boundary lubricant (gives a higher coefficient of friction than that of dry surfaces in air), while under other conditions, it lowers friction and gives a beneficial effect. Both of these situations are illustrated for heavily loaded sliding surfaces where the subsurface is undergoing gross plastic flow and an explanation is presented which appears to be consistent with all experimental facts. Carbontetrachloride is found to be more reactive chemically when the sliding surfaces are heavily strained or galled under high normal and shear stresses and containing microcracks, a situation that arises when cutting at low speed.

Introduction

At low cutting speeds, carbontetrachloride is one of the most effective cutting fluids known from the point of view of the reduction of cutting forces, and improvement of the surface finish. Yet, when a cutting tool is slid back over a surface freshly cut using CCl_4 as the cutting fluid, the coefficient of friction of the clearance surface rubbing on the freshly cut surface is higher than with sliding in air under the same conditions [1][2]. This observation suggests that CCl_4 is a good cutting fluid despite the fact it gives a high coefficient of friction as a boundary lubricant. A similar result has been reported using a friction apparatus that simulates conditions on the tool face of a cutting tool. In this test [2] a hard steel brinell ball is forced into a soft steel surface until a fully plastic zone is produced beneath the ball. The deformed specimen is then rotated relative to the ball and the coefficient of friction is measured. When such an experiment is performed on a surface wetted by carbontetrachloride, the coefficient of friction is high relative to the value obtained in air. This again suggests that CCl_4 is a negative boundary lubricant relative to air.

However, workers in Dr. Tabor's laboratory at Cambridge University have reported just the reverse role of CCl_4 [3] when a freshly turned surface was wetted with CCl_4 and a loaded slider was drawn across the surface. In these tests, CCl_4 gave a coefficient of friction *lower* than that obtained in air.

In order to further explore this paradox, a test arrangement similar to that used in reference [2] was constructed and used in further studies of CCl_4 as a boundary lubricant.

At the outset, it should be explained that this work was conducted on CCl_4 not because it represents an important boundary lubricant but because the phenomenon explored is thought to exist with other boundary lubricants but to a lesser degree. Carbontetrachloride cannot be used in practice because it is very toxic and under certain condtions, it is too active leading to chemical corrosion and reduced tool life.

Experimental Set-Up

Fig. 1 shows the principle of the test. A hard sphere is loaded against a softer test specimen until the subsurface is fully plastic. Then, with the normal load still applied, the deformed surface is rotated relative to the sphere. In order to remove the singularity at the center of the specimen, a hole of diameter d_1 is employed. It has been found that the coefficient of friction is essentially independent of the size of the hole employed in the range of $1/16$ to $1/8$ inch for a $1/2$ inch diameter sphere.

Fig. 2 is a schematic view of the arrangement used. The hard sphere is mounted in a chuck attached to a standard brinell testing machine by means of a strain gage torque dynamometer. The spheres used are standard $1/2$ inch diameter AISI 51100 bearing balls. The soft specimen is a one inch diameter cylinder of AISI 1018 steel, $1/2$ inch high. The lower cylindrical specimen is provided with a central hole ranging in diameter from $1/16$ to $1/8$ inch. After the vertical load is applied to the hardness tester and the lower specimen has been indented, the cylindrical specimen is rotated by means of the motor and gear arrangement shown in Fig. 2. Rotational speeds from 2 to 1000 rpm are possible. This arrangement enables friction to be measured when one

[1] Presently, Invitational Prof. of Eng., Arizona State University, Tempe, Arizona.

[2] Numbers in brackets designate References at end of paper.

Contributed by the Production Engineering Division for presentation at the Winter Annual Meeting, Atlanta, Ga., November 27–December 2, 1977 of THE AMERICAN SOCIETY OF MECHANICAL ENGINEERS. Manuscript received at ASME Headquarters June 22, 1977. Paper No. 77-WA/PROD-7.
Copies will be available until Aug. 1978.

of the mating surfaces is fully plastic. The frictional torque is continuously monitored using a chart recorder.

Test Results on Steel

Fig. 3 is a typical friction torque (inch pounds) versus rotational displacement (rev) curve for sliding in air following identation in air. The initial torque is a very low ($\simeq$ 8 in. lb) but rises to a value of about 50 in. lbs during one revolution. During this period of rising torque, the rotating surface is severely galled and Fig. 4 is a plan view of such a galled surface.

If dry sliding is interrupted after the initial transition to the steady state, the steady state value of friction torque is obtained immediately without any transition. This is because the surface was fully galled during the initial transition and after that, there is no further deterioration of the surface even when motion is interrupted. Fig. 5 shows the variation of the mean frictional stress (τ) versus mean normal stress (σ) when sliding in air under initial and steady state conditions.

The frictional torque (M) is converted to mean shear stress (τ) as follows [2]

$$\tau = \frac{12\,M}{\pi(d_o{}^3 - d_1{}^3)} \tag{1}$$

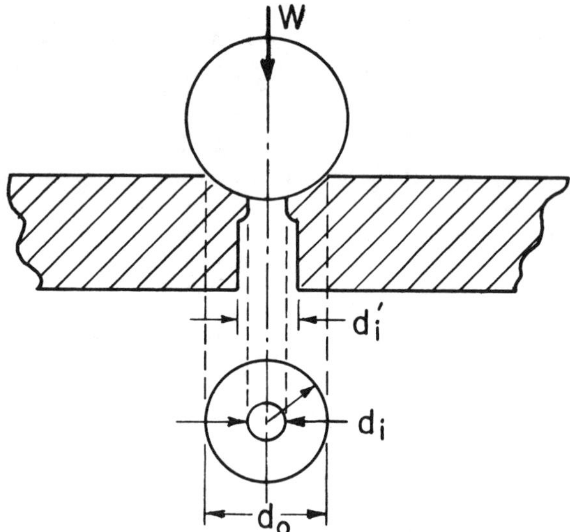

Fig. 1 Principle of Fully Plastic Friction Test

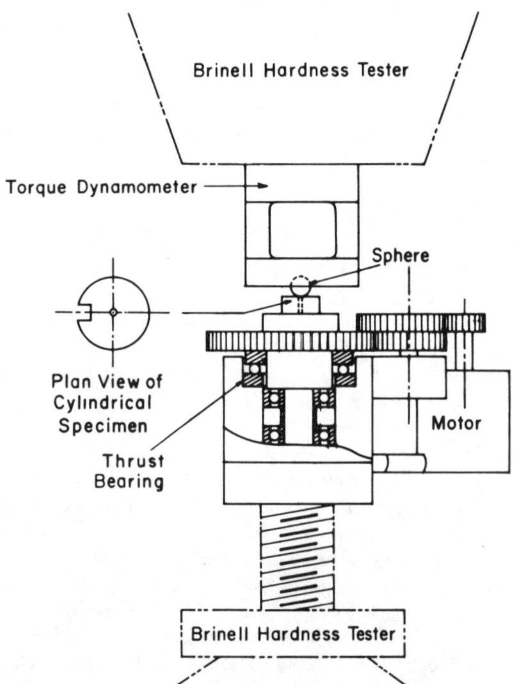

Fig. 2 Schematic Arrangement of Test Apparatus

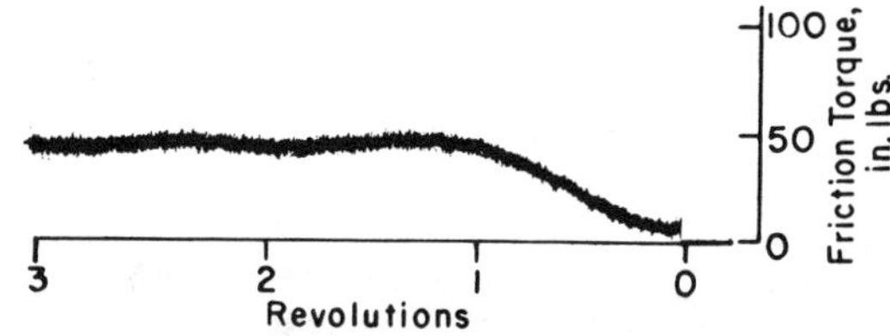

Fig. 3 Friction Torque versus Rotational Displacement for Sliding in Air. Load on sphere = 1000 kg, sliding speed = 2 rpm, hole diameter—0.125 in.

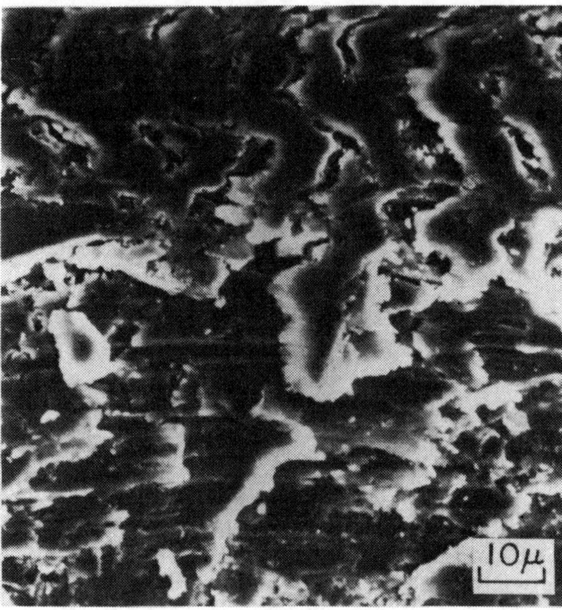

Fig. 4 Galled Surface After Three Revolutions Under Conditions of Fig. 3

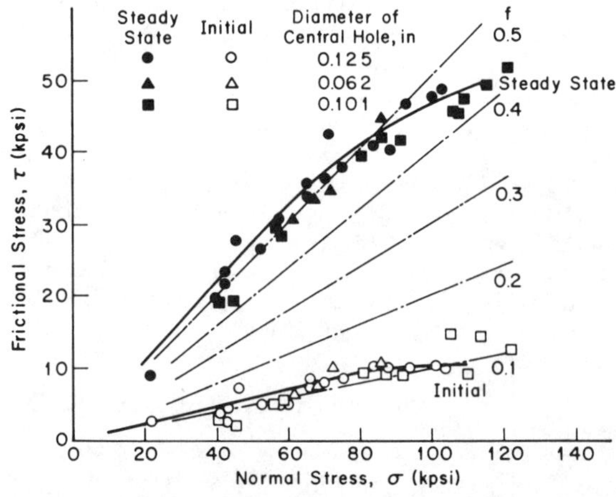

Fig. 5 Variation of Mean Shear Stress (τ) with Mean Normal Stress (σ) for Sliding in Air Under Initial (Open Points) and Steady State (Solid Points) Conditions

cent plastic zones interact. After that, there is gross plastic flow beneath the indenter instead of localized flow beneath individual asperities [5]. Fig. 12 shows schematically a rough surface before and after indentation. For a given material, the mean stress at the surface of asperities will increase as the spacing of adjacent asperities increases and hence as the surface roughness increases. The tendency for CCl_4 to react with iron will increase with pressure. For a smooth surface, the pressure will be insufficient to cause CCl_4 to react to form $FeCl_3$ and the role of CCl_4 will be to exclude oxygen which, in turn, results in high friction. This explains why the friction was so high when a tool used to cut steel or aluminum using CCl_4 was slid back over a freshly cut surface. The surface produced was so smooth that the CCl_4 did not react to form a metal chloride on the backward pass. The values of f recorded in reference [2] were initial ones rather than steady state values. This explains why sliding in air was so much lower than sliding with CCl_4.

The foregoing explanation is also consistent with the results of Fig. 11. The pressure is too low on asperities of a smooth surface to cause CCl_4 to react and hence its role is the negative one of excluding oxygen. The experiments at Cambridge University [3] were conducted on a much rougher turned surface than that of experiments described in reference [1] which were produced by a shaving cut made using CCl_4. As a result, the pressure on the Cambridge asperities was sufficient to cause CCl_4 to react and this resulted in a reduction in the coefficient of friction. The results of Fig. 10 are similar to the Cambridge ones in that a surface first rotated in air gives rise to a galled surface and pressures are sufficiently high to cause CCl_4 to react. This causes the steady state values of friction with CCl_4 to be lower than in air.

Having said all this, the role played by CCl_4 on the tool face in metal cutting is not clear. In fact, it may act as both positive and negative lubricant—positive very close to the cutting edge and negative as the chip proceeds farther up the face of the tool. From the well established concentrated shear model of cutting and the fact that strain is not uniformly distributed in a metal but occurs on relatively widely spaced planes, fracture must occur at the ends of lamellae adjacent to the tool tip (Fig. 13). The most probable direction of fracture is along AD for a ductile material but along AE for a brittle material. Fig. 13 shows the slip line field suggested by Lee and Shaffer [6] together with the corresponding Mohr's Circle diagram. This analytical approach to metal cutting arbitrarily assumes that the shear plane is in the direction of maximum shear stress (i.e. on plane S). The most probable fracture direction for a brittle material will be at an angle $\theta = 45$ deg to the plane of shear. In any case, the bottom surfaces of the lamellae will be jagged when freshly generated at the tool tip. As the chip proceeds up the tool face, its surface will be burnished. During the burnishing process, the role of CCl_4 in reducing friction will be a positive one if the pressure generated is sufficiently high to cause a reaction to occur with the metal being cut. However, as the burnished surface of the chip proceeds further along the tool face, the pressure will drop and the role of CCl_4 may shift to that of a negative lubricant. On the other hand, the metal chloride generated during the initial burnishing action may be sufficient to have a positive effect during the post burnishing contact period in which case we must conclude that the net affect of CCl_4 on the tool face is a positive one.

Test Results on Lead and Their Interpretation

Some additional tests were performed with lead as the soft cylindrical specimen. Fig. 14 shows tests performed in air. The initial value of friction coefficient was low (~0.18) but after a few revolutions, the shear stress corresponded to the bulk shear strength of lead (~1000 psi). This is due to the fact the oxide initially present tends to prevent the lead from bonding to the steel sphere, thus preventing galling. As the oxide originally present is displaced, clean lead on steel gives rise to high values of friction. To check this, tests were performed using hydrogen peroxide (H_2O_2) as the source of oxygen. The results of these tests (Fig. 15) show final values of friction equal to initial values since the oxide surface layer was replaced as rapidly as it was displaced when H_2O_2 was present.

Fig. 16 gives results for lead identations made in the presence of

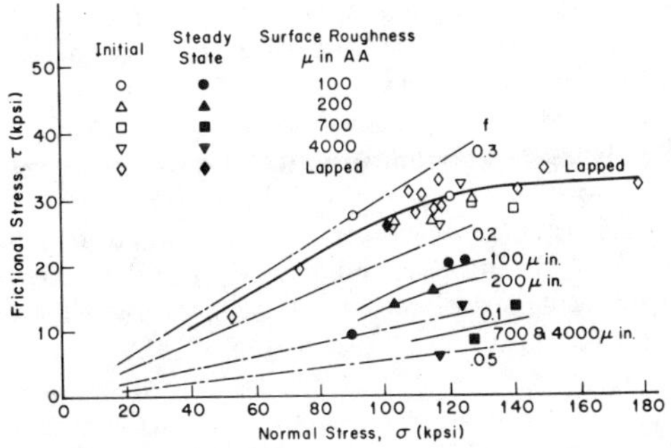

Fig. 11 Variation of Mean Shear Stress (τ) with Mean Normal Stress (σ) for Sliding with CCl_4 when the Initial Surface Roughness has Different Values in μ in, AA

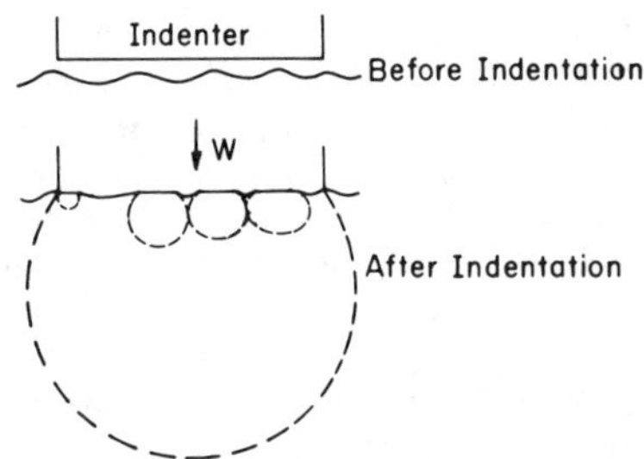

Fig. 12 Schematic View of Rough Surface Before and After Identation. The Dotted Lines are Elastic-Plastic Boundaries

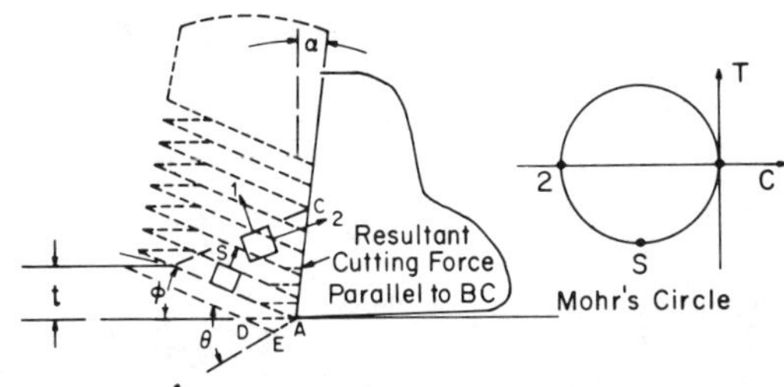

Fig. 13 Concentrated Shear Model of Cutting with Mohr's Circle Diagram for Slip Line Field ABC. Fracture Surface at Tool Tip = AD for Ductile Material but AE for Brittle Material (AE is parallel to BC)

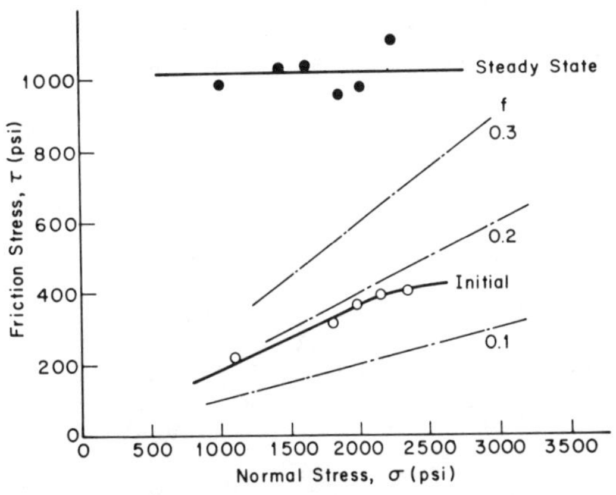

Fig. 14 Variation of Mean Shear Stress (τ) with Mean Normal Stress (σ) for a ½ in. diameter Hard Steel Sphere Sliding on Lead in Air

Lines of constant coefficient of friction ($f = \tau/\sigma$) are also shown in Fig. 5. for purposes of reference. The coefficient of friction is initially about 0.1 for the dry case but rises to about 0.5 under steady state conditions.

Fig. 6 shows the variation of τ with σ for CCl_4 and Fig. 7 shows the frictional torque versus rotational displacement for the same case. It is evident that frictional torque reaches the steady state value instantaneously in this case and consequently the initial and steady state values are the same as in Fig. 6. Also, the initial and steady state friction coefficients with CCl_4 are the same (0·28) compared with the 0·5 for the dry steady state case. Fig. 8 is plan view photomicrograph of the CCl_4 surface which is seen to be far smoother and crack free than the air surface (Fig. 4). A thin film of material was present on the surface that had been indented and rubbed with CCl_4 which was identified as $FeCl_3$.

In a further series of tests, indentation and rotation to the steady state were carried out in air followed by the application of CCl_4. Fig. 9 shows the friction torque versus displacement traces for rotation in air followed by the application of CCl_4. The friction torque is seen to drop to a very low value when CCl_4 is applied to the galled air surface. Fig. 10 shows values of τ versus σ for indentations produced in air and rotated until galling occurred (three revolutions). Then CCl_4 was applied to the cavity and values of τ were measured. The values marked initial were those measured immediately before the CCl_4 had an opportunity to react with the surface while the values marked steady state are those pertaining after three revolutions in the presence of CCl_4. The latter values are nearly an order of magnitude lower than the former ones.

Fig. 11 shows values of shear and normal stress obtained for surfaces having different values of surface roughness before indentation when lubricated by CCl_4. While there is little difference in the initial values of f with roughness, the steady state values show a large difference, rougher surfaces yielding lower values of f.

Discussion and Interpretation of Results on Steel

The foregoing results are consistent with the following explanation. Carbontetrachloride is ineffective as a boundary lubricant and is, in fact, a negative boundary lubricant if it does not reach chemically with the surface being lubricated to form a brittle compound in the case of steel ($FeCl_3$). The negative action comes from the fact that it excludes oxygen in air. Since metal oxides are somewhat beneficial in lowering frictional resistance, the presence of air is important.

It is well known that surface asperities are merely flattened by an indenter but not removed completely unless the indenter is slid across the surface [4]. This is because the plastic zones beneath each asperity grow as they are flattened under increased load, but only until adja-

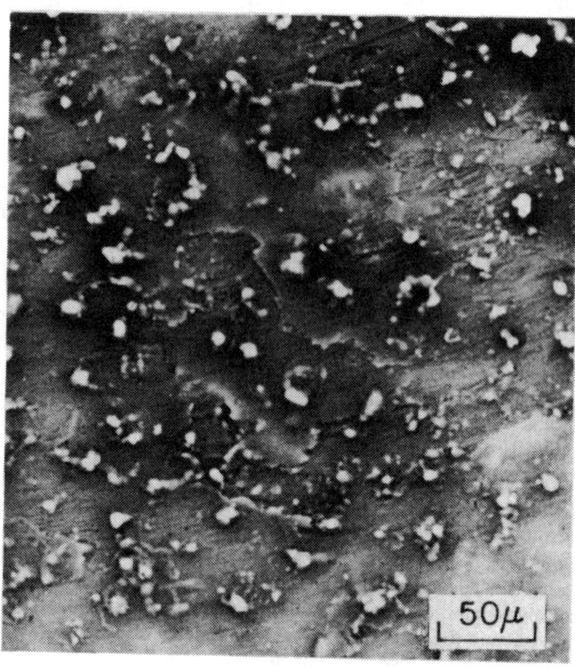

Fig. 8 Plan View Photomicrograph of Surface After three Revolutions with CCl_4 present

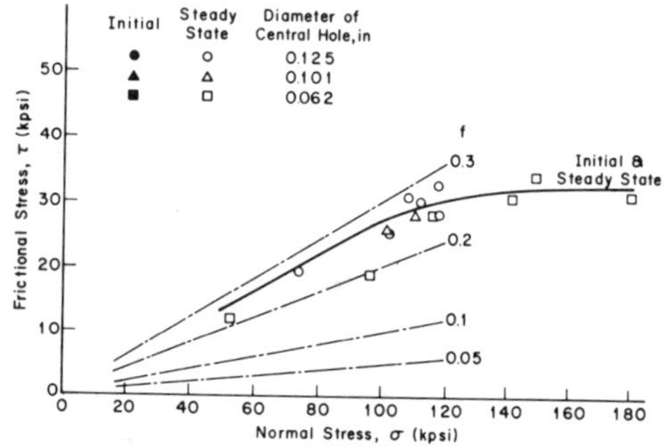

Fig. 6 Variation of Mean Shear Stress (τ) with Mean Normal Stress (σ) for Sliding with CCl_4. There is no difference between initial and Steady State Values

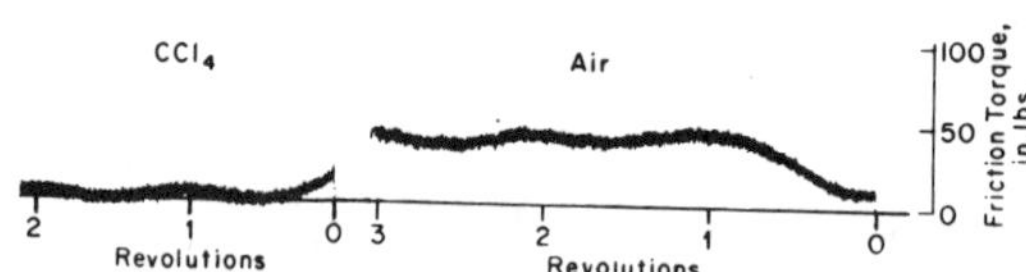

Fig. 9 Friction Torque versus Rotational Displacement for Sliding Initially on Air Followed by Application of CCl_4. Load on sphere = 1000 kg, sliding speed = rpm, hole diameter = 0.125 in.

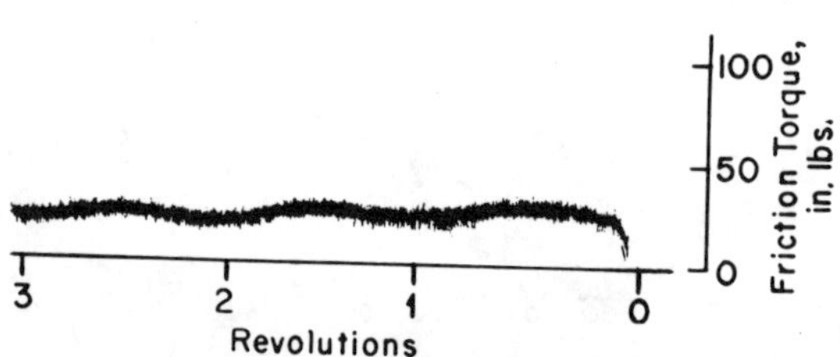

Fig. 7 Friction Torque versus Rotational Displacement for Sliding with CCl_4. Load on sphere = 1000 kg, sliding speed = rpm, hole diameter = .125 in.

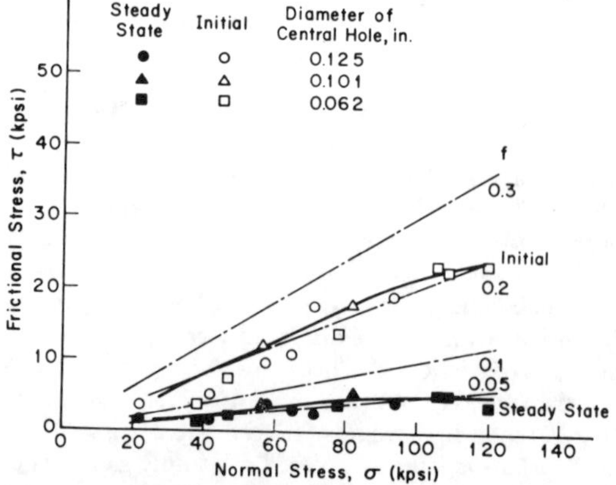

Fig. 10 Variation of Mean Shear Stress (τ) with Mean Normal Stress (σ) for Sliding in Air Followed by Sliding with CCl_4

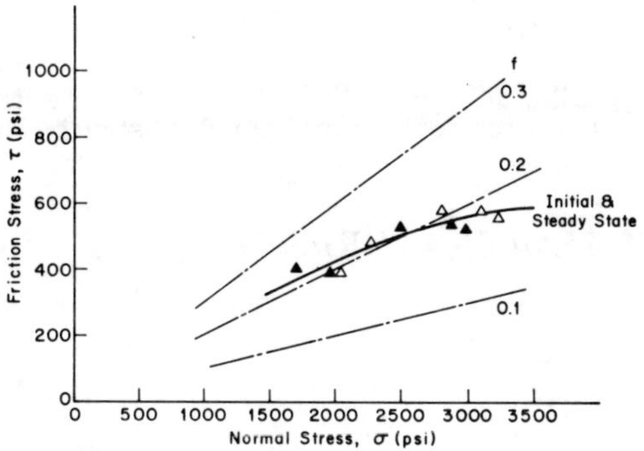

Fig. 15 Variation of Mean Shear Stress (τ) with Mean Normal Stress (σ) for a ½ inch diameter Hard Steel Sphere Sliding on Lead in the Presence of Hydrogen Peroxide

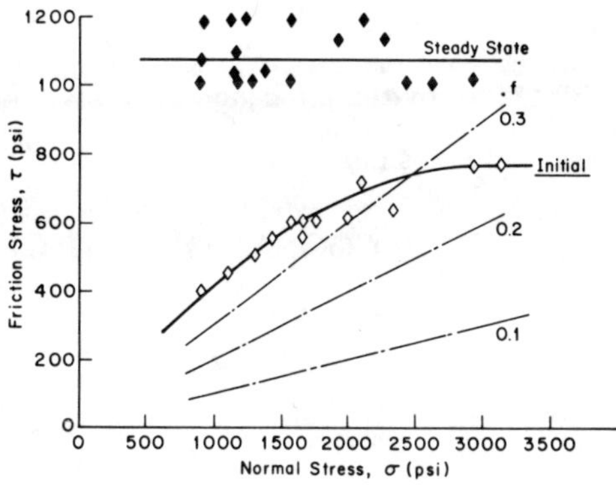

Fig. 16 Variation of Mean Shear Stress (τ) with Mean Normal Stress (σ) for a ½ inch diameter Hard Steel Sphere Sliding on Lead in the presence of CCl_4

CCl_4 followed by sliding. Initial values of friction were only slightly higher than those for the dry case since Pb Cl_3 has a slightly higher shear strength than lead oxide. However, what is more important, the Pb Cl_3 film initially produced during identation is brittle and ruptures during sliding, exposing clean lead to the steel ball. This results in final values of friction with CCl_4 that are the same as the final values for air. In both cases, the high values of *f* result from galling. Thus, CCl_4 is a negative lubricant for lead due primarily to the fact that the film formed chemically on the surface (Pb Cl_3) is brittle, exposing clean lead as it fractures.

Lead is about the only metal for which CCl_4 does not reduce cutting forces below those obtained in air at low cutting speeds. The reason for this is that lead chloride initially formed during chip formation is brittle, fractures during subsequent sliding of the chip up the face of the tool and thus expose clean lead to the tool face which results in high friction.

Concluding Remarks

This study reveals that CCl_4 can have a dual role as a boundary lubricant. It can give higher friction than air if it does not react by excluding oxygen. Or it can give low values of friction if conditions are such that it can react. The tendency for a steel surface to react with CCl_4 increases with an increase in surface roughness since the resulting increase in pressure on the asperities promotes chemical action. Carbontetrachloride is most effective on very rough galled surfaces (highly strained) and results in a surface burnishing action that is part chemical and part physical. It is believed that extreme pressure lubricants generally perform in this way although not as dramatically as CCl_4.

Carbontetrachloride is believed to be a negative cutting fluid (gives higher cutting forces than when cutting in air) for lead not so much for the reason generally stated—i.e., due to lead chloride having a higher shear strength than lead itself but because lead chloride is so brittle it ruptures on sliding, exposing clean lead which in turn leads to galling.

Acknowledgments

The authors wish to acknowlege a grant from the Office of Naval Research which has made this study possible. They also wish to acknowledge the support of Lt. R. S. Miller and Dr. K. E. Ellingsworth of ONR in connection with this work.

References

1 Shaw, M. C. "On the Action of Metal Cutting Fluids at Low Speeds," *Wear,* 2 No. 3, 1959 p. 217

2 Shaw, M. C., Ber, A., and Mamin, P. A., "Friction Characteristics of Sliding Surfaces Undergoing Subsurface Plastic Flow," TRANS. ASME Vol. 82, 342, (1960).

3 Doyle, E. D., private communication.

4 Bowden, F. P., and Tabor, D., "The Friction and Lubrication of Solids," Part I, Oxford University Press, London, 1950 p. 21.

5 Shaw, M. C. "The Role of Friction in Deformation Processing," *Wear,* 6, 1963 p. 140.

6 Lee, E. H., and Shaffer, B. W., "The Theory of Plasticity Applied to a Problem of Machining," TRANS. ASME, Vol. 72, 1951, p. 405.

Reprinted from the Proceedings of the Fifteenth International Machine Tool Design And Research Conference, S.A. Tobias and F. Koenigsberger, Editors. Copyright University of Birmingham, Published by The MacMillan Press Ltd.

COOLANTS AND CUTTING TOOL TEMPERATURES

by

E. F. SMART* and E. M. TRENT*

SUMMARY

A recently developed metallographic method of estimating temperature gradients in cutting tools has been used to demonstrate the influence of coolants on the distribution of temperature in tools used to cut nickel and iron at relatively high cutting speeds. The results demonstrate that the temperature gradients in tools used to cut these two materials are very different. In particular, there is a cool zone at the cutting edge of the tools used to cut iron which is absent when cutting nickel. The most effective coolant action is shown to be through the tool body and streams of coolant were directed to the hottest and most vulnerable, accessible faces, which were different for iron and nickel. The action of coolant only slightly reduced the temperature at the interface between tool and chip but greatly steepened the temperature gradient into the tool and at the surfaces to which the coolant was directed. It is suggested that further investigation of temperature gradients in tools would provide a more logical basis for coolant application when cutting different materials.

INTRODUCTION

A metallographic method has recently been described for determining the temperature gradients in high speed steel cutting tools[1]. This method is effective on tools used to cut higher melting point metals, such as iron and nickel and their alloys, at relatively high rates of metal removal. It depends on the fact that a heat treated tool steel is in a metastable condition and changes to a more stable condition when heated to elevated temperatures. Fully heat treated high speed steel is normally tempered at about 560°C before use and the structure and properties are practically stable up to this temperature. Over 600°C, however, the structure is modified and the hardness and other properties change to an extent dependent upon temperature and time. If the time of heating is known, the temperature can be estimated either from the observed structures or from the hardness to within an accuracy of ±25°C within the temperature range 650-900°C.

Under suitable cutting conditions this method can be used to give much more information about temperature distribution in the tools than has been obtained by other methods. In the present work the changes in metallographic structure have been used to investigate the influence of a coolant on the temperature distribution when cutting iron and nickel at relatively high speeds.

EXPERIMENTAL TECHNIQUES

All the cutting was carried out on a 13 hp lathe, the operation being continuous turning with a constant depth of cut of 1·25 mm (0·050 inch). The tools were in the form of tips clamped in a tool holder. The dimensions of the tool tip were 12·7 mm x 6·3 mm x 9·5 mm (0·5 in x 0·25 in x 0·375 in), the rake and flank surfaces being prepared to a metallographic finish. The cutting angles were:

Top rake	+6°
Side clearance	6°
End clearance	3°
Trail angle	10°
Approach angle	90°
Nose radius	0·37 mm (0·015 inch)

Tool tips were made from BM34 high speed steel (composition shown in Table 1) purchased as fully heat treated bars with a hardness of 840 HV.

The composition and hardness of the work materials are shown in Table 1. The iron was in fact a very low carbon steel in which very small areas of cementite and pearlite were visible among the ferrite grains. The nickel was a bar of commercially pure nickel which had been cast and extruded to a 2·5 inch square section.

The coolant used was a 30 : 1 dilution of Shell Dromus B soluble oil, one part oil, 30 parts water.

* Department of Industrial Metallurgy, University of Birmingham

TABLE 1 Composition of the high speed steel (BM34), iron and nickel used

	HSS (per cent)	Iron (per cent)	Nickel (per cent)
Carbon	0·87	0·04	0·15
Silicon	0·32	0·017	0·35
Manganese	< 0·30	0·14	0·35
Chromium	3·75		
Molybdenum	9·5		
Tungsten	1·65		
Vanadium	1·15		
Cobalt	8·25		
Nickel	< 0·25		99·0
Iron	Bal.	Bal.	0·4
Sulphur	< 0·025	0·010	0·01
Copper	< 0·20		0·25
Phosphorus	< 0·025	0·010	
Hardness VPN	830	83	160

The cutting time was normally 30 seconds, as experience had shown that a stable temperature gradient was established in a tool in a time considerably shorter than this and changed relatively slowly with further extended cutting times. Cutting was stopped by feeding the tool out rapidly by hand. In some cases the chip remained, adhered to the tool. To determine temperature gradients in three dimensions, at least two tools were used under each set of conditions, one for sections normal to the cutting edge and the other for sections parallel to the rake face. Metallographic sections were carefully prepared at definite locations in both directions, the ones parallel to the rake face being as near to the original surface as possible. The structures in the tool were revealed by etching in a 2 per cent solution of nitric acid in alcohol for 30 seconds.

EXPERIMENTAL RESULTS

Tools used to cut iron and nickel in air

The etched surfaces of tools used to cut the iron at 183 m/min (600 ft/min), 0·25 mm/rev (0·010 in/rev) feed for 30 seconds are shown in figures 1(a) and 2(a). Figure 1(a) is the rake face of the tool and figure 2(a) a section normal to the cutting edge at a distance of 0·62 mm (0·025 inch) from the nose, A-B in figure 1a. The corresponding temperature gradients estimated from the structures are shown in figures 1(b) and 2(b).

The main characteristics are:

(1) A zone of relatively low temperature extending along the whole cutting edge including the nose radius. In this region no structural changes were detected and the temperature was therefore less than 650°C. This zone was approximately 0·3 mm (0·012 inch) wide, i.e., somewhat wider than the feed. It is in this region that the maximum compressive stress is imposed by the cutting force. This stress falls off rapidly with distance from the edge[2,3]. There was no visible deformation of the tool edge.

(2) A very steep temperature gradient further from the cutting edge leads to a maximum temperature of 950°C at approximately 1·5 mm (0·060 inch) from the edge. At this very high temperature the yield strength of the tool steel is greatly reduced and it tends to be sheared away to form a crater[4]. The black area in the centre of the heated region (figure 1(a)) is such a crater which was filled with work material darkened by etching.

(3) The high temperature region extended to the end clearance face (C in figures 1(a) and 1(b) and this was the only part of the tool edge which was seriously weakened by heat. The compressive stress in

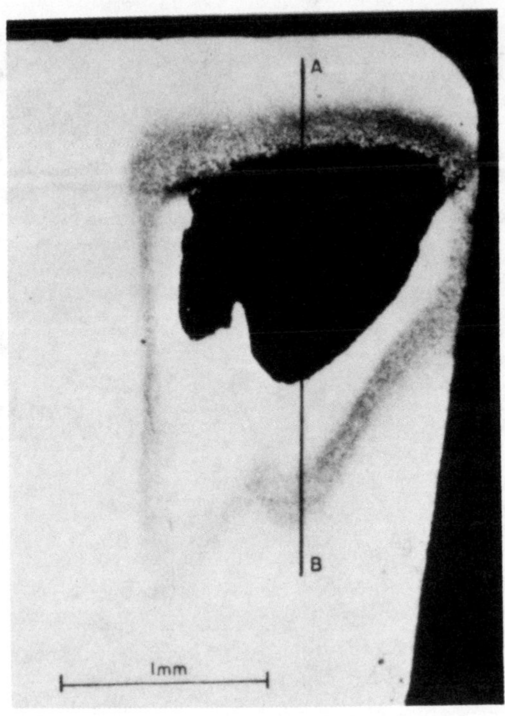

(a)

(b)

Figure 1. (a) Rake face of tool used to cut iron in normal dry atmosphere at 183 m/min (600 ft/min) 0 25 mm (0·010 in/rev) feed for 30 seconds. Etched 2 per cent Nital. (b) Isotherm diagram derived from metallurgical structures in figure 1(a).

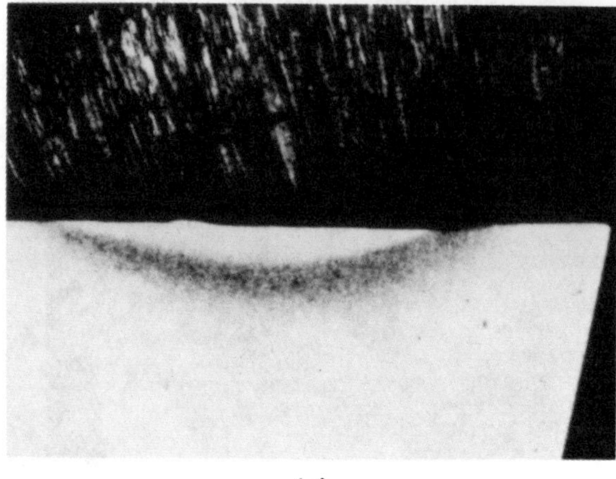

(a)

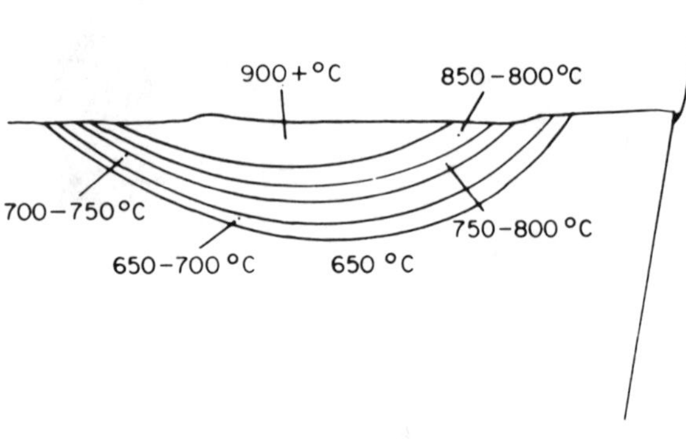

(b)

Figure 2. (a) Section normal to cutting edge of tool used to cut iron under same conditions as figure 1(a). (b) Isotherm diagram derived from metallurgical structures in figure 2(a).

this region is considerable and failure may start at this position, the tool edge being depressed and bulged out on the end clearance face.

(4) As shown in figures 2(a) and 2(b) the 'heat affected region' (i.e., above 650°C) extended into the tool to a maximum depth of approximately 0·5 mm (0·020 inch) in a crescent shaped zone.

All these features of the thermal gradients are easily reproduced. Successive cuts on the same material under the same conditions reproduce with precision the features (1) to (4) above. The least reproducible feature is the outer limit of the heat affected zone, furthest from the cutting edge, which tends to be uneven and vary from one cut to the next.

Figures 3(a) and 4(a) are the corresponding etched surfaces of tools used to cut the commercial purity nickel at 46 m/min (150 ft/min), with other conditions the same as those for cutting the iron. The temperature gradients estimated from these structures are shown in figures 3(b) and 4(b). The temperature distribution is greatly different and the characteristic features are as follows:

(1) The temperature gradients are much less steep and there is no low temperature region close to the cutting edge. As a consequence, the cutting edge was unable to support the compressive stress imposed by the cutting force and the tool edge was deformed downward as shown in figure 4(a). The layer of white material at the edge and around the nose radius in the rake section (figure 3(a)) is nickel filling the space left by the deformed edge. This tool would soon have collapsed completely.

(2) There is only a moderate increase in temperature from the cutting edge in the direction of chip flow, and in the region of maximum temperature there is a shallow crater filled with nickel (white)

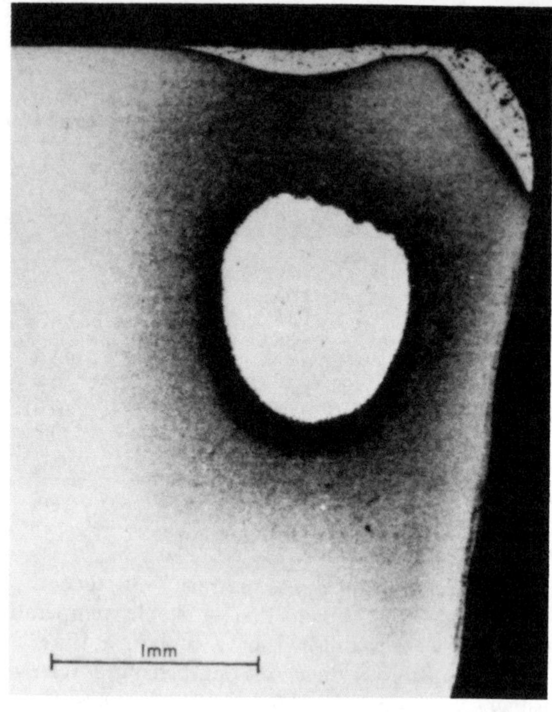

(a)

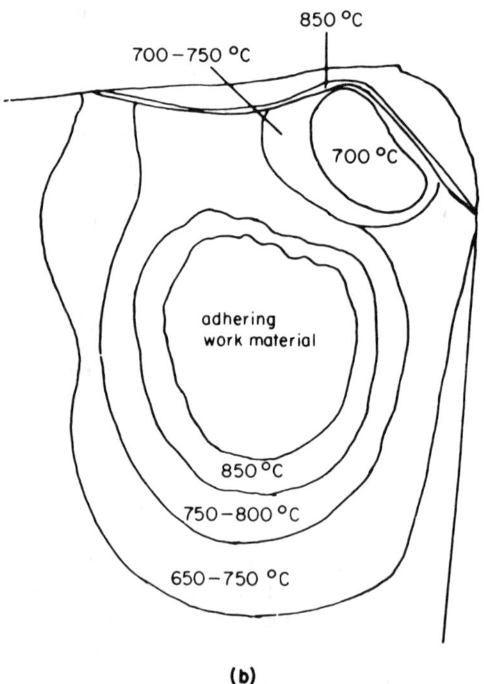

(b)

Figure 3. (a) Rake face of tool used to cut nickel in normal dry atmosphere at 46 m/min (150 ft/min) 0·25 mm/rev (0·010 in/rev) feed for 30 seconds. Etched 2 per cent Nital. (b) Isotherm diagram derived from metallurgical structures in figure 3(a).

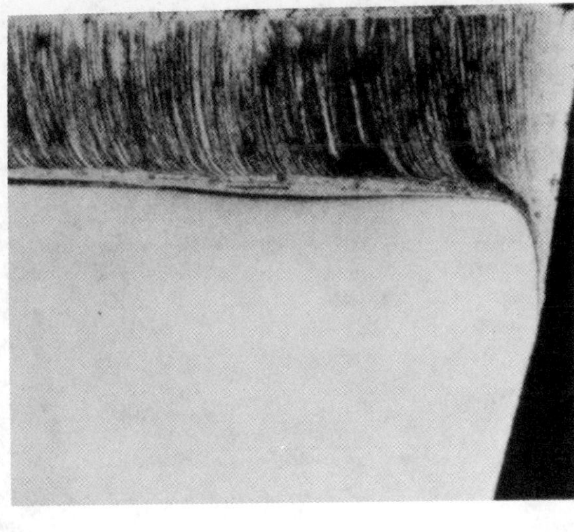

(a)

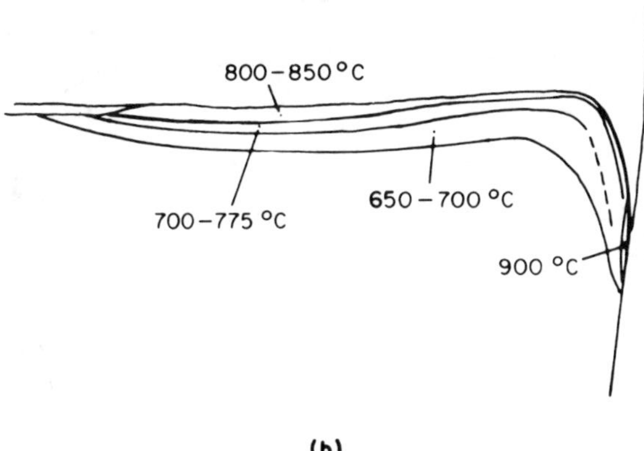

(b)

Figure 4. (a) Section normal to cutting edge of tool used to cut nickel under same conditions as figure 3(a). (b) Isotherm diagram derived from metallurgical structures in figure 4(a).

rather further from the edge than when cutting the iron.

(3) The temperature at the end clearance face was not so high as when cutting the iron, and the tool did not tend to deform and collapse at this position.

(4) As shown in figures 4(a) and (b) the heat affected zone extended into the tool all along the rake surface and including the worn and deformed cutting edges.

As when cutting the iron these major features of temperature distribution were repeatable in successive tests under the same conditions.

Temperature gradients in tool with coolants

Coolants were applied in two ways. The first was to flood the rake surface and forming chip with a stream of coolant and the second was to direct a jet of coolant at that part of the exposed tool surface which was at the highest temperature when cutting in air. In the case of the tools used to cut iron this was at the end clearance face near the nose radius, and for tools used to cut nickel the jet was directed into the side clearance and nose (figure 5).

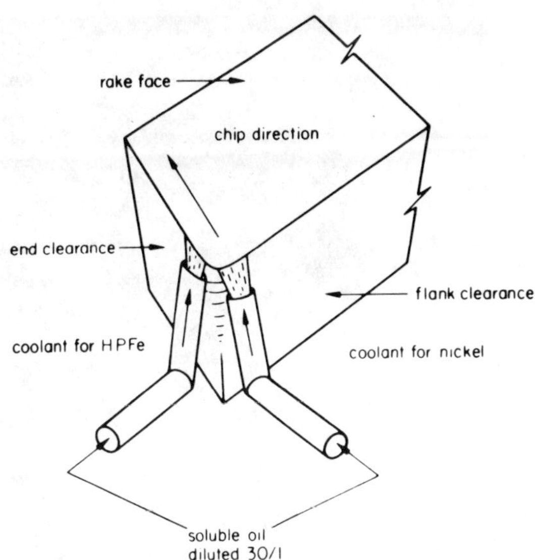

Figure 5. Illustration of selective application of coolant when cutting iron and nickel: Iron–jet applied below end clearance face; Nickel–jet applied below flank clearance face.

Figures 6(a) and (b) show the rake face of a tool used for cutting the iron, cooled by flooding on the rake face. Figures 7(a) and (b) show the tool cooled by the jet on the end clearance face. With the conventional flooding of the rake face, there is some reduction of size of the hottest region, although at one point this extends further from the cutting edge. The crater is somewhat smaller but there is no reduction in temperature at the end clearance face where the tool is beginning to deform. The low temperature region around the cutting edge is roughly the same size.

With jet cooling on the end clearance face there is a much greater effect on the temperature distribution (figures 7(a) and (b)). The central hot area is much smaller and there was no sign of a crater in this section. The coolant action was very effective in reducing the temperature at the end clearance to below 650°C and the width of the cool region all around the cutting edge was increased.

Sections through the cutting edge half way along the depth of cut (figures 8(a) and (b), 9(a) and (b)) show that the coolant has decreased the depth of the hot zone below the tool surface, and the influence of jet cooling (figure 9) is very great in this respect, although the surface temperature at the hottest position on the rake face was very little lower than when cutting in air. In figure 9 there is some shearing of the tool surface to form a shallow crater in spite of the coolant action. Figure 10 shows the influence of cooling on the temperature gradient into the tool from the hottest part of the rake surface.

The influence of coolant when cutting the nickel bar is shown in figures 11 and 12, which should be compared with figure 3. The coolant applied by the conventional flooding of the rake face (figure 11) reduced the temperature to a small extent over the whole surface. The temperature is higher than elsewhere at the cutting edge and in an area well back

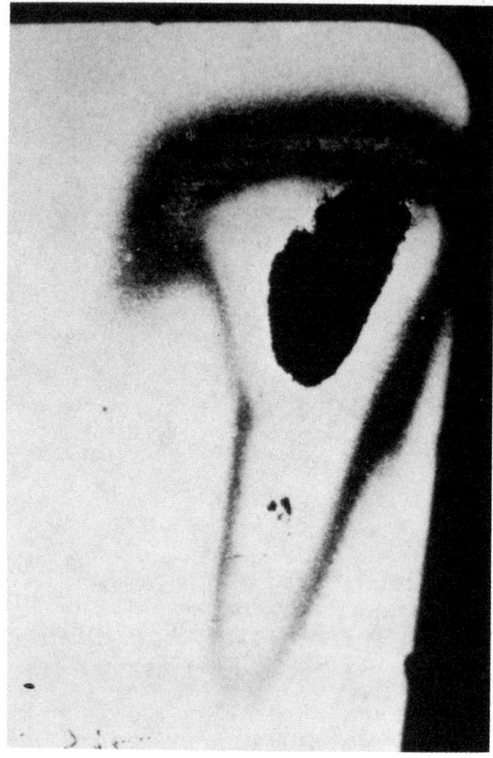

(a)

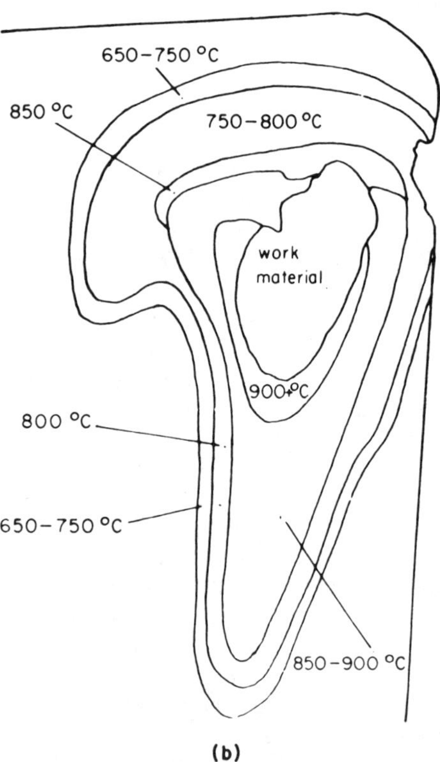

(b)

Figure 6. (a) Rake face of tool used to cut iron under a continuous flow of coolant, directed from above the rake face. Cutting conditions as figure 1(a). Etched 2 per cent Nital. (b) Isotherm diagram derived from metallurgical structures in figure 6(a).

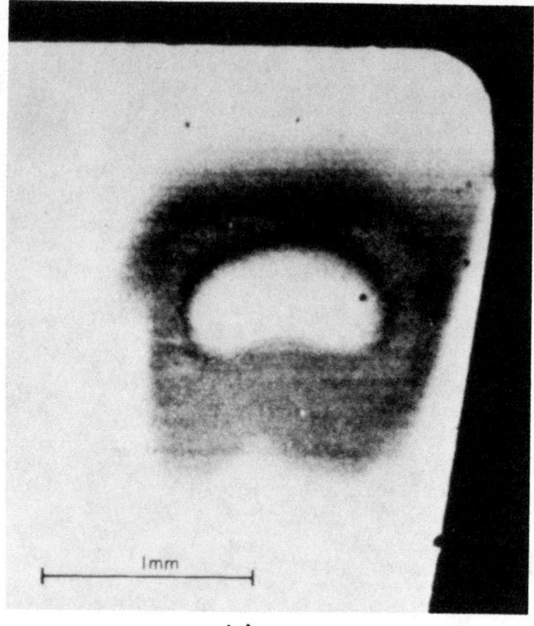

(a)

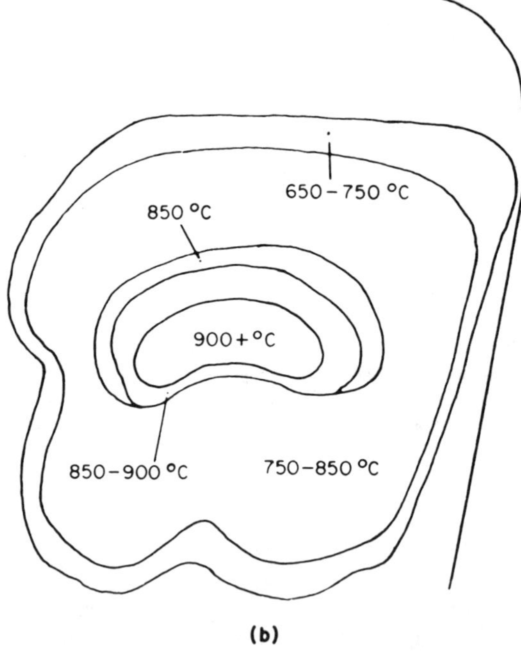

(b)

Figure 7. (a) Rake face of tool used to cut iron under a continuous flow of coolant applied from below the end clearance face (see figure 5). Cutting conditions, as for figure 1(a). (b) Isotherm diagram derived from metallurgical structures in figure 7(a).

from the edge but the temperature differences are small. No cratering was observed and, although the cutting edge and nose radius were still deformed downwards this deformation is much reduced in the cooled tool. Jet cooling in the side clearance (Figure 12) still further reduced the temperature, which was nowhere much higher than 750°C and the amount of deformation at the tool edge was much smaller. The sections through the tools (figures 13 and 14) show the considerable reduction in deformation and wear on the tool flank brought about by use of the coolant, particularly the jet in the side clearance.

DISCUSSION OF RESULTS

In considering the results the nature of the heat source must be appreciated. Under the cutting conditions used in these tests, as in most high speed

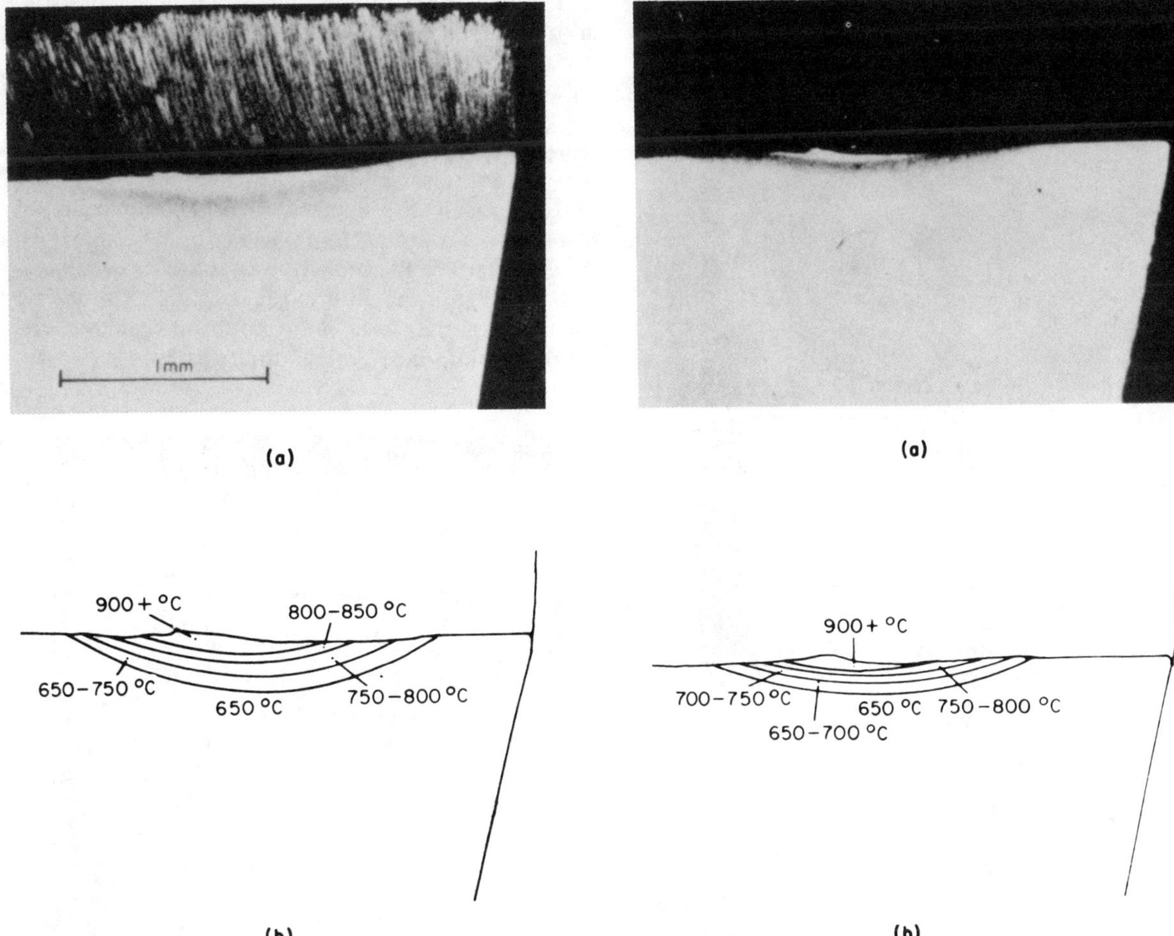

Figure 8. (a) Section normal to cutting edge of tool used to cut iron under a continuous flow of coolant directed from above the rake face. Cutting conditions, as for figure 1(a), etched 2 per cent Nital. (b) Isotherm diagram derived from metallurgical structures in figure 8(a).

Figure 9. (a) Section normal to cutting edge of tool used to cut iron under a continuous flow of coolant applied from below end clearance face (see figure 5). Cutting conditions as for figure 1(a). (b) Isotherm diagrams derived from metallurgical structures in figure 9(a).

cutting, the work material is effectively seized to the tool and movement of the chip involves intense shear in a very thin layer of the work material at the tool surface (called the 'flow zone'). This is most clearly seen in the sections through the tools used to cut the nickel (figures 4, 13 and 14). The work done on the shear plane to form the chip is converted into heat and largely carried away with the chip with little influence on the temperature of the tool. The heating of the tool is largely done by the shearing of the metal in the flow zone. Typically 20–25 per cent of the work of cutting is expended in the flow zone, but as the volume of metal in this zone is small (usually about 5 per cent of the volume of the chip) much higher temperatures are achieved. Thus in the iron chips under the cutting conditions used here, the flow zone reached a temperature of nearly 1000°C while the chip itself was never red-hot, probably about 300°C. The high temperatures reached are not readily observable but have a major influence on the performance of tools as is clear from the observations on the wear and change of shape of tools at positions of high temperature and stress.

Observations suggest that, over most of the seized interface, the bonding is metallic in character and heat is therefore readily conducted from the flow zone into the tool and the temperature in the tool at any point must be very close to that of the part of the flow zone with which it is in contact. Since there are no gaps or capillary channels at this part of the interface to permit the coolant to penetrate, the interface can only be cooled by conduction of heat through the tool or the body of the chip. The only

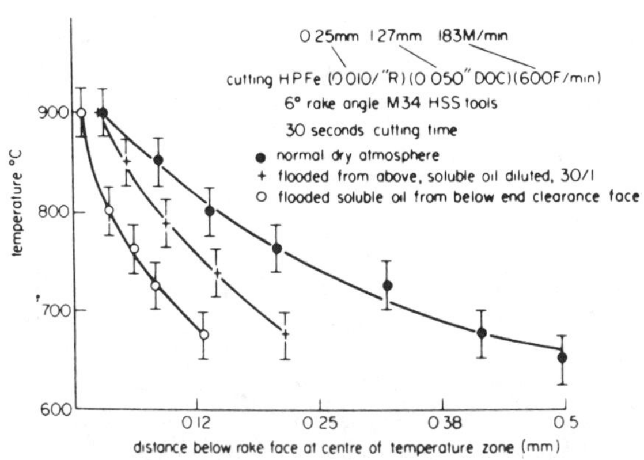

Figure 10. Influence of coolants on temperature gradients into tool from rake face.

(a)

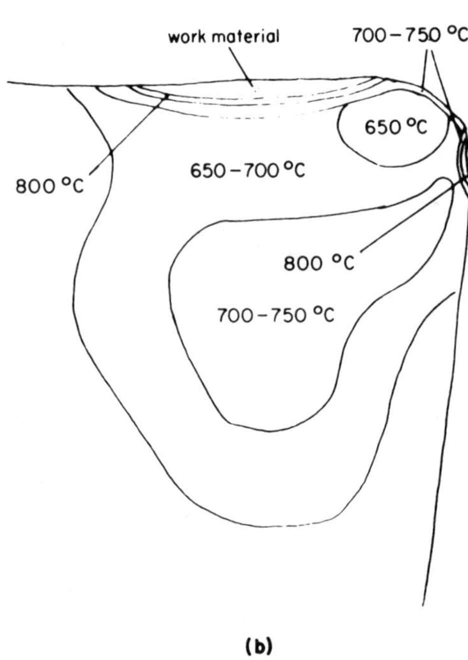

(b)

Figure 11. (a) Rake face of tool used to cut nickel under a continuous flow of coolant directed from above the rake face. Cutting conditions, 46 m/min (150 ft/min) 0·25 mm rev (0·010 in/rev) feed for 30 seconds, etched 2 per cent Nital. (b) Isotherm diagram derived from metallurgical structures in figure 11(a).

exception to this is at the edges of the chip where the edge of the flow zone is exposed to direct action by the coolant. This view of the cutting conditions existing in high speed cutting operations suggests that, in using a coolant, the aim should be a maximum reduction of temperature of those parts of the tool which suffer damage as a result of high temperature and stress.

The drastically different temperature distribution in tools used for cutting iron and nickel leads to damage to different parts of the tool. With the nickel, high temperatures were achieved at much lower cutting speeds and the tool was damaged at lower temperatures than when cutting the iron. The experimental results show that, when the rake face of the tool is flooded with coolant, there is some reduction of temperature in the tool when cutting both metals. At the interface the reduction of maximum temperature seems to be small in both cases. In the tool used for cutting the iron it is doubtful whether the maximum interface temperature was reduced at all,

(a)

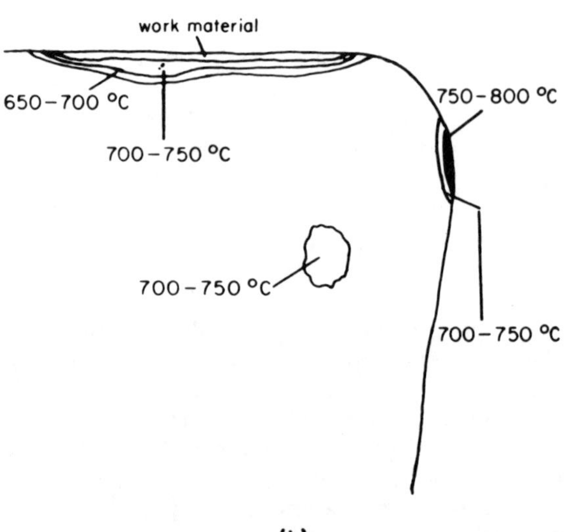

(b)

Figure 12. (a) Rake face of tool used to cut nickel under a continuous flow of coolant applied from below the flank clearance face. (see figure 5). Cutting conditions as for figure 11(a). The section is made a short distance below the level of the temperature contours in figure 14. (b) Isotherm diagram derived from metallurgical structures in figure 12(a).

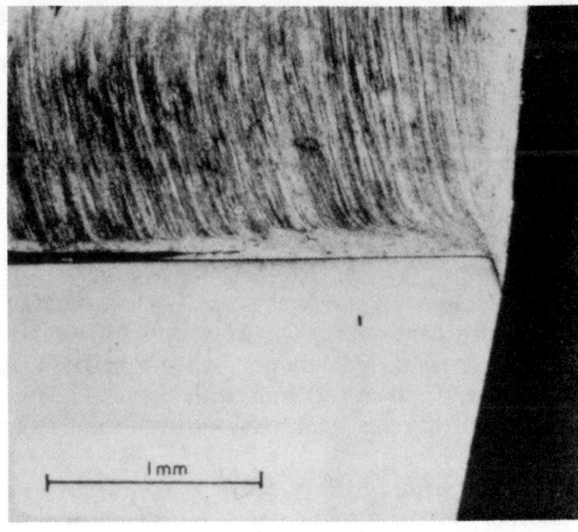

(a)

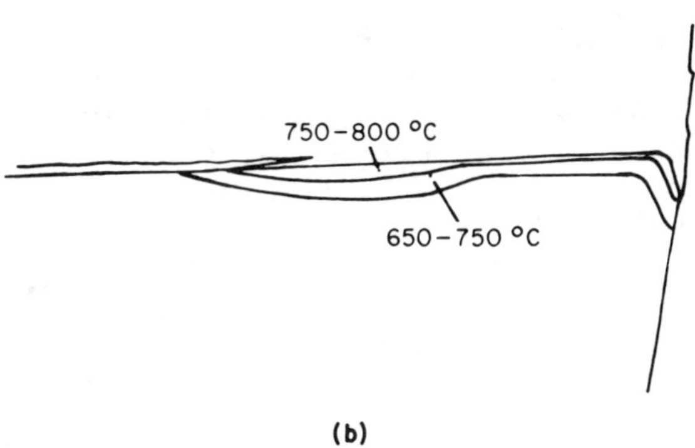

(b)

Figure 13. (a) Section normal to cutting edge of tool used to cut nickel under a continuous flow of coolant directed from above the rake face. (b) Isotherm diagram derived from metallurgical structures in figure 13(a).

although the area at high temperature may have been somewhat smaller. With the nickel the reduction of temperature at the interface was greater, no crater was observed and there was less deformation and flank wear at the edge. The greater cooling effect when cutting nickel may be due to the lower cutting speed, which gives more time for cooling through the chip body. Comparison of figure 2 and figure 8 suggests that, while the surface temperature at the rake face is relatively unchanged, the cooling of the tool body has led to a steeper temperature gradient into the tool.

Direction of the coolant to the hottest and most vulnerable parts of the tool surface appears to be considerably more effective than flooding the rake surface. In tools used to cut the nickel, the maximum interface temperature seems to have been reduced, although the fall in temperature is relatively small, and the cooling by a jet directed at the cutting edge was more effective in reducing deformation and wear of the edge than flooding the rake face with coolant (figures 3, 11 and 12). When cutting iron the results are more emphatic. While cooling of the end clearance face did not greatly reduce the maximum temperature, it restricted the area of cratering and prevented it spreading to the end clearance and nose with consequent breakdown. It greatly steepened the temperature gradient into the tool and reduced the depth to which the tool was overheated (figures 2, 8, 9 and 10). It also increased the width of the cool zone at the cutting edge. All these are important advantages which could increase tool life or permit higher cutting speeds.

These results suggest that, in high speed cutting, the most effective cooling action is obtained by promoting cooling through the tool body and that this can be achieved by use of a relatively small volume of coolant directed to the hottest accessible parts of the tool surface. The hottest part of the rake face is not available for cooling. Access to the clearance faces is restricted because the clearance angles are usually small in order to maintain the strength of the tool. Cooling might be improved by increasing the clearance angles. This would weaken the tool mechanically but the reduction of temperature could strengthen the tool and there could be optimum clearance angles which are larger than those commonly used at present.

The differences in temperature distribution in tools used to cut the iron and the nickel show that the optimum cooling arrangement (both direction of coolant and shape of tool) are likely to be dependent on the material being cut. This raises the question of

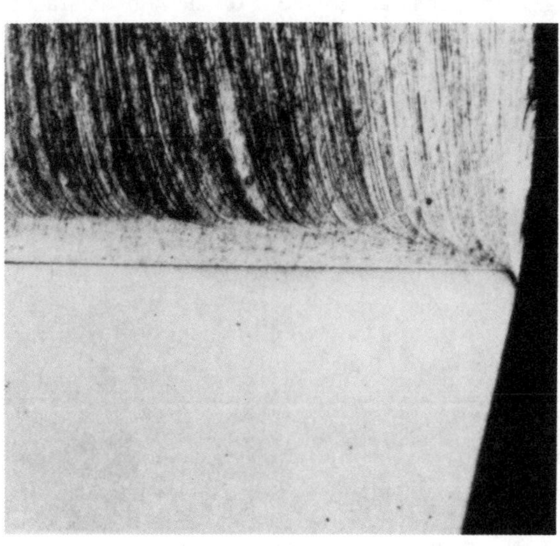

(a)

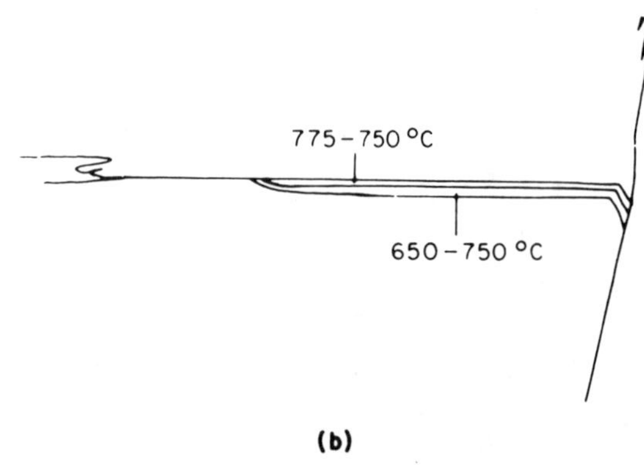

(b)

Figure 14. (a) Section normal to cutting edge of tool used to cut nickel under a continuous flow of coolant applied from below the flank clearance face (see figure 5). Cutting conditions, as for figure 11(a). (b) Isotherm diagram derived from metallurgical structures in figure 14(a).

the significance of the tests carried out on these two materials. Work done so far on steels, including carbon and austenitic stainless steels, has shown that the temperature gradients in tools used to cut these steels at high cutting speeds are similar in character to those in tools used to cut the iron demonstrated here. The addition of alloying elements greatly reduces the cutting speeds at which high temperatures appear but the temperature pattern remains much the same. Work on nickel based alloys has been more limited, but so far this suggests that tools used on the main nickel-based alloys show temperature gradients like those of tools used on the nickel with high temperatures near the cutting edge and shallow temperature gradients. Why iron and nickel and their alloys should behave in such fundamentally different ways during cutting is not clear and is the subject of continuing research.

The temperature gradients and cooling effects demonstrated here are likely to apply to steels and nickel-based alloys under conditions where the heat source is a flow zone on the rake face of the tool. In general this is in the high speed cutting range. It has not been possible to carry out similar tests using carbide tools because they do not undergo structural changes like those of high speed steel which can be used to monitor temperatures. However, observations suggest that the heat source is of the same character when machining with carbide tools as high speed steel[5] and the general character of the temperature distribution should therefore be the same. The values of temperature must vary with the thermal conductivity of the tool, i.e. with the rate at which heat is conducted from its source in the flow zone. The present experiments suggest that the influence of the thermal conductivity of the tool material on the interface temperature might not be very great. It has been shown that greatly increasing the temperature gradient in the tool by cooling it has had relatively little effect on the temperatures at the interface.

Where cutting conditions are such that a flow zone on the tool rake face is not present, the general character of the temperature gradient in the tool could be greatly different from those demonstrated here. In particular the present experiments do not relate to conditions where a built-up-edge exists or where, as at very low speeds, the work material may slide over the tool surface. Conclusions regarding coolant application would be very different for these cutting conditions.

Because of the great variety of conditions used in industrial machining operations, no simple experiments can produce results directly applicable to all conditions. What has been put forward here is the proposal that the study of temperature gradients in tools can provide a more logical basis for the application of coolants in high speed cutting. This cannot be a substitute for long time tool wear testing and industrial experience, but it could provide a guide to development of cooling techniques. The simple experimental work recorded here should be supplemented by long time tests and further studies of temperature gradients with a variety of different work materials, tool shapes, cutting conditions and coolant arrangements.

CONCLUSIONS

The results of tests in which the temperature gradients have been estimated in high speed steel tools used for cutting iron and nickel with and without coolant, have led to the following conclusions for relatively high speed cutting conditions where the heat source is a flow zone at the tool rake surface.

(1) The temperature gradients in tools used for cutting iron and nickel were very different in character, the temperature gradients being much steeper when cutting iron, with a cool zone near the cutting edge which was absent when cutting nickel.

(2) The major cooling effect was through the body of the tool and the coolant acted most effectively when directed to the hottest and most vulnerable exposed part of the tool surface.

(3) Effective use of the coolant reduced only slightly the maximum interface temperature between tool and workpiece but greatly increased the temperature gradient into the tool and near the cooled surface.

(4) Coolant directed to the most vulnerable part of the tool surface which is accessible reduced the damage to the tool.

It is suggested that further investigation of temperature gradients could lead to a more logical guide to the applications of coolants when cutting different materials at high cutting speeds.

ACKNOWLEDGMENTS

The authors gratefully acknowledge the financial support given for the work by the Science Research Council. The work was made possible by the use of the facilities at the Department of Industrial Metallurgy at the University of Birmingham. We should like to thank Mr G. J. Gunnell for his work in preparing the illustrations. The authors are grateful to the International Nickel Co. Ltd. for supplying the nickel bar used in the machining tests.

REFERENCES

1. P. K. Wright and E. M. Trent. *Journal of the Iron and Steel Institute* (May, 1973) **211**, Part 5, 364-68.
2. N. N. Zorev. *International Research in Production Engineering* (1963) ASME, New York 42-9.
3. G. W. Rowe and A. B. Wilcox. *Journal of the Iron and Steel Institute* (March, 1971) **209**, no. 3, 231-2.
4. P. K. Wright and E. M. Trent. Metallurgical appraisal of wear mechanisms *Metals Technology* (January, 1974) **1**, no. 1.
5. E. M. Trent. Conditions of seizure at the tool work interface *I.S.I. Special report* **94**, 11.

CHAPTER 2

CLASSIFICATION OF FLUID TYPES

Presented at SME's Westec Engineering Conference, March 1974

Cutting Fluid Applications For Today's Materials

By. A. Dennis Frazier
Rust-Lick Inc.

INTRODUCTION

There are many facets to effective and efficient metal cutting. Any change which will improve one of these areas, improves the total operation. Often overlooked, but none-the-less an important facet of any machining operation is the cutting fluid. Production can be increased and costs reduced through the judicious use of better fluids.

Cutting fluid results can best be measured in terms of tool life. Longer tool life results in maximum volume of metal removed per tool grind, minimum shut-downs for tool change, and minimum number of tool grinds per unit of production. Cutting speeds can often be increased, surface finishes improved, and dimensions stabilized with correct application of better fluids.

All of the above lend themselves to increased production while minimizing costs. In the overall view, company profits will be dependent on the savings that can be made on metal, labor, tool, operating, maintenance, and overhead costs.

NATURE OF CUTTING FLUIDS

Although alterations have been made in the formulation of cutting fluids since their introduction, the basic roles have remained unchanged. They are two-fold: (1) head dissipation, or cooling, to prolong tool life and cool workpiece and chips and to permit higher cutting speeds; and (2) lubrication to reduce friction, to improve the finish on the metal cut, and to prevent rusting. By being applied from directly over the cutting zone, the fluids also serve to flush away metal chips.

Briefly, there are three major types: insoluble, or straight oils that are derived from animal (including marine), vegetable, or mineral (petroleum) sources; soluble, or emulsifying oils that are derived from similar sources but are emulsified with water; and alkaline, or synthetic, fluids that have little or no oil and are water-based coolants. The term "cutting fluids" as applied to both cutting oils and (soluble) coolants has gained wide acceptance. Although an alternate term, lubricoolants, (3) is descriptive, it is not ordinarily used.

Emulsifiers (e.g., soaps, petroleum sulfonates), germicides or bactericides (e.g., orthophenyl phenol, formalin), rust inhibitors (e.g., hydroxyl amines, inorganic and organic nitrites, and nitrates), and extreme pressure (EP) agents (e.g. sulfur and/or chlorine, and phosphorous) are added in varying proportions to these basic fluid types. Other additives sometimes used are: anti-foaming agents (e.g., silicones), water softeners (e.g., carbowax, polyethylene glycol), water conditioners (e.g., phosphates and borates), dyes (e.g., fluorescein) and perfumes (e.g., mint oil).

The original cutting fluid was water; then fatty oils (which often become rancid) were added to form oil-in-water emulsions; next straight insoluble oils were used; and, most recently, synthetic, chemically (non-oily) soluble, water-based coolants. This latter group now comprises from one-half to two-thirds of the cutting fluid market.

PURPOSE OF CUTTING FLUIDS

Satisfactory performance using cutting fluids is conditional upon the following:

(1) Lubricating tool and chip to reduce heat of friction, tool wear, and power consumption.
(2) Rendering antiwelding properties to the tool and work surface. This action prevents galling of metal particles on the tool point and finished work surfaces and has a vital effect of the finished product.
(3) Cooling the tool to reduce abrasive wear at high temperature. The work must also be cooled to prevent distortion resulting from residual heat. This procedure is important in order to produce final work pieces with accurate dimensions.
(4) Protecting the tools and the finished work surfaces from the effects of rust.
(5) Preventing the work surface from becoming discolored. Of importance here, is the proper selection of a fluid for a specific job with consideration for the composition of the material involved.
(6) Being free from smoking and fogging when used.
(7) Maintaining a moderately mild or non-existant odor.
(8) Remaining stable during the usable period and not forming deteriorating products which are offensive.
(9) Possessing no toxic properties or other chemical of pyhsical properties which would react negatively upon the psychological responses of the operators.
(10) Having no residual contamination of the workpiece either at ambient shop temperatures of probable elevated temperatures to which the part may be subjected in service.
(11) Having no effect on painted or machined surfaces of the machine tool.

PERFORMANCE

Factors Affecting Water Soluble Oil Performance

Water soluble cutting fluids are generally carefully balanced compounds of chemicals and oils. This careful balance is easily destroyed by various contaminants. Soaps, chemicals, rust preventatives, hand cleaners, and straight cutting oils must not be allowed to mix with water solubles.

Although there may be no outward appearances of change in a contaminated soluble oil, rust inhibition and germicidal properties may be seriously affected. The extreme pressure additives present in heavy-duty water solubles are frequently reneered ineffective by contaminants. When this occurs, tool life and part finish suffer and machines and parts become susceptible to rust. Regularly scheduled changes of water soluble cutting fluids helps to assure troublefree operations. The small cost of keeping fresh, clean coolant in a machine far outweighs the cost of poor tool life and rust parts and machines caused by contaminated coolants.

Rapid bacterial growth and foul smells also develop if water solubles are not changed regualrly. Rancid water solubles are injurious to operator health and can cause serious skin irritations if continued contact is allowed.

Factors Affecting Performance of Straight Cutting Oils

Straight cutting oils are generally highly refined mineral oils to which various elements lide sulfur, chlorine, and fatty materials have been added. These additives are present to help the oil to resist the extreme temperatures and pressures found in cutting operations. They also aid in preventing chip welding and controlling chip formation.

These oils are also affected by contaminants but perhaps not to such a degree as water solubles. Straight cutting oils should be changed or filtered at intervals to prevent the fluid-up of "fines" in the oil which tend to irritate the pores of the skin as well as scratch and mar the finished parts.

Water must be kept out of straight cutting oils as it reacts with the sulfur and chlorine in the oil and eventually forms corrosive acids which tend to attack both the machine and the parts. Water contaminated cutting oils should be discarded and no attempt made towards reclamation.

TYPES OF CUTTING FLUIDS

While there are numerous cutting fluids on today's market, almost all of them fall into one of three categories:

(1) Cutting Oils
(2) Emulsifiable Oils
(3) Chemical Fluids (often referred to as synthetic)

There are many possible breakdowns. These will become evident later on in the paper as individual formulations are discussed. Listed here are the three types with one exemplary breakdown:

(1) Straight Cutting Oils (not mixed with water)
- A. Inactive Oils
  1. Straight Mineral Oils
  2. Fatty Oils
  3. Fatty Oil-Mineral Blends
  4. Sulfurized Fatty-Mineral Oil Blends
- B. Active Oils
  1. Sulfurized Mineral Oils
  2. Sulfo-Chlorinated Mineral Oils
  3. Sulfo- or Sulfo-Chlorinated Fatty Oil Blends

(2) Emulsifiable Oils (Soluble Oils)
- A. Emulsifiable Mineral Oils
- B. Super-Fatty Emulsifiable Oils
- C. Extreme-Pressure Emulsifiable Oils

(3) Chemical (synthetic) Cutting Fluids
- A. True-Solution Type
- B. Wetting-Agent Type
- C. Wetting-Agent Type with Extreme-Pressure Lubricant

The composition of emulsifiable mineral oils is comprised of a light-bodied (usually 100 SUV at 100°F) mineral oil which becomes emulsified with water by the addition of either petroleum sulfonates, amine fatty acids, resin condensates, chromium oleate, and coupling agents such as glycol.

The low costs of these emulsifiable mineral oils accounts primarily for their wide use. They have the desirable characteristics of a good rust inhibition and adequate lubricity for the ordinary cutting applications. The customary dilution would be approximately one part oil to twenty parts water.

Similar to the emulsifiable oils outlined above are the super-fatty emulsifiable oils. With the addition of fatty oils (lard, rapseed, and sperm oils), they become oilier and, therefore, suitable for tougher machine operations. Quite often these oils are sucessfully used for machining aluminum and for certain tapping operations.

The dilutions usually vary from one part oil to eight-fifteen parts water.

The extreme-pressure emulsifiable oils, in addition to fatty, contain sulfur, chlorine, or phosphorus. As a result, they give off boundary-lubricating properties for handling machining operations even more difficult than those handled by the emulsifiable oils mentioned above. The usual dilution would be one part oil to five-forty parts water.

These oils are, in everyday usage, known as heavy-duty soluble oils. Occasionally they have taken the place of the the oil-type cutting fluids for broaching, gear hobbing, shaping, shaving, and turning operations. As water-based coolants they cool better than oils. Smoking and fogging, so often characterize heavy cuts, are either brought to a minimum or eliminated entirely with the use of the soluble oils.

Specific formulas of these emulsifiable oils are prepared for grinding operations. The concentration would be one part oil to twenty-five to sixty (sometimes even higher) parts water. The solution fluids are made up, for the most part, of amines, borates, glycols or ethylene or propylene oxide condensates, phosphates, rust inhibitors (inorganic and organic nitrites), and sequestreing agents. Their optimun usage as grinding solutions inhibits rust and allows for the rapid dismissal of heat.

It should be taken into consideration, however, that these fluids can interfere with the operation of chucks, turrets, slides, or other moving parts as they are apt to leave a residue of hard or crystalline deposits formed by water evaporation. These agents form clear solutions in water (a dye is often used to color water): one part agent to fifty-two hundred and fifty parts water. The addition of these chemicals to emulsifiable oils or other chemical coolants provides more effective rust-inhibiting properties.

Wetting-agent types (using one or more wetting agent) augment the wetting action of the water. Their application lends greater uniformity of heat dissipation and of anti-rust action. Also, this kind of fluid may include anti-foaming agents, humectants, mild lubricants (organic or inorganic) and water softeners.

Versatility is a strong characteristic of this wetting-type of chemical fluid. It is ideally suited for the lubrication of moving machine parts and alloys very rapid heat dissipation. It has supplanted cutting oil on numerous operations and is effective with high-speed steel as well as carbide tools. The usual mixture would be one part of wetting-type chemical fluid to ten-thirty parts water.

There are two types of wetting-agent fluids--one plain and the other with extreme-pressure lubricants, such as chlorine, sulfur, or phosphorus, added. The purpose of the latter is to impart extreme-pressure or boundary lubrication effects. It serves excellently for tough machining jobs and is usualbe with both high-speed and carbide tools. Usual dilution would be one part wetting-agent to five-thirty parts water.

Mineral oil is usually solvent extracted, highly refined organic oil, derived from crude oil; normally oils of from 100 SUV at 100°F to 150 SUV at 100°F are used as a base for most soluble oils and cutting oils.

Surfactants are agents added to cutting oils and soluble oils to reduce surface tension. The oldest and most commonly used have been sperm oils, animal fats, and fatty acids. More recently, the synthetic types consisting of phosphorus compounds and alkylphenoxy polyethoxy ethanols, (a common comparison would be washing detergents) are being used. By lowering the surface tension of the oil or water emulsion, these materials permit the active ingredients to reach the work area more rapidly giving a cooling effect. With the addition of E.P. (Extreme Pressure) agents, this occurs more effectively.

Coupling agents are usually alcohols and glycols. Being soluble in both oil and water, these agents assist in complete emulsifications of the soluble oils.

Additives, such as sulfur, chlorine and phosphorus, are used for "anti-weld" purposes and are known as E.P. (Extreme Pressure) additives. These additives react chemically with the metals being cut to help prevent "built-up" on edges of cutting tools. The polar additives cover the same materials as surfactants.

Bactericides are self-explanatory. They are usually chlorinated phenols, mercury compounds or quaternary ammonias.

Rust Inhibitors used are either Nitrites or Phosphates.

USING AND HANDLING CUTTING FLUIDS

Of real importance to the life of a tool in particular, or machining operations in general, is the method of applying the cutting fluid. Special equipment is not usually a requisite for good results, although there are numerous extremely effective devices and systems on the market for supplying fluids to the cutting areas. Increased production and tool life, a better finish, and lower consumption of power are possible with the customary low-pressure application methods used provided these basic rules are precticed:

(1) Completely envelop the cutting tool edge and the work by applying a copious stream of fluid. This application has a two fold purpose. One, to supply adequate fluid to the cutting zone and, two, to provide a cooling action that prevents an undue rise in temperature. An empirical rule developed for the operation with lathe-type tools is that the inside diameter of the coolant supply nozzle should be at least three-fourths of the width of the cutting tool.

(2) Direct the fluid at the zone of chip formation and again completely cover the portion of the tool producing the chip. Another nozzle which supplies fluid from below (along the flank of the tool) is desirable in operations involving heavy-duty turning and boring.

(3) Apply fluid through hollow tools for horizontal operations. This type of application is preferable to an external one as it provides an adequate flow of fluid at the cutting edges and flushes chips out of the way.

(4) Apply a full flow of liquid at low pressure for grinding operations. This method usually proves very effective. If there is undue splashing from a large volume of liquid, a precautionary method would be the installation of splash guards on the machine rather that the reduction of flow.

(5) Apply the cutting fluid for most milling operations in a copious quantity to both the incoming and the outgoing sides of the cutter. This is sometimes accomplished with fan-shaped nozzles at least three-fourths of the width of the milling cutter.

(6) For face milling, direct the fluid at all cutting edges. Keep the cutter completely enveloped in fluid by using a ring-type distributor.

MAINTAINING CUTTING FLUIDS

Cutting fluids, like any other fluids that are used over and over again, must be cared for properly. There are several precautions that should be observed.

Cutting Oils. Oil-type fluids perform satisfactorily if applied in full flow to the tools and the work, and if sufficient volume is maintained in the system to hold oil temperature around 70°-75°F.

Cutting and grinding oils become contaminated rapidly during use. Extraneous materials, chips, dirt, etc., should be removed continously or at periodic intervals by filters, strainers, centrifuges, or settling tanks. The mechanical edge-type filter, incorporating metal strips or disks as the filter element, acts principally as a strainer. The absorbent type filter uses paper disks, cotton waste, or cloth bags as the filtering element. Magnetic filters are suitable for seperation of ferrous particles.

Centrifuging is used for removing heavy contaminants and particles from oil. Centrifuging, together with a heating unit and settling tanks, is often used for extracting oil from chips.

All cutting oil systems should be drained at intervals, manually cleaned, flushed and replenished with filtered or new cutting oil. Frequency of cleaning depends upon individual conditions.

Emulsion Fluids. Emulsions generally require more maintenance and care than cutting oils do.

In preparing emulsions, always add the oil to the water. This initial mix should be agitated thoroughly while the oil is being added, otherwise soap forms which combines with the water causing the mineral oil to separate out. If not enough water is used, or if the water is added to the oil, an invert emulsion will result in which water particles are dispersed in the oil phase. Though suitable for certain drawing operations, invert emulsions are undesirable for metal cutting.

Water used in preparing emulsions is very important. Soft waters, where available, present no problems. But hard waters containing various minerals and salts often hinder or impede emulsification. It is not uncommon for emulsions made with hard water to "break" readily, that is, to separate into a stratified condition with a layer of oil or creamy emulsion floating on the surface. Such separation is detrimental.

Modern metal-forming lubrication

Reprinted from Tooling & Production, October 1981

In many of today's metal-forming operations two factors are often overlooked as a means of increasing process efficiency and productivity—type of die lubricant and its method of application. Recent innovations regarding both permit manufacturing engineers to reduce operational cost while improving performance, and to comply with increasingly stringent government regulations. To better understand the state of the art concerning this technology let's begin by examining its development.

By Ron Newhouse
Manager Technical Services
Franklin Oil Corp
Bedford, OH

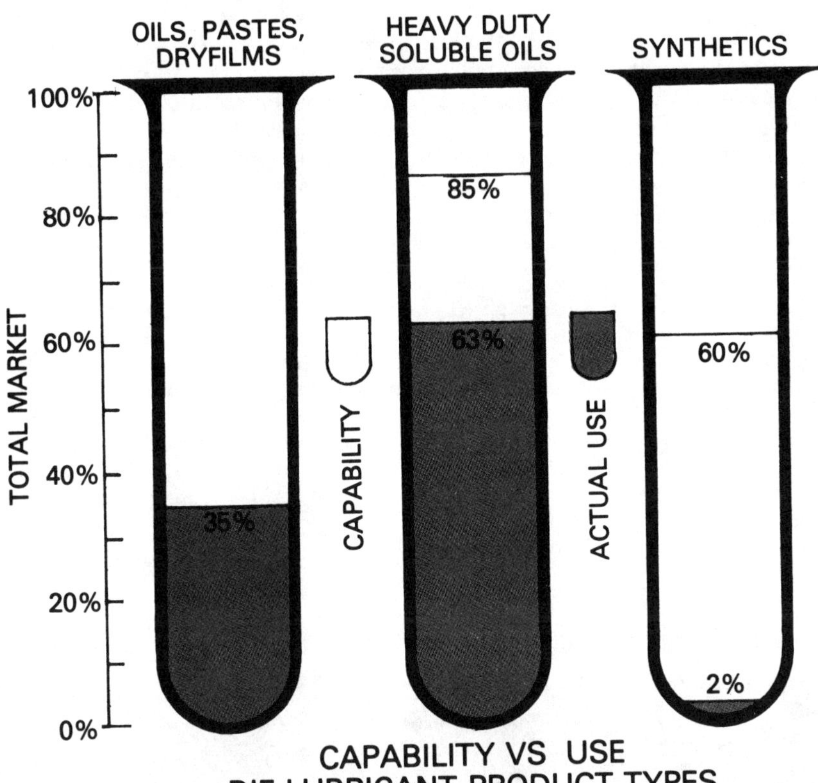

1. Graph shows estimated capability compared to actual use of several metal-forming lubricant groups. Only the severest drawing and extrusion operations (15 percent of the industry) are beyond the performance levels of current heavy-duty, water-soluble oils. Current state-of-the-art synthetics can handle 60 percent of present market needs. This figure will improve as technology increases. Further, both soluble oils and synthetics are much less expensive to use than straight oils. They are easier to clean and with proper treatment can be disposed of at less cost.

Evolution of metal-forming lubricants

Early in this century, straight petroleum-based oil (sometimes referred to as mineral oil) was recognized by those involved in pressworking as a way to prevent punches from prematurely wearing and breaking. The heftier the part to be stamped, the more viscous the oil used. At that time, viscous or boundary lubrication, as it is called today, was the way to go. It wasn't long, however, before it was realized that die life was poor, and the shape and depth that metal could be drawn was extremely limited.

A breakthrough came by reacting soaps with animal fats. This product, labeled a soap-fat paste, offered more lubricity or "slip," thereby allowing metal to flow easier under drawing pressure. One of its first applications was on solid roofs for automobiles. Previously, all auto roofs had a canvas center because cutting out the middle of the blank allowed the metal to be formed without breaking. Soap-fat pastes permitted using a solid blank, thus improving the structural integrity of the design while reducing cost.

During this same period it was discovered that soaps and fats were not only slippery, but had higher film strengths than straight petroleum-based oil. This led to the addition of fats to oil and consequently the development of lard oils. These two products, paste for drawing and lard oil for stamping, dominated pressworking during the '20s, '30s and '40s.

Early in the development of soap-fat pastes a major modification was made—solids in powder form were added to dramatically improve the deep draw capability. Materials such as mica, talc, whiting, calcium carbonate and aluminum sterate were used to enhance the mechanical extreme pressure characteristics of the lubricant. This type of product became known as a pigmented paste.

Difficulty in cleaning, plus buildup on presses and dies, were major disadvantages when using this type of material. Moreover, recent government-enforced disposal has reared its head, via the Resource Conservation and Recovery Act, as another major problem concerning the use of these lubricants. In fact, many companies still using pigmented pastes are being forced to change due to the cost penalties connected with disposing of the solids, and difficulty in locating a firm to do the work.

Performance of these original pigmented pastes was surpassed in the '50s by the surface reactive group of lubricants containing sulfur, chlorine and phosphate. Such additives do not depend on viscous lubrication or mechanical film strength to furnish tool and work separation. Rather, they react with the metallic surface on the tool and workpiece to form a metallic oxide which then becomes the bearing surface. Because these oxides have high resistance to frictional welding, they prevent punches from prematurely dulling through buildup and wear. Moreover, they preclude welding in draw dies, which is the major cause of galling and breakage.

The popular metal-forming compounds during the '50s were those petroleum-based oils containing fatty additives, sulfur, chlorine and/or phosphate in varied amounts and combinations. During the late '50s and early '60s, coolants were borrowed from the machining side of metalworking and adapted to metal forming. The advantage of these products was the ability to be mixed with water. Even though some of these solutions contained small percentages of fat and/or the reactive materials group to provide lubricity, they were generally used only as coolants for high-speed stamping. Today's highly sophisticated water-soluble drawing and stamping oils evolved from these early products.

The breakthrough precipitating this evolution occurred in the mid '60s. Until then, 8 to 12 percent was the maximum amount of chlorine that could be stably held in the emulsion. Combining several newly developed emulsifiers allowed an increase to 20 percent, thereby expanding the capabilities of water-soluble chlorinated oil for difficult deep-draw jobs.

Presently, other additives for increasing anti-wipe, lubricity and polarity characteristics, along with up to a 40 percent chlorine content, have advanced these products to the point where they successfully compete with all but the most potent straight oils and pigmented pastes. Such lubricants, in varying mix ratios, can do 85 percent of all the drawing and stamping jobs, **Figure 1**. They are the current leader in the die lubrication field and are rapidly replacing the small market share still held by the old paste and straight petroleum-based oil compounds.

Modern metal-forming soluble oils contain all the lubricity and extreme pressure additives normally found in straight oils. Diluent mineral oil is used only as the vehicle for holding the chemical package together until water is added by the user. (Note that the percentage of additives in a 4:1 mix is the same as a medium grade blended straight oil.)

The new kid on the block

Heavy-duty water-soluble stamping and drawing oils will be with us through the '80s and possibly the '90s. However, a new family of products, called water-extendable synthetics, is threatening to steal some of the application market.

...continued

2. Two test pins are shown after Falex testing. This test is performed on a machine that rotates a carbon steel pin between two V-block jaws submerged in a cup of the lubricant being tested. Each steel pin has a brass shear pin designed to break if the test pin and jaws weld together. Lubricants are rated by the amount of pressure that can be applied before the shear pin breaks, or the degree of scarring (wear) on the pin if a break does not occur. (Note: Maximum pressure that can be applied is 4500 lb.) The pin on the left was tested using a medium-duty chlorinated soluble oil, the pin on the right with a synthetic lubricant. Pressure was increased to the maximum over a 10-min period and then maintained for 1 min. It can be seen that the pin at left is more elongated and the scarring is deeper. In addition, its shear pin broke just as the pressure reached 4500 lb. In comparison, the pin at right continued to run through the entire test cycle. Although this test has little direct correlation to the drawing capabilities of a lubricant, it does give a good indication of the amount of load the lubricant will carry before failing.

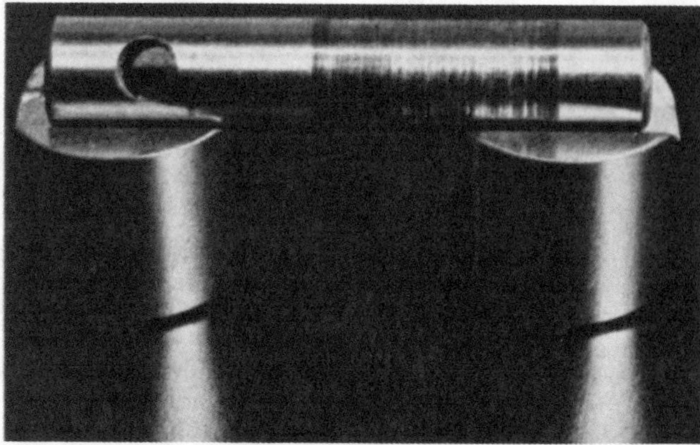

3. Cup drawing test conducted at a major engineering university compares the capability of several metal-forming lubricants. Top photo shows a 5.2" draw using a pigmented paste. Middle photo shows a 5" draw using a water-extendable synthetic lube mixed at 4:1. Bottom photo shows a 4.8" draw that failed when using a heavy-duty chlorinated soluble oil mixed at 3:1. Even though the pigmented paste produced the best results, problems regarding disposal make it an unfavorable choice.

They are usually soap-polymer-based solutions containing numerous high technology additives. Their performance range is still being probed, but initial tests indicate they do have muscle. For example, during a standard Falex test to determine film strength and ability to reduce friction under load, a synthetic drawing compound (mixed 5:1 with water) outperformed a 5:1 solution of chlorinated soluble oil, **Figure 2**.

Further verification of the capability of water-extendable synthetics in metal forming involved a standard cup draw test, **Figure 3**. Here, it outperformed a well-known heavily chlorinated water-soluble oil. The tests were conducted on 409 muffler stainless. Not only did the synthetic draw a larger blank, it did so at a 4:1 mix, compared to the standard 3:1 ratio for the chlorinated product. These test results indicate that synthetics meet the requirements for general-purpose stamping, forming and light-to-medium drawing. Successes to date include the drawing of kitchen range tops, flashlight battery cans, vacuum cleaner tanks and auto suspension components.

Energy conservation and pollution control regulations are adding impetus to the dissemination of synthetic lubricants in metal forming. For example, parts produced with a synthetic can usually be cleaned in a mild alkaline solution at less than 100 F. In some cases, they do not have to be cleaned at all because the synthetic may not interfere with subsequent processing (i.e., mastic bonding, welding, painting etc).

Furthermore, synthetic lubricants can offer additional cost savings when complying with pollution control regulations. In many cases industrial users are paying more to have spent water-soluble products hauled away for treating than the original cost of the compound. On the other hand, some synthetics carry a very good biodegradability rating, thus requiring little (if any) prerelease treating. Care should be taken, however, if there is a plan to dump directly into city sewers. Biodegradability ratings furnished by the manufacturer should be submitted to the water treatment department for approval. Also, care should be taken to separate tramp oils before dumping. The newest synthetics are formulated to kick tramp oil to the surface where it can be easily skimmed and drummed, **Figure 4**. Refer to the accompanying box entitled *Waste treatment of synthetic lubricants* for details regarding their disposal.

When considering the use of water-extendable synthetics, be sure to check the following:

- Contact a qualified supplier with the capabilities to do the job. The manufacturing and control of synthetics are not easy. These products require a high degree of technology and very close control to assure consistent performance and quality, **Figure 5**.
- The mating of the right product to operational requirements is half the battle. Field experience with synthetic lubricants is invaluable.
- Use the bottom-up approach, not top-down. There is a tendency to try a new product on the toughest job in the shop. This approach could cast an undeserved negative light on the product. First of all, it might not be the fault of the lubricant, but tooling, design or metal inadequacies. Second, the synthetic might not do the two or three tough jobs, but it could save time, trouble and money on the 15 or 20 jobs that are the bread and butter of the operation.
- Check the lubricant's compatibility with nonferrous materials because staining can occur if the pH is more than 8.5.
- Check carbide compatibility because cobalt leaching can occur if the pH is also greater than 8.5.
- Watch for foaming in heavily agitated recirculation systems.

Application is equally important

Another important factor in metal-forming lubrication is the application method. This is evident when considering how it relates to the way the lubricant performs. Use too little and performance suffers; use too much and costs become disproportionately high. Cleaning, disposal and housekeeping costs are also related to application.

To review the evolution of application equipment, let's start with swabs, rags and brushes. This type of hand application was improved with the introduction of the paint roller. Although originally designed for applying paint to wall surfaces, it was quickly adopted by industry for lubricant application and is still the

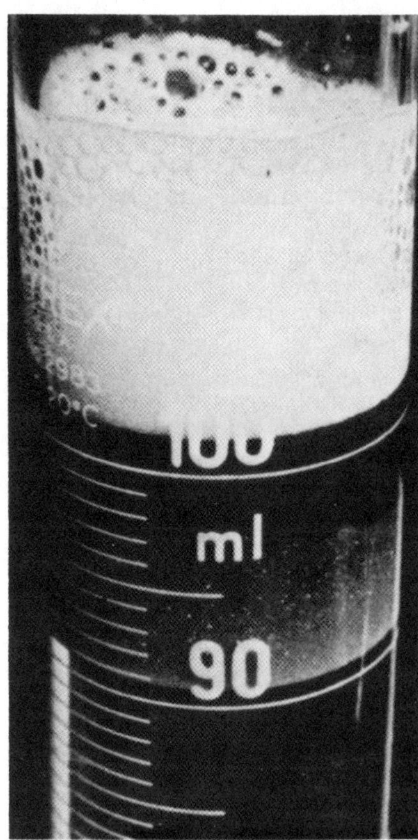

4. Illustration of tramp oil kicked out of a synthetic lubricant. The foam cap results from mixing the oil and the synthetic. Separation was so fast (5 min) that this cap did not have time to dissipate. In production, the separated tramp oil can be easily skimmed and drummed for disposal or recycling.

best hand tool for this purpose; however, hand application in medium- to high-production plants is the most expensive method available. When factors associated with hand application, i.e., over usage (waste), under usage (less die life), open containers (contamination), dependence on operators (Murphy's law) and housekeeping (mess), are considered, rather than just initial costs, this becomes apparent.

Improvement came with the introduction of the drip method. This was probably started when an enterprising operator found out he could beat the piecework rate by poking a hole in the bottom of a can, thus allowing the lubricant to continuously drip on the strip stock. The drip applicator obviously has come a long way since then. It has been combined with various accessories such as felt wiper pads and rollers to improve coverage, but they still have obvious shortcomings. Their small reservoirs must be refilled frequently by an operator. They are hard to control due to equipment vibration. They are mounted before the stock feed, causing work slippage during an in-feed, and they only provide single-point application.

Next came the development of roller coating, ranging from the inexpensive drip-feed type to the well-engineered precision applicator that applies liquids in controlled mill thicknesses. The recirculating types are a significant improvement in single-point application. The range is broad in both price and performance and is proportional to the size, automation and film control required.

For years the machining side of metalworking has used flood application of coolants/lubricants on high-production machines. Until recently, only a few specialty press manufacturers, such as Waterbury Farrel, offered flood systems as an option on their equipment. Now, other press builders are offering this option or are considering it for the near future. Flooding of the die area in high-production stamping or drawing offers one of the best methods of positive lubrication, heat removal and sliver or metal-fine flushing available.

Deep-drawn cup jobs, such as battery cans or lipstick tubes, that literally smoke on exiting a die when using a straight oil with single-point application are cool enough to hold after changing to a water-soluble product applied by flooding. Note, however, that when the water-soluble lubricant was tried with a single-point application system it was wiped off at the second station, resulting in tool galling.

Retrofitting with a flood system can be difficult, especially on a large press. Even though this is the case, numerous large automotive plants have adapted the flood system to operations used in making trim, hardware and suspension parts. They range from a few isolated presses using individual sumps to 27 presses operating from a large central system. Pay back on installation cost has ranged from 9 months to 18 months, based on average annual five-figure savings. These savings include eliminating

5. Here, a liquid chromatograph is used to analyze the chemical constituents of a synthetic lube to ensure quality and consistency.

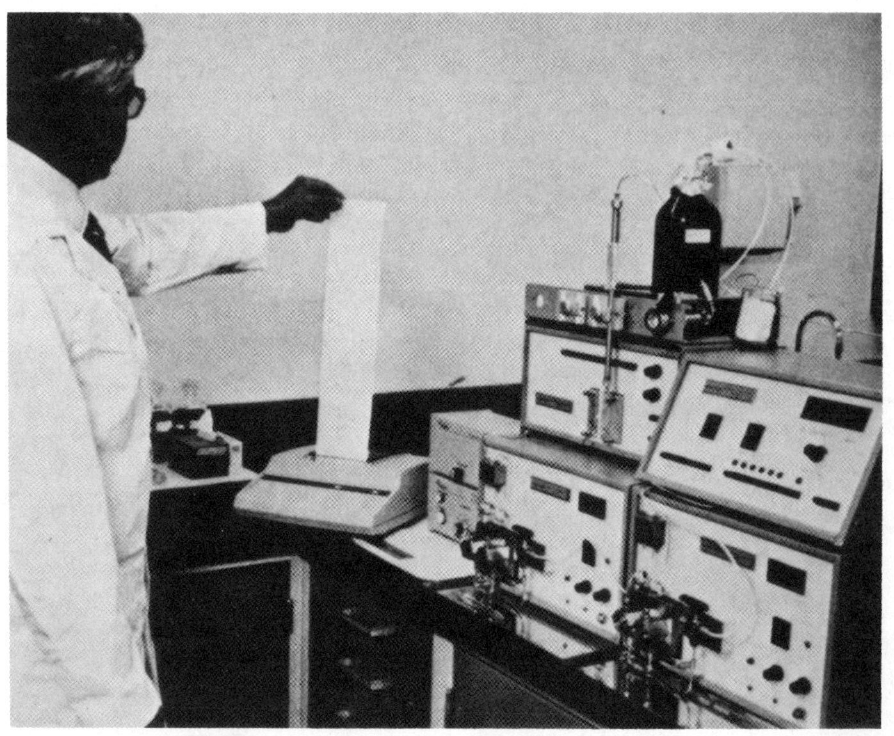

6. Photo at left shows an airless spray die-lubrication system from Pax Machine Co, Selina, OH. Each nozzle or jet has its own pump. The pump works similar to a diesel fuel injector, literally injecting the lubricant through the line and out the jet in a sequence synchronized with the press cycle. Timing can either be via a rotary cam switch mounted on the press crank or by an automatic device mounted in the unit. Filtration is also important to any kind of spray equipment. This unit uses a 75 micron bag-type filter with 300 sq in of surface. Photo above shows how positive die lubrication can be consistently achieved. Here, the jets have been permanently mounted in the die. Quick disconnects to the lubricator make the setup simple, plus assuring proper location of the spray pattern.

cleaning operations, reducing storage and distribution costs, and reducing direct lubricant usage. Although flooding is an ideal method of applying die lubricants at multiple points or die stations, it is definitely not for everyone.

There is a versatile lower-cost method of multistation application available. It is the airless spray system. Spray application has been available for many years, but in the past has created several problems. For example, when air is mixed with the lubricant in a venturi-type system, fogging and bounce occur, resulting in contamination of the surrounding area. Plant personnel complain and in some cases union problems develop. With the advent of the airless system these problems were solved.

Airless spraying was first developed and used for applying paint. These systems were large and expensive. In addition, they were designed to apply the finely controlled spray patterns necessary to achieve a good finish. It was quickly realized that this control was not necessary when applying an adequate lubricant film, thus opening the door for an off-the-shelf-type airless spray system designed for cost-effective application, **Figure 6.**

With today's high-speed presses, rate of lubricant delivery is definitely a significant factor to consider. The speeds at which the airless spray applicator units operate varies from 350 spm to 900 spm. At the high end of the range they appear to be putting out a continuous spray. Maintenance, especially at medium to high speeds, is also a factor that has caused problems in the past. For example, O-ring swelling and wear would produce pump failure. This was solved when certain systems incorporated an automatic oiler that fed air-line oil to the pumps for lubrication rather than depending on the die lubricant.

Of the types available, hand, drip, roller coat, flood and spray, only the last three should be considered in modern metal-forming operations (e.g., recirculating roller coat for single-point application and flood or spray for multipoint application). Moreover, lubricant and application method should be established during the job planning stage by tooling or production engineering. Unfortunately, this is not usually the case, and the responsibility falls on the shoulders of operations personnel.

In summary, consider the following points:

- Heavy-duty, water-soluble drawing and stamping compounds have been the most popular metal-forming lubricants in the '70s.
- Water-extendable synthetics will be the way to go in the '80s.
- No matter what lubricant is used, it is only as good as the application method selected.
- The ideal time to plan for both the lubricant type and application method is in the early stages of engineering.
- After the lubricant has done its job, there is still responsibility for disposal.

Solid lubricants can solve metalforming problems

Reprinted from Precision Metal, April 1977

Solid lubricants fulfill a number of functions in various metalworking processes, but their main job is the prevention of abrasive wear, galling and seizing.

By JAMES HEFFEL
Dow Corning Corp., Technical Service & Development Department

Most metal forming operations present classic illustrations of boundary friction. As a result, unless the contact area between die and workpiece is lubricated effectively, the result can be galling and tearing of the part and damage to the die.

Boundary friction is caused by the inherent roughness of any metal surface. Although they may appear to be smooth and polished, all metal surfaces consist of minute peaks and valleys. When two such surfaces first come into contact, the load is supported by only a few points. The actual area of contact is just a small fraction of the apparent area of contact and the true pressure at the points in actual contact is many times the apparent pressure. As a result the contacting peaks of the two surfaces may cold weld. Also, the mating peaks and valleys tend to interlock.

Both of these actions resist motion. The welded junctions tear apart and the peaks of the harder material plow through the softer material. The result is metal transfer from one surface to the other, abrasive wear, galling and seizing.

The resistance to motion can be reduced by lubrication. The lubricant reduces friction by separating the two surfaces with a thin film with little resistance to shear. In the case of oils and greases, this film is fluid. It is established by hydrodynamic forces as a function of pressure and relative velocity between the bearing surfaces.

With hydrodynamic lubrication when the pressure is too great, and/or the relative velocity between the two surfaces is too low, a hydrodynamic film cannot form and boundary friction occurs. To reduce friction, and the resulting wear, a boundary lubricant is required under these conditions.

Boundary lubricants

Most boundary lubricants are dry solids, such as graphite, molybdenum disulfide and waxes. These materials resist being squeezed out from between the bearing surfaces and maintain relatively low coefficients of friction at high loads and low relative velocities.

Moybdenum disulfide (MoS_2) functions to reduce boundary friction, as tiny flat plates adhere to a metal surface but slide freely against each other. When they have been distributed to form a smooth film on the entire surface, the layers of solid lubricant form a thin, slippery coating which protects the surface from wear.

Molybdenum disulfide also has a load-carrying capacity to withstand pressures beyond the yield point of most metals. In addition, its coefficient of friction actually decreases as load increases.

Solid lubricants

In many metalforming operations, pressures are almost always too high and relative velocities too low to establish a hydrodynamic film between the contacting surfaces. Liquid lubricants may be squeezed out to make contact, which can lead to metal pickup on dies, galling, scoring and tearing of the workpiece, and short die life.

Solid lubricants are also heat-resistant. Heat-resistance is important in hot forming, of course, but frictional heat may be high even in cold forming operations.

Molybdenum disulfide powder is stable to 750F (399C) in air, to 1500F (816C) in the absence of air. Dry lubricant coatings which will withstand 1200F (649C) service temperatures are commercially available. Graphite has an upper temperature limit of 1500F (816C).

Types of solid lubricants

Solid lubricants are available in a variety of forms: powders, pastes, greases, dispersions and coatings.

Properly applied solid powders, such as graphite, molybdenum disulfide and fluorocarbon, have excellent lubricating properties. However, initial application may be difficult. Also, powder lubricants have a finite life and reapplication may be difficult, although this is seldom a factor in forming operations.

Pastes are viscous compounds usually containing 60% or more of lubricating solids by weight. The fluid component simplifies application and is usually selected for its secondary characteristics, such as temperature, stability or compatibility with other materials.

Solid lubricant coatings can be described as paints with the pigment

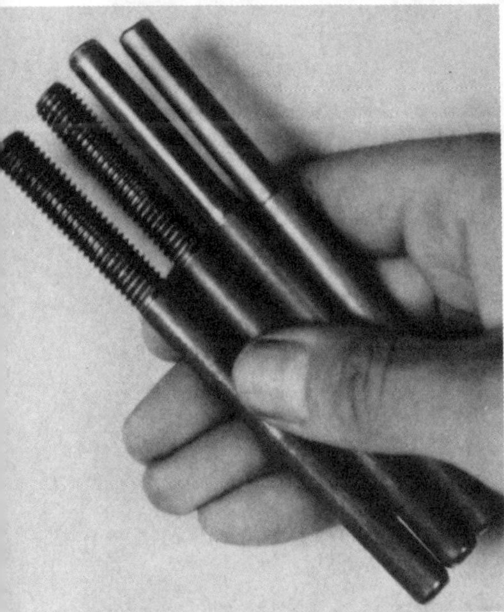

Fig. 1—Use of MoS_2 powder increased die life from 5000 to 100,000 parts in a cold heading operation of these Type 430 stainless steel parts. Wire is reduced from 0.375 in. to 0.358 in. diameter.

Fig. 2—Lubricated with Molykote® 557, the edges of aluminum and steel strip no longer are abraded by the hardened steel guide dies of this roll forming machine.

replaced by a lubricating powder. They contain a binder in a solvent system, and are available in either baking or air-dry systems. Some solid lubricant coatings have the added advantage of being clear and colorless as well as dry, eliminating the need to clean parts after forming.

Generally, the types of solid lubricants most useful in metalforming are the powders and coatings. However, dispersions in oil are sometimes used as additives in conventional drawing oils and both dispersions and greases frequently improve lubrication of presses and other equipment.

Boundary lubricants have improved results in a variety of metalforming operations, including drawing, stamping, heading, bending, forging and die casting. They have increased die life and improved surface finish. In one deep drawing operation, for example, the addition of Molykote® M dispersion to the drawing oil increased die life from 15,000 to 60,000 pieces.

Aluminum extrusion

A number of aluminum extruders have found that Molykote® 321 bonded lubricant coating increases the number of billets which can be extruded before a die must be polished. It also improves the surface finish of the extrusions. For example, Trim Alloys, Inc., Randolph, Massachusetts, has doubled the life of extrusion dies with this lubricant while operating at higher press speeds.

Fig. 3—A dry coating eliminated cleaning these extruded aluminum window frames after forming.

Mideast Aluminum Industries also uses the dry lubricant coating at their Mountaintop, Pennsylvania extrusion plant. It has been effective on bridge dies, porthole dies and lightweight flat dies. The lubricant improves surface finish and permits more billets to be run before the die must be pulled and cleaned. It also reduces breakthrough pressure on flat dies and helps strip bridge dies.

Extrusion dies are coated before being heated in the die oven. The coating is also sprayed on the front end of each billet before it is heated. This eliminates the need to relubricate the dies during the run and improves stripping efficiency, reducing dead cycle time by about 15 %.

Cold heading

Stainless steel is difficult to form. MoS_2 powder is one lubricant used by a number of shops to cold head stainless steel.

For example, Vico Corp. produces a part in 0.375-in. (9.5 mm) dia. Type 430 stainless steel wire pieces which is reduced to 0.358-in. (9.1 mm) dia. by heading. (See figure 1) The company tried a number of lubricants, including coating the wire with copper, but maximum die life was only 5,000 pieces. With MoS_2 powder, die life exceeds 100,000 pieces. The powder is applied to the wire in a draw box mounted on the draw bench.

Roll forming

Ro-La-Lume Corp. of America roll forms coated aluminum and steel strip into slate for awnings, window shutters, patio covers, carports and steel buildings. The company discovered that the hardened steel guide dies tended to abrade slivers of metal from the edges of the strip stock passing through them at speeds up to 100 fpm (30.5 mpm). (See figure 2) Carried into the forming dies, the slivers produced surface nicks and dents and damage to the coating on the strip. Frequently, less than 300 ft. (91.4 m) of strip could be formed before the machine would have to be shut down to clean out the accumulated shavings.

By spraying Molykote 557® on both edges of the coil of stock, the problem was eliminated and a complete 600-lb. (272 kg) coil could be formed continously.

Fig. 4—Colorless lubricant has doubled the life of steel punches used on aluminum bus bars.

Bending, punching & broaching

The clear solid film lubricant is also widely used in bending, punching and broaching operations, especially with aluminum. It has extended tool life and does not need to be cleaned from the finished part.

The Adams & Westlake Co. conservatively estimates that it has saved $7,500 annually by applying this lubricant on rollers and dies of hydraulic aluminum benders. The company fabricates windows for motor homes using complex aluminum extrusions as the frames, as shown in figure 3. The extrusions are bent as much as 120° with radii of 2 to 3 in. (5.08 to 7.62 cm). A lubricant is necessary to prevent marring or scratching of the 0.002-in. (0.005 mm) anodized finish. Clean, solid lubricant did the job without cleaning operations.

A (solid) lubricant has more than doubled the life of punches used to pierce ¼-in. (6.35 mm) silver- and copper-plated aluminum bars used as bus bars in electrical switchgear made by I-T-E Imperial, South Holland, Illinois, shown in figure 4. The press operator merely sprays the lubricant on the punches at the first sign of metal pickup, without cleaning the bars.

Success in the punching operation has led to use of the transparent lubricant in a number of other operations at I-T-E. It has eliminated scraping of the silver plating from the surface of the bars during bending. Taps require sharpening half as often and lubrication of self-tapping screws greatly speeds their assembly.

Lima Register Co. uses Molykote® 557 solid film lubricant in broaching aluminum tubing. The company assembles heating and ventilating grilles, registers and diffusers of anodized 6063T5 aluminum bars using aluminum tubing mullions. A broach expands the 0.312-in. (7.92 mm) O.D. mullions into 0.316-in. (8.02 mm) holes in each grid bar to hold the bars in place.

At first broaching tools broke so often that the tool room had trouble keeping up. Dipping the tubes into the clear, dry lubricant before broaching has solved this problem. The coating has other advantages, too. Lubricated mullions can be stored and need not be relubricated later, and the clear coating need not be cleaned from the assembled grids.

Fig. 5—Molykote® 557 prevented wax buildup and metal pickup on dies and core pins in a sizing operation.

Fig. 6—Forging dies which produce this connecting rod produced 36,000 parts without rework when lubricated with a eutectic-graphite blend.

Die casting

A small amount of Molykote® M-30 dispersion has eliminated sticking and breaking of die casting ejector pins at Valley Die Cast, Detroit, Michigan. The lubricant is also applied to guide and core pins. The liquid carrier vaporizes at high temperatures without forming any gummy residue. The coating of molybdenum disulfide which remains is stable at all zinc die casting temperatures.

One zinc die caster has found that applying a Molykote® 321 coating to the dies as a release agent greatly accelerates blooming the die. Previously it took several days to obtain a hardware finish on certain die cast parts; parts from coated dies give a hardware finish right away. The coating permits up to 400,000 parts to be cast before any buildup is noted on the die.

P/M sizing

Sizing, or coining, P/M parts increases their density and improves dimensional tolerances. However, sizing calls for a lubricant capable of easing ejection and preventing scoring and metal pickup at extreme pressures. If the lubricant is inadequate, parts are difficult to eject from the die and may even strip core pins right out of the press. These problems are especially severe with intricate parts and parts with large surface areas, such as gears.

Several powder metal producers have found Molykote® 557 clear coating to be effective as a sizing lubricant. At Sinterbond Corp., Gloucester, Massachusetts., it has prevented metal pickup and eliminated wax buildup on tungsten carbide dies and core rods; see figure 5. Parts are dipped to apply an even coating on every surface of the most intricate parts. The coating does not need to be cleaned from the finished part.

Hammer forging

Recently forging lubricants have been developed which use nonpolluting, nonflammable eutectic salt blends as the graphite carrier. The eutectic salts melt at forging temperatures to form a highly viscous fluid film between the heated die and hot workpiece. The graphite provides additional protection, should metal-to-metal contact occur. Supplied as a dry powder, it is diluted with water and applied by spray or swab.

The connecting rod shown in figure 6 is forged in a three-cavity die with six strokes of a 6000-lb. (53.4 MN) Erie hammer. Salt (NaC1) was the primary lubricant, although the die was occasionally swabbed with heavy oil. Use of the eutectic salt-graphite lubricant increased die life from 20,000 to 36,000 parts.

The eutectic salt-graphite lubricant was also tested for forging tractor track rollers made of SAE-1536 steel. The rollers are 10-in. diameter, 6-in. (15.2 cm) long and weigh approximately 17 lbs. (7.72 kg). The part is formed in a 3000-lb. (26.7MN) Erie steam hammer with three to five blows. Die temperature is 300F (149C) and billet temperature 2200F (1205C). The standard lubricant was No. 6 fuel oil applied by swabbing.

The salt eutectic-graphite lubricant sprayed on the die gave excellent forming and release.

Conclusion

Metal forming presents a classic case of boundary friction. Pressures are so high and relative motion so slow that a hydrodynamic film cannot be established to separate the surfaces of the workpiece and the tools. The result is galling and tearing of the work and metal pickup on punches and dies. These problems can usually be solved by using a solid boundary lubricant. The slippery solids form a protective film between the work and the tooling, protecting them from metal-to-metal contact, reducing rejects and extending tool life. Solid lubricants also lubricate effectively at higher temperatures than conventional oils and greases.

Solid lubricants come in a variety of forms. Dry powders and bonded coatings are the ones which are usually most useful in metal forming. Most forms are nonflammable and nonpolluting and some are also nonstaining and need not be cleaned from the finished part. **PM**

Reprinted from Tooling & Production, November 1981

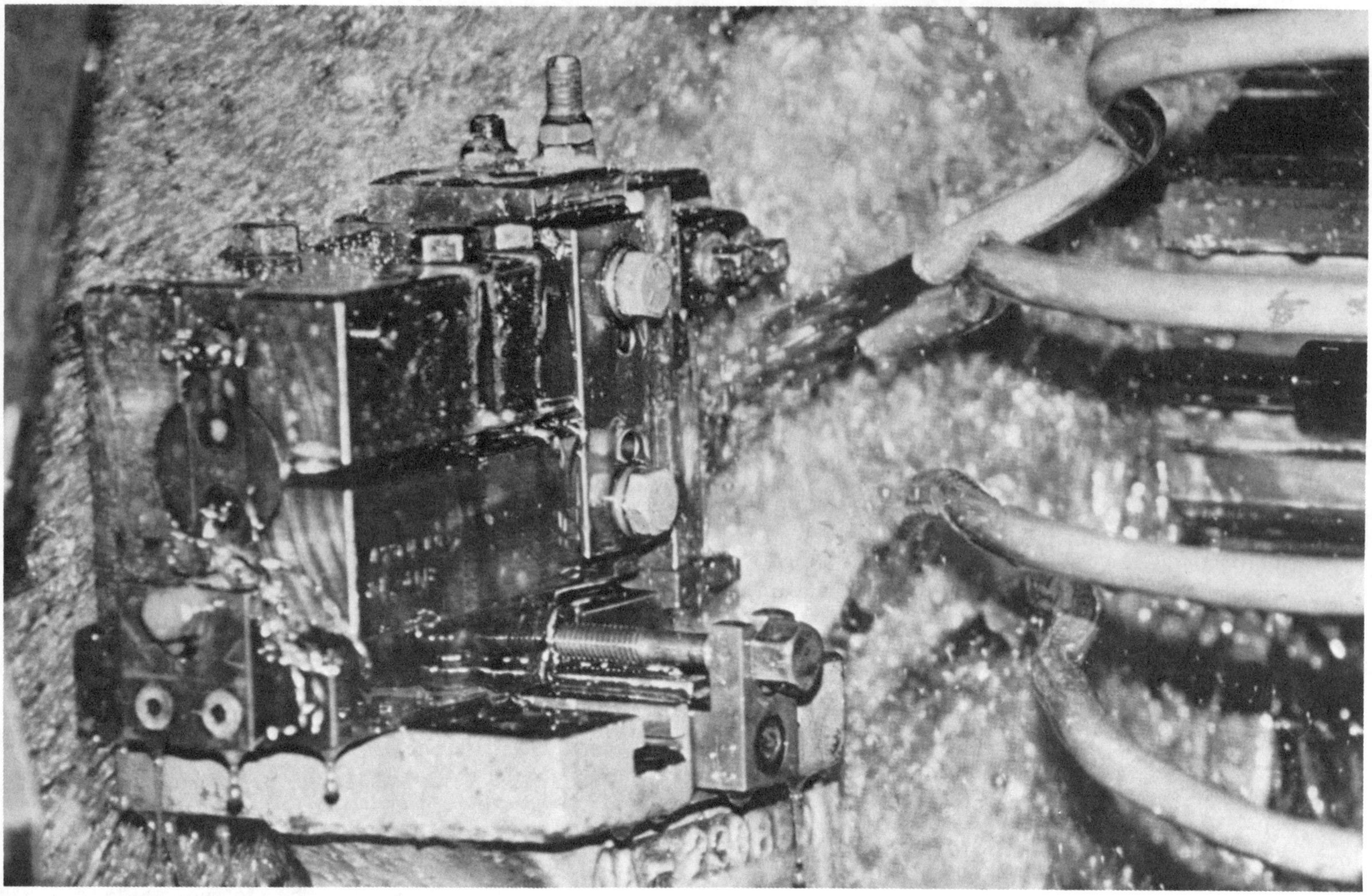

Turning

Elements of selecting and using metal-cutting fluids

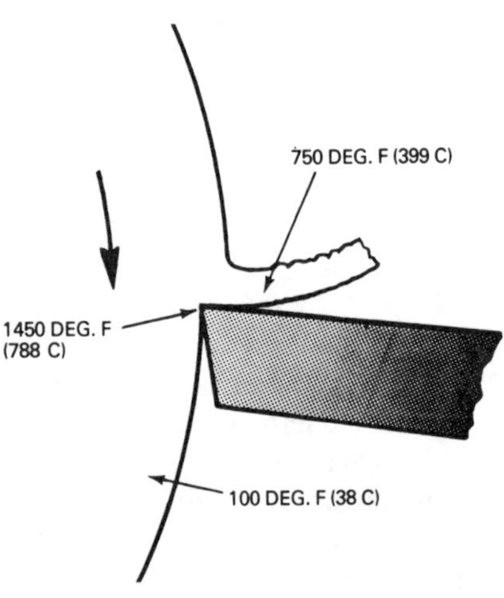

1. Temperature is highest at the tool tip during machining. These are approximate temperatures resulting from turning mild steel without coolant at 400 sfm.

One of the most prevalent opportunities for increasing productivity in metalworking—one that is often overlooked though it seldom requires additional capital investment—is the selection of the best cutting fluid for the job. Using the proper fluid can increase production by permitting increased feeds or speeds or both. Moreover, the right fluid can reduce the number of machining operations by improving workpiece surface finish. It can also lower tooling costs by increasing the number of parts per tool grind, thereby reducing the quantity of tool purchases as well as downtime for tool changes.

by H F Weindel
Senior Engineer
Mobil Oil Corp
Fairfax, VA

Why they are used

The functions of cutting fluids are to remove the heat generated in a metal-cutting operation, reduce friction and tool wear, prevent chip welding and control the built-up edge on tools. They also protect work against rust and wash away chips from the cutting area.

The major cause of the high temperatures associated with metal cutting relates to deformation of the metal by the cutting tool, **Figure 1.** This accounts for 60 to 70 percent of the generated heat. When cutting fluids flow over the workpiece, heat is transferred to the fluid and carried away. The remainder of the heat is generated by the friction between the tool and the chip or the tool and the workpiece. Cutting fluids reduce this by lubricating the surfaces. Reducing heat can have an immediate beneficial effect

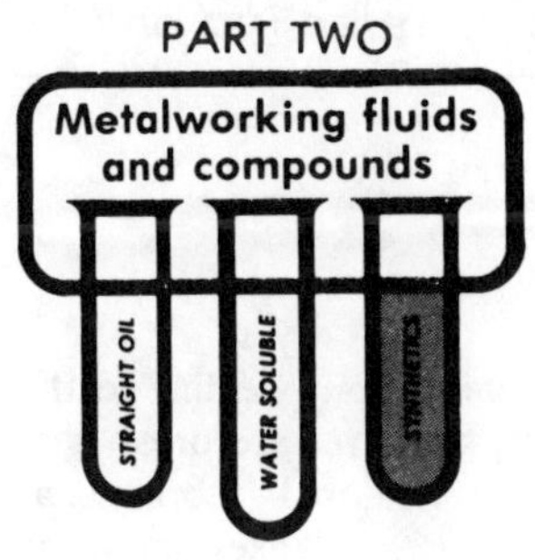

on tool life. In some operations, reducing tool-tip temperature by as little as 25 F can result in a 100 percent increase in tool life.

In addition to reducing friction and controlling heat, the cutting fluid must also control the site and mobility of the built-up edge on tools, especially when ductile metals are machined. The built-up edge is a small deposit of chip material riding behind the tool's cutting edge, **Figure 2**. Sufficient buildup must be maintained to focus the rubbing action of the chip an appreciable distance from the cutting edge. When too much buildup occurs, however, some of it may slough off between the tool and the workpiece, thus marring the finish.

The built-up edge is controlled through additives in the cutting fluid which, at the temperature encountered at the cutting edge, react chemically with the metal to interfere with the welding action that forms the buildup. A by-product of reducing the welding between the chip and tool is reduced cratering at the tool's cutting edge and increased tool life.

The cutting fluid should also lubricate boundary areas at the tool nose, side of the tool and just below the cutting edge. While these areas are not subject to the same rubbing as the tool tip, inadequate lubrication can result in excessive wear. Lubricity additives will protect these boundary areas by adhering to them, either chemically or by polar attraction to form a protective film, even where there is no fluid.

In addition, since the freshly machined surfaces of ferrous metals are particularly susceptible to oxidation, the cutting fluid should provide protection against rust at least while the parts are being processed. Also, in many cases the fluid must not stain or corrode the metal being machined.

Another function of the cutting fluid is to flush away chips that could clog the tool, damage the workpiece or otherwise interfere with metal cutting. This is especially important in operations where chips tend to clog such as deep-hole drilling. When this happens, the drill may be forced off center, and chips can wedge between the work and the drill tip, thus preventing cutting.

Furthermore, removal of the chip is a significant factor in reducing heat buildup. This is because much of the generated heat is located in the chip. When chips are not removed immediately, they can transfer this heat to the tool, resulting in greatly reduced tool life.

Cutting fluid types

All cutting fluids are of two basic types: mineral oil based (cutting oils) and water miscible. Cutting oils are petroleum products of various viscosities that contain suitable additives. These oils are of two general types, staining or nonstaining, determined by whether they stain copper and other nonferrous metals. This staining is due to the activity of certain anti-weld additives. Water-miscible fluids are of two general types—emulsifiable (soluble) oils and synthetic or aqueous coolants. (Note: The term semisynthetic covers the gray area between these types). Emulsifiable oils are supplied as concentrates that are mixed in-plant with water to form oil-in-water emulsions. Synthetics are true chemical solutions that contain no petroleum and are also supplied as concentrates.

Mineral oils contain high degrees of anti-weld and lubricity additives. They are used in moderate-speed operations with high-speed steel or carbide tools where finish, accuracy and tolerance are important. Water-miscible fluids are used in high-speed operations with carbide or ceramic tools, where high metal removal rates are encountered and cooling is the major consideration.

Additives are used in cutting fluids to improve performance. For example, typical anti-weld additives include chlorine, phosphorus and sulfur. These are used in a variety of compounds, either singly or in combination, to achieve the proper anti-weld activity, plus control built-up edge. The additives react with the metal to form metallic compounds

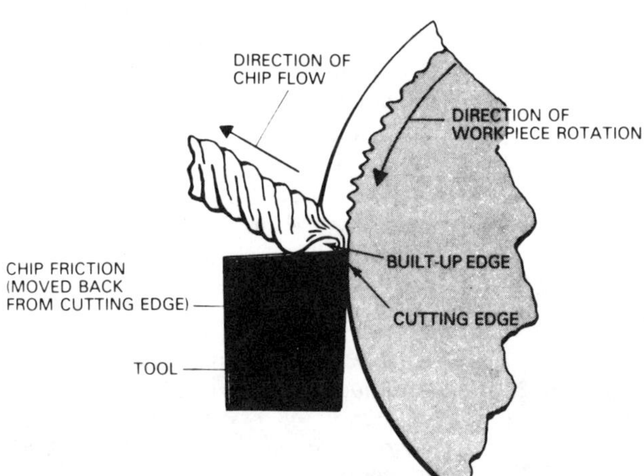

2. Illustration shows the position of the built-up edge slightly behind the tool's cutting edge.

...continued

that prevent chips from welding to the tool. Lubricity additives are often fats. These offer exceptional lubricity at lower temperatures. Waxes may be substituted in fluids subjected to higher temperatures.

Some cutting oils also have additives that reduce the misting of flowing oils, and some water-miscible fluids contain additives to retard the growth of bacteria. Refer to the box entitled *Classifying cutting fluids* for further detail regarding cutting fluid subcategories.

Factors affecting selection

The selection of a cutting fluid is determined primarily by the properties of metal being worked and the machining operation. Another factor is cutting speed, which depends on the metal and operation.

Ductility is an important property in determining the kind of chip formed and the amount of built-up edge. Brittle metals form discontinuous chips while ductile and semiductile metals form continuous chips, **Figure 3**. As ductility increases, the amount of built-up edge increases. This requires a fluid with better anti-weld properties.

The hardness of the metal is also an important property. As hardness increases so does tool wear. Here, lubricity properties are required.

Almost as important as the type of metal is the severity of the machining operation in cutting fluid selection, **Figure 4**. Turning is the least severe operation, even though the tool edge is almost continuously exposed to the generated heat. In this operation, heat is removed easily by the cutting fluid and the unconfined flow of the chip. Under commercial manufacturing conditions, where high speeds, feeds or depth of cut are required, or where finish is critical, cutting fluids are required for cooling, lubrication and good tool life.

Milling is slightly more severe than turning, because of chatter, shock loading or finish. Consequently, the major concern is also cooling, but with some need for lubricity additives.

Drilling is an even more severe operation because the tool edges are buried in the workpiece. There is little opportunity for cooling by radiation or a large volume of cutting fluid. In addition, chips are forced to move along the shank of the drill and transfer much of the heat to the tool. Here, fluids with good lubricity and anti-weld properties are required.

In deep-hole drilling, when the fluid is delivered under pressure, through the drill stem, more effective cooling and flushing are possible. The most suitable fluid is a light cutting oil with good lubricity.

Boring is a turning operation, but is about as severe as drilling. Somewhat stronger lubricity and friction-reducing properties are needed in the cutting fluid than for turning.

In gear cutting, the tool is subjected to great shock as it comes in contact with the gear blank. In addition, escape of the chip and control of the built-up edge are difficult. Consequently, the cutting fluid must have strong anti-weld properties to control built-up edge, and high lubricity characteristics to protect the cutting edge from rapid wear. These same properties also work to insure accuracy and proper surface finish.

Cutting fluids used in thread-cutting operations, i.e., both tapping and chasing, should have especially strong anti-weld and lubricity characteristics because of the severity of the process. The many small cutting teeth with their sharp points are difficult to protect from ex-

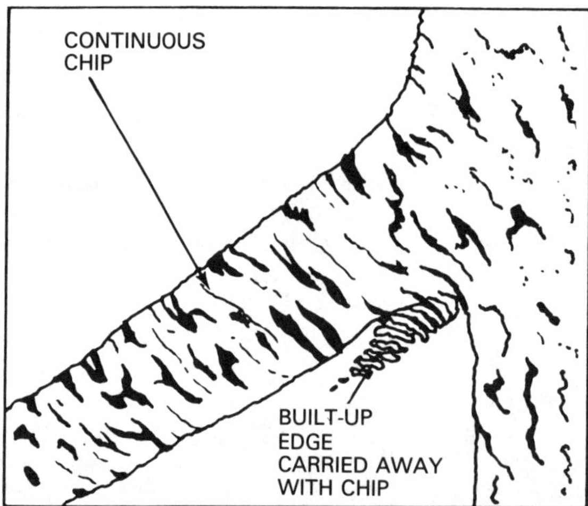

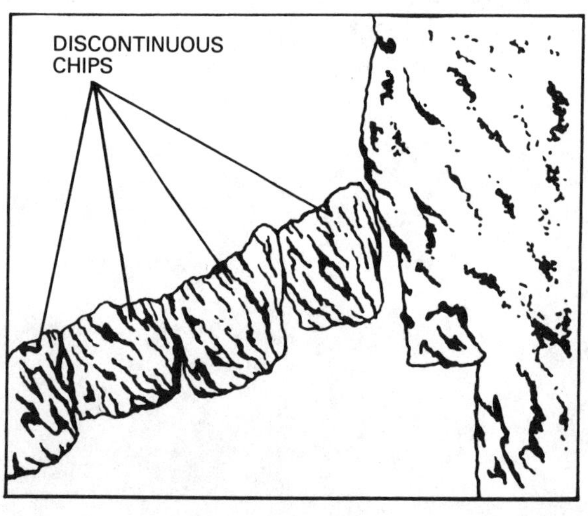

3. Type of chip formed is determined by the metal. Brittle metals form discontinuous chips with little or no built-up edge, while more ductile metals form continuous and increasing built-up edge.

cessive wear, burning or destruction. Tapping operations are especially difficult because the tool is buried in fairly hot metal with little opportunity for cooling.

Internal broaching is the most severe metalworking operation. Each cutting tooth is in contact with the workpiece during the entire cut, and each succeeding tooth is buried deeper in the cut. This makes it difficult for the cutting fluid to reach a tooth during machining. In addition, the chip is confined between the work and the tool.

Cutting fluids used in broaching must have maximum anti-weld and lubricity properties to protect cutting teeth and insure proper finish and accuracy. Because of the difficulty in getting fluid to the teeth, more viscous fluids are used so they will cling to the teeth, providing lubrication during the entire cut.

Cutting speed, while dependent on the metal and the machining operation, is also important in the selection of a cutting fluid because of its effect on the rate of heat generation. At low to moderate speeds the anti-weld and lubricity properties of a cutting fluid are most important. Cutting oils have been shown to be the best in these applications.

These speeds can also be affected by the type of composition of the cutting oil used. Some oils permit high cutting speeds with little or no effect on tool life or finish, permitting higher production.

At high speeds, heat builds up faster and cooling becomes more important. The built-up edge is smaller because it is carried away by the chip, **Figure 5**. There is also little time for the cutting fluid to reach the tool tip and form anti-weld materials. Further, high-speed machining operations generally use carbide or ceramic tools. In these operations, use of water-miscible fluids permits better heat removal.

The type of machine tools in a shop also plays a role in cutting fluid selection. For example, with some equipment it is difficult to prevent lubricating oil from leaking into the cutting fluid system, or cutting fluid from leaking into the lubrication system. In situations such as these, dual-purpose or tri-purpose cutting oils have been developed that can be used in both systems. These are noncorrosive and have high friction-reducing properties. Leakage does not then adversely affect either machining operations or machinery operation.

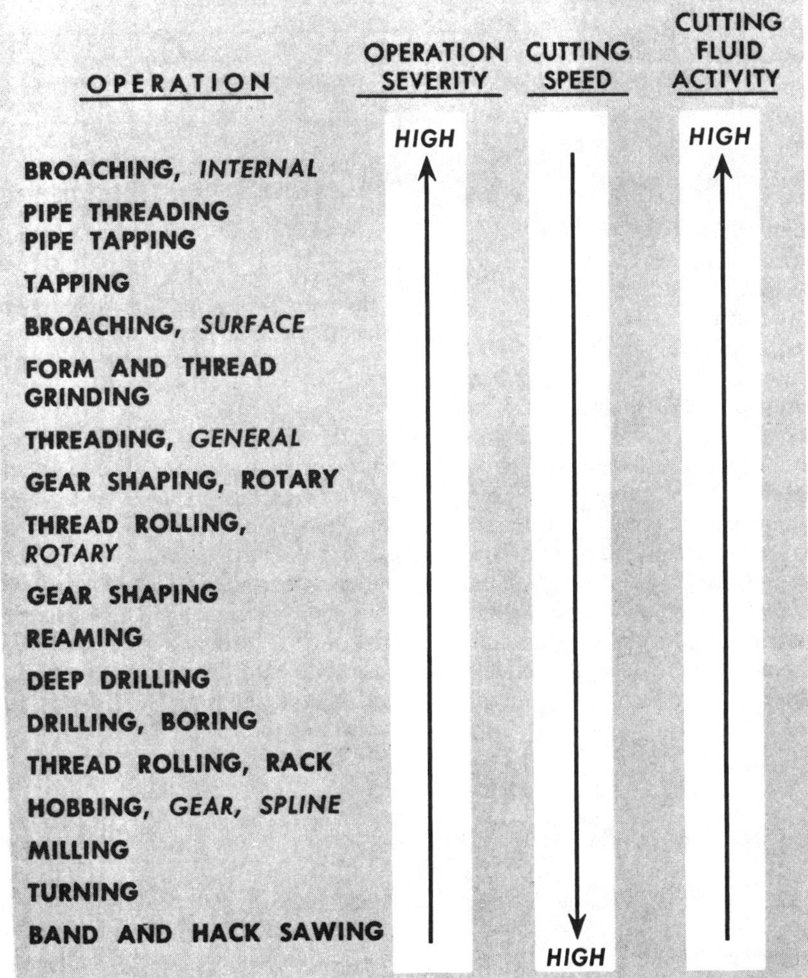

4. The above chart shows the relative severity of several machining operations and their effect on cutting speed and cutting oil activity.

To achieve this compatibility, however, some compromise in the anti-weld properties must be made, so these oils are generally not as active as the staining-type oils. They can be used as lubricating oils in severe operations where the problem is lubricating oil leaking into highly active cutting oil because they do not significantly reduce the machining efficiency of the more active oil.

Proper handling

Improved handling of cutting fluids can have a profound impact on their performance. Indeed, improper handling can conceivably negate potential benefits expected from the careful selection of a fluid.

Proper handling can provide several benefits:

. . . continued

• Decreased labor for application and makeup.
• Better control of fluid condition.
• Cooler cutting fluid which improves heat control.
• Reduced incidences of dermatitis.
• Lessened chances of applying the wrong fluid.
• Longer cutting fluid service life.
In addition, a good cutting fluid handling program should include fluid application and fluid cleanliness.

To be effective, cutting fluids should flood both the tool and the workpiece. In most cases one or more low-pressure streams are preferable because they provide large volumes of fluid without excessive splashing. A good rule of thumb is to supply the same number of gallons of fluid per minute as the maximum horsepower required for the cutting application.

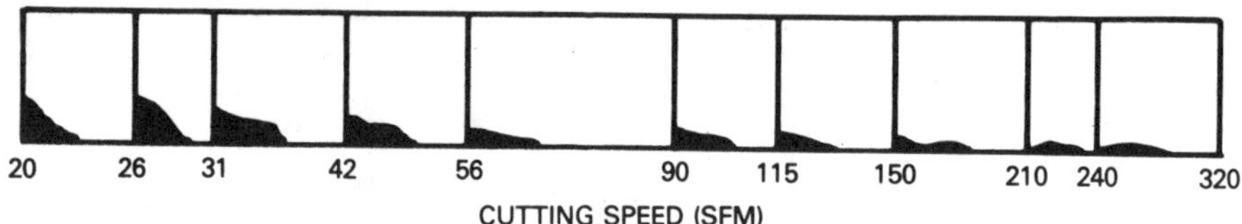

5. The size of the built-up edge that forms when machining ductile metals decreases as cutting speed increases. At high speeds buildup disappears. This is why increasing cutting speed and decreasing feed often results in improved finish.

Just as important as the amount of fluid supplied is where it is applied, **Figure 6**. In turning, the largest amount of fluid should be directed to where the metal is deformed and the chip formed. In milling, the stream should be directed at the point of cutting and on each side of the tool. In drilling, tapping and internal broaching, the stream should be directed into the hole. If the hole is large and fairly deep, a second stream should be directed on the other surface of the workpiece.

Cutting fluids, as they drain back to machine reservoirs, pick up and carry a considerable amount of metal fines. When these contaminants are recirculated back to the work, they can scratch the workpiece, thereby causing a poor finish. They also increase tool and pump

Classifying cutting fluids

The two general categories of cutting fluids can be broken down further, depending on type of additive and application.

Cutting oils

Within the broad classification of cutting oils are the following types:

Straight, nonadditive mineral oils are occasionally used for light-duty machining of free-cutting ferrous metals and nonferrous metals where requirements are not severe.

Mineral-fatty oils are commonly employed in automatic screw machines where operations are not especially tough, where high surface finish is desired and where staining cannot be tolerated.

Chlorinated mineral oils are particularly suited to machining soft, draggy metals with low machinability ratings, and in operations such as drilling, turning and milling stainless steel.

Sulfurized mineral oils are useful for mild to severe machining (e.g., drilling, turning, threading, broaching) on tough metals of high ductility.

Sulfurized-fatty mineral oils possessing both anti-weld and lubricity properties will provide good finishes on metals that are hard to cut and where tool vibration is normal. Parts made of tough, draggy materials causing high cutting pressure characterize such an application.

Sulfo-chlorinated mineral oils with high anti-weld action are required for broaching and threading soft, draggy steels that have a great tendency to develop built-up edges.

Phospho-sulfurized mineral oils comprise a group of corrosive and noncorrosive oils possessing a wide range of properties. These multimetal oils have supplanted many of the previously mentioned oils that have a comparatively narrow application range.

Synthesized additive oils are a group of oils incorporating a new additive technology. These fluids have additive packages that act synergistically; i.e., each additive increases the effectiveness of the others. Consequently, conditions at the cutting edge are dramatically modified to reduce friction between the tool tip and the workpiece. Improved performance is claimed when machining all types of metal under a wide range of pressures, temperatures and operations. Tools last longer with the new oils because they remain cooler with less burn and wear. Machining rates may be increased, thereby converting the benefit of longer tool life to greater productivity. Their increased lubricity also reduces the energy consumption in the machining operation.

Water solubles

The water-miscible fluids fall into four general types:

General-purpose soluble oils are emulsifiable oils that contain limited amounts of lubricity and anti-weld additives for use in light-to-moderate applications.

Heavy-duty soluble oils contain substantially higher levels of anti-weld and lubricity additives for heavy-duty machining operations.

General-purpose aqueous coolants are water solutions with limited amounts of wetting and lubricity additives, used largely in honing and grinding operations.

Heavy-duty aqueous coolants contain larger amounts of anti-weld and lubricity additives, along with some mineral oil to enhance machining performance.

wear, and can shorten fluid life, especially in soluble oils.

These contaminants can be removed by several means. Filtration is the most common, and easiest. However, unless two stages of filtering are used, the filters need ample volume to avoid too frequent replacement or cleaning.

Coarser metal particles can be removed by settling. This can be speeded up through the use of centrifuges, which can also remove small particles that are not readily removed by settling alone.

Oil reclamation

For shops using cutting oils, a considerable quantity is lost each day in the chips shipped out the door. Depending on the viscosity of the oil used, this lost fluid can be as much as 50 gal of oil for every 1000 lb of chips. Reclaiming this oil can significantly reduce costs, plus removing it usually results in a better price for the chips from scrap dealers.

The first step to reclaiming is getting the chips from the machinery to the reclamation area. This can be done by simply loading them into a wheelbarrow. Larger shops may use a centralized conveyor system. Where different types of cutting oils are used, it is important to keep the chips separated to prevent mixing the oils. An alternative is to use only one fluid in-plant.

The most effective means of extracting oil from chips is with a centrifuge. While the oil can be reclaimed simply by allowing the oil to drain off by gravity, a centrifuge permits quicker and more complete oil removal. Oil removal may also be hastened by breaking up the chips in a crusher before settling or centrifuging. In addition to reducing the volume, the crusher also breaks up long turnings so they will be easier to handle.

A good reclamation program can also reduce oil costs by cutting the amount of fluid lost through leaks in equipment. This oil can be collected and reused after it has been cleaned up, along with the oil reclaimed from chips. It is most important, however, that all reclaimed oil be cleaned before it is returned to service. The same measures used to keep oil clean in machines can also be used on reclaimed oil.

Reclaimed oil cannot be reused indefinitely because additives are continually lost through service or during reclamation. A fluid analysis program will provide valuable information on the fluid condition and additive strength. Often, reclaimed fluid with lower additive levels can be used in less exacting operations. In many cases, the addition of a new fluid as make up is adequate to restore reclaimed oil to its original condition.

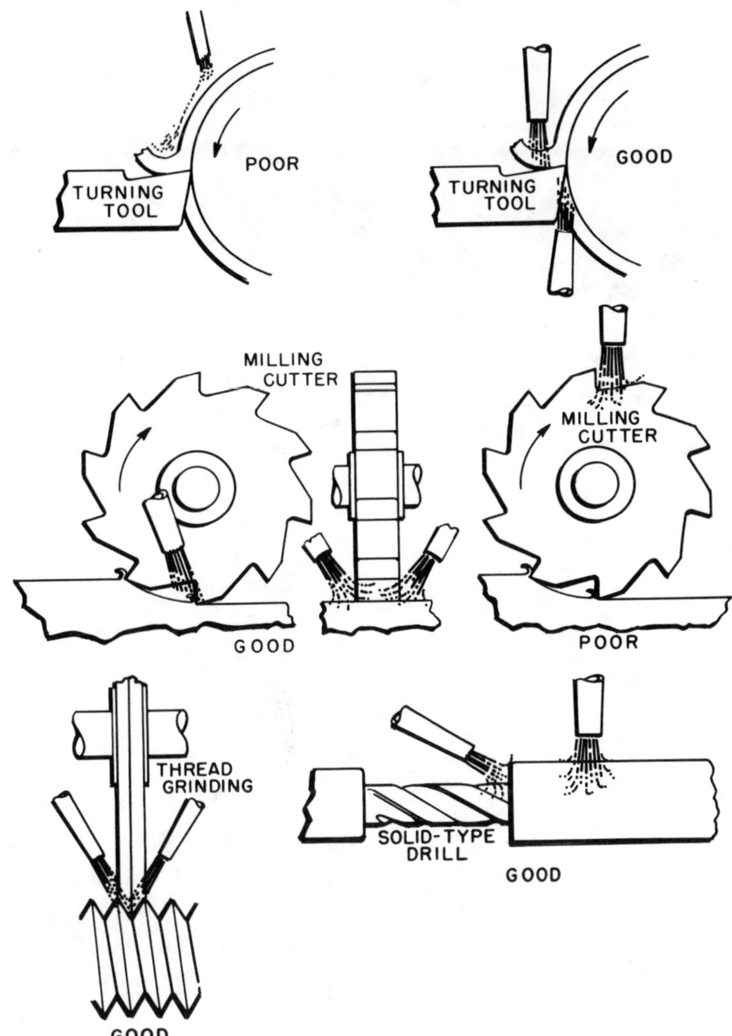

6. This illustration shows proper and improper cutting fluid applications. Cutting fluids should be applied so they will reach the areas to be lubricated and cool the hottest areas most effectively without causing undue splashing.

A comprehensive cutting fluid program, including proper selection, handling and reclamation, can be a vital influence in increasing profits by reducing machining costs. Such a program can reduce the cost per piece by permitting increased feeds and speeds, decreased tool wear, improved surface finish and fewer rejects.

Reprinted from Tooling & Production, February 1982

Synthetic metalworking fluids: A CLOSER LOOK

Synthetic coolants/lubricants are not snake oil. In fact, they contain no oil. Significant advancements have been made concerning this fluid technology and a closer look should dispel the myths.

Synthetics are water-based fluids consisting of chemical lubricating agents, corrosion preventatives and other additives. They are distinguishable from soluble oils in that they range from clear to translucent solutions rather than opaque emulsions. When using a synthetic an operator can actually see the cut; vision is not obscured as with an emulsion.

by Harris R Vahle
VP Operations
Pillsbury Chemical & Oil Inc
Detroit, MI

The fluids can be used for all metal-cutting operations: drilling, milling, turning, broaching, reaming, grinding etc, **Figures 1** and **2**. They are also being used in high-speed stamping and deep drawing. Further, they can be formulated for use with a wide range of metals (both nonferrous and ferrous) such as aluminum, stainless, titanium, brass, nickel, Monel, magnesium etc. Any material that can be cut with an oil can be cut with a synthetic.

Unfortunately, there are no industry-wide specs for synthetic fluids. The technology is new and manufacturers are keeping their formulations secret to maintain a competitive edge. The fluids are supplied as concentrates that are mixed with water in the user's plant. Dilution depends on the application and typically ranges from 10:1 to 60:1. Dilution does affect the extreme-pressure lubricity of the fluid, so heavier cuts require a higher concentration. Refer to the dilution chart for typical ratios used for metal-cutting operations.

Syn versus sol

Synthetics generally exhibit good stability under all water conditions, whether hard or soft. Such metalworking fluids are true chemical solutions—all of the elements are uniformly distributed. Because they are one-part fluids, lubri-

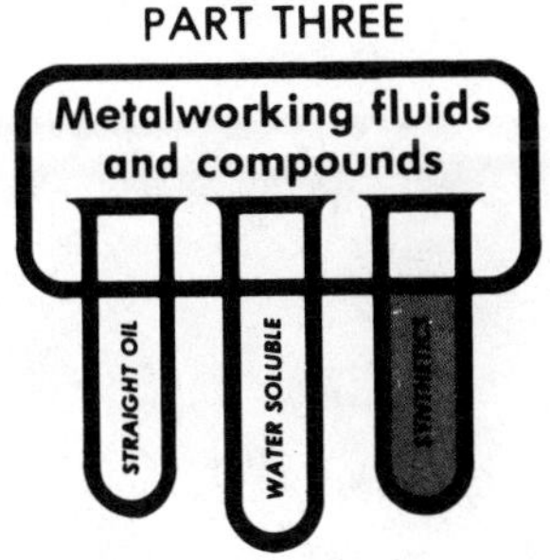

cation and cooling occur simultaneously. This improves heat transfer from both workpiece and tooling, while helping to produce a better cut.

In contrast, soluble oil (an emulsion) is a two-part fluid consisting of oil droplets suspended in water. When an oil droplet passes by the tool's cutting edge, it lubricates; when water passes by, it cools. No matter how fine the droplet structure, there is always this slower two-phase action.

Some synthetics have better extreme-pressure (EP) lubrication capabilities than most oils (i.e., higher film strength and lower coefficient of friction). Sulfur and chlorine are typical EP additives found in oil and are not appropriate for synthetic fluids. Rather, the EP additives used are generally phosphoric compounds, e.g., phosphate esters.

Because of superior cooling and lubricity, tool life can typically be increased from 20 percent to 250 percent in machining applications by using a synthetic fluid.

These fluids also provide detergent (self-cleaning) action. They wet and penetrate very well to break up and dislodge dirt. This helps to keep the coolant system clean. In some cases this penetrating characteristic can lift paint from machine tools. Usually this problem is caused by either inadequate surface preparation (e.g., dirt under the paint) or improper paint selection. In most cases, an epoxy-based paint is recommended. A good-quality paint placed on a clean surface will not exhibit peeling or bubbling when exposed to synthetic fluids.

Although detergent action suggests sudsing or foaming, synthetics can have excellent foam resistance. However, not all synthetics are foam resistant. Antifoam additives (usually silicone-based materials) offer only temporary control requiring that they be added on a regular basis. The newer synthetics are formulated so they have inherent foam resistance. Even for high-pressure or high-agitation operations such as gundrilling, these fluids provide better control over aeration than a petroleum oil. This is because it takes less time, generally, for an air bubble to escape from such a synthetic than it does from an oil.

Another potential drawback regarding detergency is the washing away of petroleum oils used to lubricate machine slides and ways. Although this is true, with good synthetic products, it is not a problem because they can provide multi-purpose capability. Several manufacturers offer dual-purpose fluids that may be used as both a coolant and a high-water-base hydraulic fluid. More significant is the recent development of a *quad-purpose* fluid that can be used as a coolant, a gearbox lubricant, a slide-and-way lubricant, and a high-water-base hydraulic fluid. As a coolant and hydraulic fluid, the concentrate is reduced with water—95 percent water and 5 percent solution. As a gearbox or slide-and-way lubricant, it is used undiluted. In fact, one plant is using only a synthetic to provide lubrication for slides carrying 2600 lb fixtures from one machining station to

1. Although in the early stages of production, this transmission valve body is machined with a synthetic fluid. The material is SAE 308 aluminum. There are 15 multiple-diameter step bores gun reamed to within 0.001" tir. They range from two diameters up to six diameters per bore; microfinish is 80 rms maximum. In the past, the industry-standard for gun reaming this part was mineral seal oil. The fact that it can be successfully machined when using a synthetic is an outstanding technical achievement.

2. This input carrier (made of 1010 steel) is manufactured by a major automotive company. The component is turned, broached, reamed, drilled, bored and burnished, all with a synthetic fluid. Note: 1010 steel is very soft and, thus, difficult to machine.

...continued

Typical dilutions for synthetic metalworking fluids

Material	Application	Dilution
Cast iron	Drilling, tapping	10:1 to 20:1
	Turning, milling	20:1 to 40:1
	Centerless grinding	15:1 to 30:1
	Surface grinding	25:1 to 40:1
Steel	Tapping, broaching, sawing	10:1 to 40:1
	Drilling, milling, turning,	20:1 to 50:1
	Centerless, ID and surface grinding	30:1 to 60:1
Aluminum	Tapping, broaching, sawing	10:1 to 20:1
	Drilling, milling, turning	15:1 to 30:1
	Centerless, ID and surface grinding	20:1 to 40:1

another. Such an all-purpose fluid helps to uncomplicate the machine tool user's life.

Cobalt leaching is a problem associated with using synthetic fluids. The result: degeneration of carbide tooling and carbide products. The problem can be avoided by inclusion of an inhibitor in the fluid to preclude ionic exchange.

Also note that metals susceptible to white corrosion will sometimes stain when using water-based fluids. For example, zinc or aluminum will react with water and spotting will occur. If a synthetic fluid is formulated properly, however, it deposits a light protective coating that prevents water from reacting with (and staining) the metal. Another type of staining is caused by chemical selection. Again, if the fluid is properly formulated for the application, staining will not occur.

Because of advancements in synthetic fluid technology, there are few applications where petroleum-based fluids are preferred. Currently, straight oils may offer an advantage in extremely severe operations such as broaching. Inroads have been made, however, and synthetics are being successfully used in some broaching jobs previously performed with heavy-duty straight oils.

Finally, synthetic fluids eliminate smoking during machining and reduce coolant misting. Since oil is an allergenic irritant, misting can promote respiratory problems and dermatitis. In some plants air filtration equipment must be used. In contrast, a synthetic fluid doesn't become airborne as readily as an oil because it has a heavier specific gravity. If misting should occur it is more acceptable because the fluid is not an irritant.

Long life, less cost

The biggest factor causing hesitation about using synthetic fluid is erroneously perceived high cost. The initial price per gallon for a synthetic is higher than the price per gallon for oil. However, when considering the long run—the cost of the fluid over its useful life—synthetics become extremely cost effective.

A comparison by a major automotive company shows that the longer life of synthetics actually provides a cost advantage over oils. In this study, the average service life of synthetics was determined to be four times longer than for oils—two years versus six months (although with proper maintenance synthetics can last indefinitely).

Since the synthetic lasted two years, the annual cost to fill the system was one-half initial cost. On the other hand, since the oil lasted six months, the system was refilled twice per year. Therefore, annual cost to fill the system was two times initial cost. Also, recharge/cleaning costs for the system using the synthetic fluid was one-quarter that of the system using soluble oil, and only one-quarter as much fluid had to be waste treated. The net result: synthetics cut yearly coolant system costs by 24 percent.

The main reason for synthetic cutting fluid longevity is resistance to tramp oil contamination. Tramp oils will alter the droplet structure of a soluble oil. This depletes the fluid's emulsifying ability, making the droplets larger: The coolant will then operate less effectively.

On the other hand, all oils entering a system charged with a synthetic fluid are completely rejected by the solution, thus preventing contamination. Here, tramp oils float to the surface, where through use of a simple skimming process, they can be recovered and reused.

Also, it's interesting to note that a synthetic does not provide a food source for bacteria while the emulsifiers, fats etc, in soluble oil do. Bactericides will control bacteria growth; however, these agents in soluble oils have to work im-

3. One of the services a synthetic coolant manufacturer can provide is microbiological testing to help a user monitor and control bacteria and fungus levels.

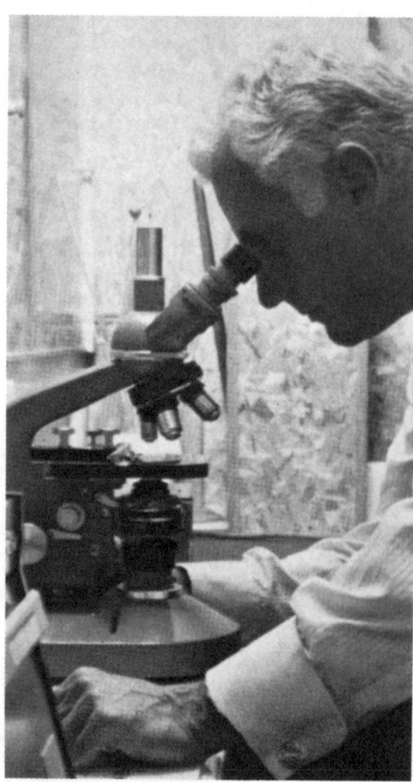

mediately to maintain the product. So, the longer soluble oil is in storage, the less effective the bactericide will be in service because it has already been working. A synthetic does not place this demand on its bactericide, therefore it is much more effective when used. This means the fluid is less likely to sour and produce odors.

Related to bacteria control is control of fungus-type organisms (i.e., fungus, mold and yeast). A common misconception regarding synthetics is that they promote the growth of such organisms. They do not thrive in oil, but may in a non-oil-based fluid. A synthetic concentrate, is not a food source; water and contaminants are (the box *Fine filtering for synthetic fluids* details a new system for removing contaminants). At any rate, their growth can be easily controlled with a fungicide.

Coolant control procedures

	Test procedure	Normal condition	Deviation	Corrective action
Daily tests	Coolant concentration	Depends on application. Consult manufacturer.	Higher or lower than normal.	If too high, add water. If too low, add synthetic concentrate.
	Operating pH	Range from 8.5 to 9.8.	Higher or lower than normal.	Check for unusual contaminant or high contaminant levels.
	Tramp oil contamination	Approximately 1 to 3 percent.	Higher than normal.	Remove excessive tramp oil. If skimmer is used, check to see if it is operating properly.
	Suspended solids contamination	Approximately 0.1 to 0.2 percent.	Higher than normal.	Check to see if filters are clean and filtration equipment is operating properly.
Weekly tests	Biological contamination	Depends on application. Consult manufacturer.	Higher than normal.	Add biocide.
	Corrosion control	Depends on application. Consult manufacturer.	Lower than normal.	Add more corrosion control additive.

Changing over

When converting from an oil to a synthetic fluid, major equipment changes normally are not required. The synthetic may simply be placed into the coolant/lubricant system, although it is recommended that it be thoroughly cleaned.

First, the system should be totally drained; then a purging fluid, containing an effective cleaner and a strong bactericide/fungicide, should be added and circulated throughout. The purgant must be capable of dislodging sludge, tars, waxes, oils and greases; loosening bacterial nests and fungus growths; and suspending these particulates so they can be flushed out. After the purgant is drained, the system should be flushed again with a light solution of the synthetic.

No matter how thorough the cleaning, it is not practical to expect that 100 percent of the sludge etc will be removed. Thus, the first charge will contain residual material picked up by the synthetic's detergent action. These contaminants should be subsequently removed.

It is advisable to use better-quality seals when changing over. Viton and neoprene seals are usually recommended, although some synthetics are compatible with Buna N.

The most important part of a successful synthetic coolant system program is fluid maintenance and control. If it is properly maintained and controlled, there is no reason why a synthetic fluid should ever be dumped.

There are procedures necessary for effective coolant control. The fluid manufacturer can suggest methods suited to each user's needs. Typically, maintenance consists of daily and weekly tests. The daily tests involve monitoring coolant concentration, pH level, tramp oil contamination and suspended solids contamination. The weekly coolant maintenance tests include biological contamination control and corrosion control, **Figure 3**. Refer to the *Coolant control procedures* chart for recommendations on maintaining synthetic fluids.

What if you must dump?

The majority of modern synthetic fluids are organic chemical solutions. Being organic, they are usually biodegradable. So, because of EPA regulations, biodegradability is an advantage.

There are a couple of catchy acronyms associated with waste control: BOD (biological oxygen demand) and COD (chemical oxygen demand). BOD refers to the chemical changes or oxidation brought about by biological activity. COD refers to the chemical changes or oxidation brought about by chemical activity encouraged by a strong chemical oxidant. BOD and COD are important because a combination of biological and chemical oxidation is required for waste disposal.

...continued

Synthetic fluids exhibit a higher BOD: COD ratio than do oils. Some engineers believe this is a problem. Not so. The reason the petroleum oil BOD:COD ratio is low is that it cannot be degraded. Any biodegradable product will always have a higher BOD:COD ratio than one that is not biodegradable.

However, BOD:COD must be carefully controlled in the waste treatment process. If wastewater were to be discharged with excessively high BOD:COD levels, it could harm the environment by depleting oxygen in the fresh water supply. The result: fish kills. For this reason BOD:COD have to be reduced prior to discharge.

It is widely known that organic matter can be treated (oxidized) by using acids. This is used in most industrial waste disposal facilities. Organic materials like synthetics can be handled by such systems with little modification.

Some of the modifications necessary to handle a synthetic in a conventional acid-caustic waste treatment system include using more acid, more caustic and more retention time. This additional material and longer time are offset by the fact that synthetic fluids last considerably longer (if not indefinitely) compared to petroleum-based fluids. Therefore, much less fluid has to go to waste treatment.

Since all metalworking fluids must be treated before release into the environment, the less waste there is to handle the better. Overloading a waste disposal system obviously is not advisable. Slowing the quantity of discharged waste through extended fluid life is a key to successful waste control in the future.

Fine filtering for synthetic fluids

Diatomaceous earth (DE) filtration is becoming a practical method in metalworking filter technology, allowing extra-fine filtering. Such filtration keeps fluids cleaner, enabling the user to get better surface finish, longer fluid life and increased productivity.

DE is a light siliceous or silica-type material, derived from the skeletal remains of minute planktonic algae, that forms a soft earthy rock. It is easily crumbled into a powder used to coat filter media. In typical operation, the filter is first filled with fluid, then the DE is added and circulated through the closed system, coating the filter tube to form a filter cake. When the fluid flows through the filter cake, suspended solids are trapped. A regenerative process allows the DE to be used over and over until its maximum holding capacity is reached. Then the filter can be backflushed (via air pressure) into a sludge receiver to clean it out. DE filtration systems are designed as polishing filters; a rough filter should be used ahead of them to remove large particulates.

Recently, United States Filter Fluid Systems Corp, Whittier, CA, introduced a new DE system that filters fluids down to 1 micron—even bacteria can be filtered out. Called Econo-Flo™, it is a regenerative low-pressure filter in a complete packaged system.

Conventional synthetic fluids can only be filtered down to 20 microns before they begin being stripped of key components, thereby losing their lubricating qualities. The photo shows a special synthetic fluid being successfully run through a DE filtration tester.

Reprinted by the Society of Manufacturing Engineers from Iron Age, October 26, 1981

CAN SYNTHETIC LUBES OVERCOME FRICTION IN THE MARKETPLACE?

Synthetic lubricants have gotten rave reviews in performance. But they still have a long way to go to achieve their projected goals.

By John J. Obrzut

"Though synthetic lubricants now account for hardly 2 pct of industry's consumption of lubricating oils and greases, their use will probably double every three or four years during the 1980s." So say two scientists, Dr. Russell Murphy and Dr. David Neiswender of the Mobil Research & Development Corp., Paulsboro, N.J.

They cite two chief reasons for this projected growth in synthetic lubricants—they conserve energy; they reduce oil consumption.

A survey by the C. H. Kline & Co., Fairfield, N.J. agrees that synthetic lubricants account for about 2 pct by volume of the lubricant market, but contends that because of their higher price, they represent about 10 pct of the dollar value.

The Kline survey projects that synthetic lubricants will represent about 15 pct of the total lubricant market by dollar value by 1985.

Other forecasters predict that the 1980s will be a period of strong growth for the synthetic lubricants, doubling in volume over that of the past ten-year period. Even these estimates are described as being conservative. However, the base from which these estimates are made was quite low 10 years ago.

Whether the various surveys agree or disagree on the outlook regarding the extent to which synthetic lubri-

Some synthetic cutting fluids foam when highly agitated. Pillsbury fluid (left) produces little foam.

Legend: Excellent; Very Good; Good; Fair; Poor

	Soda Ash (1924)	Sodium Nitrite	Amine Nitrites	Amine Borates	Amine Borate with Synthetic Lubes	1980 Generation
Foam Resistance	Very Good	Very Good	Excellent	Fair	Excellent	Excellent
Corrosion Protection, Ferrous	Good	Very Good	Excellent	Excellent	Excellent	Excellent
Corrosion Protection, Nonferrous	Poor	Fair	Poor	Good	Good	Excellent
Tramp Oil Rejection	Good	Excellent	Excellent	Excellent	Excellent	Excellent
Service Life	Fair	Fair	Good	Excellent	Excellent	Excellent
Extreme Pressure (E/P) Lubricity	Fair	Fair	Good	Good	Excellent	Excellent
Residual Film	Poor	Poor	Fair	Good	Very Good	Excellent
Co-Solvent (Cleaning) Capability	Fair	Poor	Fair	Good	Very Good	Excellent
Resistance to Atomization (Mist/Smoke)	Poor	Poor	Fair	Fair	Very Good	Excellent
Multi-Process Capability	Poor	Poor	Poor	Poor	Good	Excellent
Biosphere (Microorganism Control)	Poor	Poor	Fair	Very Good	Very Good	Excellent
Biodegradability	Very Good	Poor	Poor	Very Good	Very Good	Excellent

Source: Pittsburgh Chemical & Oil, Inc.

The Long and Short of Synthetic Lubricant Properties

Synthetic	Advantages vs. Mineral Oil	Limiting Properties
SHF	High Temperature Stability	Solvency/Detergency*
	Long Life	Seal Compatibility*
	Low Temperature Fluidity	
	High Viscosity Index	
	Improved Wear Protection	
	Low Volatility, Oil Economy	
	Compatibility with Mineral Oils and Paints	
	No Wax	
Organic Esters	High Temperature Stability	Seal Compatibility*
	Long Life	Mineral Oil Compatibility*
	Low Temperature Fluidity	Antirust*
		Antiwear and Extreme Pressure*
	Solvency/Detergency	Hydrolytic Stability
		Paint Compatibility
Phosphate Esters	Fire Resistant	Seal Compatibility
	Lubricating Ability	Low Viscosity Index
		Paint Compatibility
		Metal Corrosion*
		Hydrolytic Stability
Polyglycols	Water Versatility	Mineral Oil Compatibility
	High Viscosity Index	Paint Compatibility
	Low Temperature Fluidity	Oxidation Stability*
	Antirust	
	No Wax	

*Limiting properties of synthetic base fluids which can be overcome by formulation chemistry.
Source: Mobil Oil Corp.

done by Mobil, improvement in power transmission efficiency ranged from 2 pct to 8.8 pct. High oil temperature, an indication of wasted energy, was about 20°F cooler with the synthetic lubricant.

Mobil's Delvac 1, an engine oil made from synthesized hydrocarbons in the polyolefin family and esters, plus additives, is finding greater use in truck fleets, off-the-highway equipment and industrial trucks.

In tests with over-the-road matched vehicles, trucking associations, along with the Department of Transportation (DOT), showed that fuel savings were 3 to 4 pct by using the synthetic engine oil, along with a Mobil synthetic transmission and rear axle oil.

With medium- and heavy-duty industrial equipment, synthetic lubricants have given designers a degree of freedom they haven't had before. They can now increase machine power and speed output within limits without increasing machine size or weight, often with gears and shafts of the same sizes.

One gear maker, for instance, was able to increase the rating of a gear reducer from 18 to 19½ hp by simply switching from a mineral oil lubricant to a synthetic lubricant. Another gear maker was able to increase input horsepower to a worm gear by 15 pct.

Longer lubricant life is another plus for the synthetics. In some cases, in fact, it's a major advantage. It can cut costs two ways, first on lubricant cost and, secondly, on the cost of changing it. And with today's complex machines, it's more of a problem because gears and bearings are not always readily accessible.

Anyone with a sludge problem? Again, synthetic lubricants may be the answer. They resist oxidation very well and don't contain the compounds that produce sludge in mineral oils. Although their cost per gallon is much higher, their longer life may give them an economical trade-off.

Commonwealth Edison changed the oil in its nuclear plant coolant pumps twice a year—each time at a cost of $500,000 in lost power. It now uses a synthetic gear oil that lasts a full year between shutdowns.

But gear boxes and bearings are not the only applications for synthetics. They're used as cutting fluids for machining a great variety of materials.

"Even tough-to-cut metals, such as SAE 308 aluminum, can be machined successfully with synthetic cutting fluids," says Robert K. Rauth, president, Pillsbury Chemical & Oil, Inc., Detroit. "Synthetics have advantages over oils in a number of areas, an important one being tool life. They can increase tool life greatly because they have excellent extreme-pressure lubricity."

Also, synthetics reject tramp oil. Tramp oil in an oil coolant system reduces its effectiveness and results in premature dumping. Not so with synthetic cutting fluid. "Synthetics are true chemical solutions," Mr. Rauth notes, "and totally reject oils. Tramp

How Synthetic Coolants Compare with Soluble Cutting Fluids on Costs

10,000 gal Transfer Machine System

Item	Average of Four Synthetic Coolants	Average of Low and Medium-Priced Soluble Cutting Fluids
Purchase price per gallon	$4.30	$1.71
Recommended dilution	1:26 (3.8%)	1:20 (5%)
Cost/mixed gallon	$0.162	$0.081
Cost to fill system	1,620	$810
Expected fluid life	2 years	6 months
Replenishment ratio	1:83	1:25
Replenishment, cost/day	$33.10	$33.00
Waste treatment, gallons/year	5000	20,000
Annual Costs		
Fill system	$810	$1,620
Make up (replenishment)	8,275	8,250
Recharge cleaner and labor	525	2,100
Waste treatment	250	1,000
Total Annual Cost	**$9,860**	**$12,970**

Source: Pillsbury Chemical & Oil, Inc.

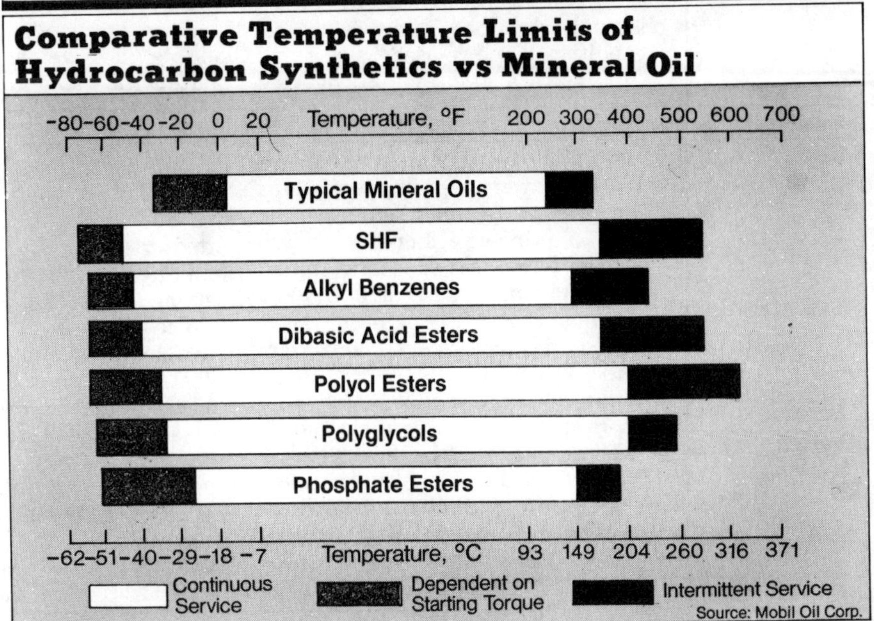

cants will grow through the 1980s, one thing is certain—they're expanding their niche in the market.

First, how does one define a synthetic lubricant? J. George Wills, chief technical editor of the Mobil Oil Corp. says that "synthetic-base fluids are man-made and tailored to have a controlled molecular structure with predictable lubricating properties."

Among the materials he classifies as synthetic-base fluids are the synthesized hydrocarbons, organic esters, polyglycols, phosphate esters, and other synthetic lubricating fluids. And he points out that among the synthetic lubricant bases now in use, the first four account for more than 90 pct of the volume. A number of high-cost materials make up the last category.

Synthetic lubricants do more than save energy and oil, Dr. Murphy contends. "The ability of synthetic lubricants to transmit power more efficiently through gears and other mechanisms gives equipment designers greater opportunities in one of their perennial goals. By specifying synthetic lubricants they can use smaller and more compact components and they can sometimes eliminate lubricant coolers. This will eventually reduce equipment costs."

Dr. Neiswender points out other advantages, one example being that synthetics are often used as answers to design problems—but after the fact. He notes that equipment intended for intermittent use but then used continuously or equipment with a poorly designed lubrication system, may break down quite often. Synthetic lubricants can often solve such problems and minimize breakdowns.

Synthetic lubricants are also being used where temperatures reach extremes. In fact, the first synthetics were used for just such applications for jet aircraft. Development work continues on synthetic lubricants that will give jet engines higher output and make them more fuel-efficient at even higher temperatures and pressures.

How can the metalworking and other industries benefit from synthetic lubricants? A study by the American Society of Mechanical Engineers (ASME) indicates that industry's use of synthetic lubricants can save as much as 11 pct in energy consumption by simply reducing friction in the machinery it operates.

It's an immediate and economical way to reduce energy losses in machines in three different ways: Synthetic lubricants maintain their viscosity better than mineral oils over wide temperature ranges, enabling better efficiency at lower temperatures; they're more easily formulated to give them polar films with a strong affinity for metals and thus better protection; and thirdly, their molecular structure is such that their traction coefficient (similar to the coefficient of friction for solids) may be made lower than that in mineral oils.

Dr. Murphy explains that the film of lubricant that separates two machine parts must be sheared to allow them to move, and as such, the greater the coefficient of traction, the more energy it takes to shear the film. High coefficients of traction account for much of the inefficiency in power transmission.

Synthetic lubricants can be formulated with built-in low coefficients. The energy savings in many applications can be significant.

For example, in a test of a mineral-base gear oil and a synthetic gear oil, both having similar operating characteristics, the synthetic gear oil gave a worm gear 7.3 pct higher efficiency in power transmission.

On the other hand, because synthetic lubricants can be formulated more readily to meet specific requirements, they hold promise as traction-drive fluids where the higher coefficient of traction would be desirable.

The energy savings possible through the use of synthetic lubricants are real, say the scientists. They point out that in 10 tests with synthetic gear oil, three of which were

Where Synthetic Lubricants Found Markets

Field of Service	Synthetic Fluids Used
Industrial	
Circulating Oils	Polyglycol, SHF, Organic Ester
Gear Lubricants	Polyglycol, SHF
Hydraulic Fluid (Fire Resistant)	Phosphate Ester, Polyglycol
Compressor Oils	Polyglycol, Organic Ester, SHF
Gas Turbine Oils	SHF, Organic Ester
Greases	SHF
Automotive	
Passenger Car Engine Oil	SHF, Organic Ester
Commercial Engine Oil	SHF, Organic Ester
Gear Lubricant	SHF
Brake Fluids	Polyglycols
Aviation	
Gas Turbines	Organic Ester
Hydraulic Fluids	Phosphate Ester, Silicones, SHF
Greases	Silicones, Organic Ester, SHF

Source: Mobil Oil Corp.

oils can be skimmed from the surface, reconditioned and re-used."

Mr. Rauth also cites the longevity in the life of a synthetic cutting fluid—four times longer than oil. Thus, refill costs are lower; make-up costs are lower; cleaning costs are lower; and waste treatment costs are lower—making the more expensive synthetic fluids less expensive to use in the long run. "One central coolant system using a synthetic hasn't been dumped in three years," says Mr. Rauth.

A large bakery that had problems with gear-drives and transmissions as well as with its chain-and-sprocket drives switched from petroleum-base lubricants to synthetic diester-base lubricants, made by the Keystone Division of Pennwalt Corp., at costs running three to four times higher than it had been paying per gallon for the petroleum lubricants.

In spite of the much higher unit costs, the bakery received a life span from the synthetics that was ten times that of the conventional lubricants—without nearly the number of problems, considerable saving in downtime and less chance of error. In addition, it saved sustantially on energy as a bonus.

Synthetic fluids often serve where oil-base products give up. Such was the case at Kelsey-Hayes' Detroit plant which went from hot to cold cleaning of machined parts. The oil-based additives separated from the washer solution and plugged up the system, according to Kurt Vollers, the plant's general manager.

A search led to the selection of a water-base synthetic in Pillsbury's 5000 series, one that was specially formulated for cold cleaning. "It supplied a detergent action that continuously cleaned the washer system, eliminating the plugging. It left the parts clean and rust-free," says Mr. Villars. With these and other advantages, he estimated that last year's savings at about $250,000.

The endorsements for the synthetic lubricants seem to be coming from the producers and their satisfied users. But some potential users don't seem to be ready to try them.

Some just can't justify the difference in costs under any circumstances. Still others may be waiting for another oil crunch.

"At that time (1973-74)," says G. J. Colucci, service engineer, Fluid Power Products Dept., E. F. Houghton & Co., Valley Forge, Pa., "85 pct or more of the 300 million gallons of hydraulic fluids consumed were oil-base.

"Since then, rapid strides have been made in developing synthetic water-soluble-additive concentrates which dilute with water to make fluids containing 95 pct water and 5 pct additives.

"Some are single-purpose fluids for hydraulic systems only. Others are dual-purpose fluids used undiluted or diluted in the hydraulic system of a machine tool, or diluted as a cutting or grinding coolant."

What these synthetic concentrates have done was simply to save users a substantial amount on hydraulic oil consumption. In concentrated form, these fluids sell for several dollars per gallon.

But when they're diluted to a ration of 19:1 with water, their cost drops to 25¢ or 35¢ a gallon, plus the labor cost for mixing. It's still a very sharp drop from the per gallon cost for hydraulic oil.

Although the cost differential favors the high water-base fluid, direct conversion of a system from oil to the synthetic high-water-base fluid is seldom possible. Some modification of the system is usually necessary to make it compatible with these fluids. Also, dilution of the concentrate requires special handling before it's used.

Aside from cost, another major advantage of these fluids is that they're biodegradable, and as such, any leakage can be handled on the spot or usually through a municipal waste water treatment facility. This treatment differs from that for a soluble oil-type hydraulic fluid in that the oil component is not biodegradable, says Mr. Colucci, and must be separated from the fluid so that the water can be dumped into a sewage system.

Tests performed by Exxon showed that its Synesstic 32 synthetic industrial lubricant, formulated from high-quality diester base stocks, can reduce energy significantly. By using a conventional industrial-type contact bearing, lubricated with the Synesstic 32 lubricant, it consumed 13.9 pct less power than a bearing using a petroleum-base oil.

In full scale compressor tests on reciprocating and centrifugal compressors in the field, net savings per year for fuel, lubricants, electricity and maintenance amounted to $100,000 on the reciprocating compressor and $2000 per year on the centrifugal compressor. □

Reprinted from Tooling & Production, December 1980

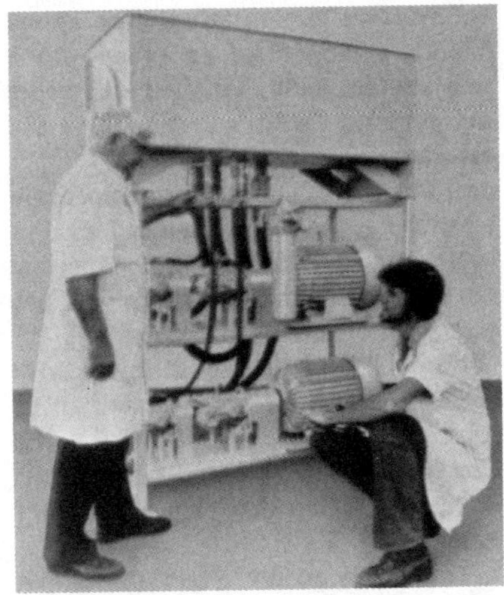

Sperry Vickers has designed power units for stacked modules with the reservoir elevated above the pump and motor. The modular concept is compatible with both HWBF and oil hydraulic systems.

This compact lever-operated, four-way directional valve is capable of handling 10 gpm and operating at 5000 psi. It can be subplate or manifold mounted.

Wet-armature directional valves are designed to effectively reduce hydraulic fluid leaks. This cutaway shows a Sperry Vickers DG4S401 wet-armature valve for HWBF applications.

High-water-base fluid

NEW TECHNOLOGY SAVES OIL

by Robert N Thayer
Market Development Specialist
Sperry Vickers Div
Sperry Corp
Troy, MI

Petroleum's soaring prices and uncertain supply are persuading many companies, particularly the automakers, to consider high-water-base fluid (HWBF) as an alternate hydraulic fluid. Some of the reasons include low fluid cost, excellent fire resistance, lower transportation and storage costs, easier cleanup and disposal (biodegradable), cooler operation and uniform viscosity—lesser need for machine warmup. Also, it is nonpolluting of coolants and air.

Optimum results are obtained when hydraulic systems are *designed* for use with HWBF. Since HWBF contains 95 percent water and 5 percent soluble oil additives, it has a higher vaporization pressure than hydraulic oil. Thus, it vaporizes more easily. This could cause a problem at the inlet to the hydraulic pump where pressure is reduced by the pump's action. Vapor bubbles going through the pump can cause cavitation damage, noise and wear.

To prevent vaporization, the fluid should be gravity-forced into the pump from an elevated reservoir, a pressurized tank or a supercharge pump. The most popular method in current use is an elevated reservoir. This arrangement provides several advantages for either HWBF or oil hydraulic systems:

- Longer pump life.
- Quieter operation.
- Elimination of aeration on inlet side of pump.
- Lower fluid temperature, because of improved air circulation around tank.

Stack plan

Elevating the reservoir also facilitates a new design idea: *stacking modules.* The stacking-module power-unit concept for HWBF had its inception about five years ago in the automobile industry. Since that time, many machine-tool users have recognized the advantages derived from the module's flooded-inlet system for hydraulic pumps, plus the added savings of reduced floor space.

The Sperry Vickers stacked hydraulic power units, for example, employ an overhead reservoir that keeps the pump's inlet continually flooded. The overhead reservoir also minimizes cavitation and eliminates the need for air-bleed valves, thus reducing the number of system components. Furthermore, it virtually eliminates the need for pump priming.

Power units for HWBF or oil are available with reservoir capacities of 20, 30, 60, 100, 150 and 200 gal. The key benefit from this design is that a reservoir can feed one or more power-unit modules, with separate inlet lines for each pump. Each module can accommodate an electric motor up to 30 hp and single or tandem piston pumps.

Overhead stacking reduces the number and size of individual power units as well as saving plant floor space. Multiple pump combinations and multiple modules may be arranged so that the most practical and cost-effective system can be worked out. By merely bolting together the frame of each module, a flexible, versatile system can be designed to meet specific requirements.

Tandem pumps using one motor provide cost savings by reducing the total number of motors and space required.

Improving capability

HWBF capability involves more than merely elevating the reservoir. Naturally, suitable system components are required. For example, Sperry Vickers standard B-Series in-line piston pumps and standard valves have been tested and proved for use with HWBF at 1200 rpm, 1000 psi, 40 F to 120 F fluid temperature, and a positive pressure on the pump inlet.

For applications requiring higher pressures and speeds, such as die-casting machines and plastic-injection-molding machines, there are special pumps capable of operating at 1800 rpm and up to 2000 psi. Pumps to run at higher pressures are currently under development. While piston pumps operate best with HWBF, vane pumps can serve if a special pump cartridge is installed.

As a cautionary note, HWBF is not compatible with certain materials such as cork, leather, paper, cadmium, zinc, lead, aluminum, magnesium and most paints. Thus, it is recommended that the inside surface of the reservoir be left bare or covered with a paint known to be resistant to HWBF.

Because standard Sperry Vickers in-line piston pumps and standard hydraulic valves can be used with HWBF, the only added cost for hardware is for elevation of the reservoir. This increase is modest for one-module power units, and for two- or three-module power units, there is an actual cost reduction. Since hydraulic systems designed for HWBF are also compatible with oil, these systems should be specified for suitable applications on all new machinery and production equipment.

Manufacturing experience with HWBF should be acquired gradually to learn its characteristics and proper application. It is essential that the fluid be properly prepared and the HWBF system be maintained by knowledgeable service personnel. While HWBF is not a panacea for the cost crunch of hydraulic oil that faces US industry, there are many overall benefits to its use. But to realize these, it is vital to adhere to recommended practices for design, installation, operational limitations and maintenance of equipment.

At the Sperry Vickers, Omaha, NB, manufacturing facility, Harold E Bargar, retired plant engineer, reports that engineers set up a multiple-station part-clamping system to prove the effectiveness of HWBF. The clamps (see photos) hold steel cam rings for machining on a Turchan NC contour milling machine. Originally, the clamps used 100 percent oil, a 10-gal reservoir and a 5-gpm pressure-compensated variable-volume pump set at 700 psi. Now, a Sperry Vickers Model F6-25V12 balanced-vane constant-displacement pump serves, operating at 1200 rpm and 1000 psi. Engineers had to redesign the power unit to accept the pump and yet not overpower the fixture-holding clamps on the machine.

In good design, the reservoir should equal three times the pump capacity—roughly 40 gal in this case. To provide for a flooded pump inlet, the reservoir's fluid surface was elevated 4 ft above the centerline of the pump. Fluid velocities in the lines were calculated at 4 fps in the pump inlet line, 15 fps in the working fluid lines and 7 to 8 fps in the exhaust lines. Line diameters were set at ½" ID for the working lines. A 149 micron (nominal) strainer filters the inlet, and a 10-micron unit fits in the outlet.

Pressure-compensated variable-volume pumps can reduce gpm output while maintaining pressure at system requirements, thus using minimum horsepower. But the constant-displacement pump chosen for this setup cannot cut its fluid output. Thus pressure on the pump must be reduced without dropping overall system pressure. To do this, an accumulator and an unloading relief valve are employed. Maximum system pressure is limited by the valve, which unloads the pump (back to the reservoir) when the accumulator is charged to full pressure. An internal check function of the valve prevents the system fluid from unloading to the reservoir. This setup cuts power consumption, increases pump life and reduces the need for coolers. For info, circle E37.

Machine operator positions vane-pump cam ring for milling operation.

Vane-pump cam ring in position.

CHAPTER 3

SELECTION CRITERIA

Presented at the 31st Annual Meeting of the American Society of Lubrication Engineers, May 1976. Updated for Improving Production With Coolants and Lubricants.

A Drilling Test for the Evaluation of Cutting Fluid Performance

G. W. SKELLS, JR. and S. C. COHEN (Member, ASLE)
Van Straaten Chemical Company Chicago, Illinois 60606

Various tests have been established and used to evaluate or rate the performance of cutting fluids. This paper describes a drilling test based on tool wear for the evaluation of the performance of cutting fluids. This test is easily established in the laboratory or plant. Data obtained over several years use indicate that it can be of value to industry where a method of screening, or of rating the performance of new cutting fluids is required.

INTRODUCTION

A large number of tests have been proposed and used to evaluate the performance of cutting and grinding fluids. These tests can be broadly separated into two categories of simulative and machining tests.

Nonmachining tests such as the Falex (*1*), (*2*), (*3*) and Four Ball (*4*), (*5*), (*6*) have been utilized to evaluate extreme pressure additives, antiweld or antiwear agents of various lubricants, and cutting fluids. Although these tests are useful for other types of lubricants, there has been no actual correlation shown between these tests and the performance of cutting fluids.

Machining tests can be subdivided into two types:

One, tests which measure energy input or cutting forces (*7*), (*8*). These tests are primarily used to rate the machinability of metals. Exceptions are the tapping torque test and the drilling torque test, which have been used for evaluation of cutting fluids (*9*), (*10*). These tests do show differences in cutting fluids and there appears to be a correlation to field experience in some specific operations.

Two, tests which measure tool life or tool wear as a direct indication of machinability or of cutting fluid performance (*11*), (*12*).

For the purpose of developing high performance cutting fluids, the authors felt that it was essential to have a test that correlated with actual performance. Since the users of cutting fluids are directly concerned with tool life, the authors selected the second type of test, a machining operation, where it is possible to accurately measure the tool wear.

After a study of varied machining operations, the authors decided against machine tools with multiple cutting edges because of the problems in accurate measurement of wear. This primarily limited them to turning, shaping, planing, and drilling. Turning was rejected because of the cost of the equipment and test materials; planing and shaping were rejected because of the variation in speed of the cutting tool through the work material, and also because these machines are not normally equipped for cutting fluids. The drilling test was selected because of its economy and the availability of fairly small, rigid machines with variable speed drive and an adjustable feed. Also, tooling was available in a large number of materials, both high speed steel and carbide, and it was possible both to reproduce point geometry and to measure wear accurately on the land of the drill with a quality drillpoint grinder.

DESCRIPTION OF TEST APPARATUS

Drill Press

A heavy duty, rigid drill press is powered by a 3 horsepower motor, supplying a continuous variable speed to the spindle between 0–3,500 rpm. The drill press is equipped with the standard positive feed mechanisms, allowing a wide variation in feed rates (Fig. 1). The work material is held in place by a heavy duty vise, attached to the work table by a rigid indexing table that can be manually moved in the X, Y plane. Coolant is supplied from an external tank and pumped through one large rigid nozzle, insuring adequate flow of fluid to the point of cut, and the return is through the normal machine drain.

Fig. 1—Laboratory drill press

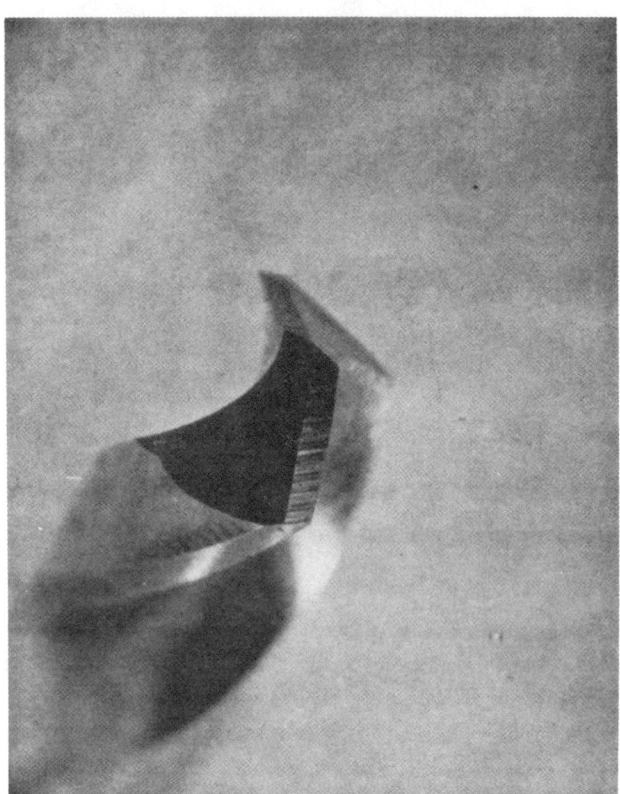

Fig. 2—Four facet prismatic drill point

Drills

To maintain rigidity, the authors use only screw machine length (2½ in long) drills. Our standard drill is a ¼ in M-1 twist drill. These are used for our everyday tests. However, for special materials, the authors may use drills made from other high speed steels or carbide.

The drills used are ground to a prismatic point with a four facet point geometry using a precision drill grinder (Fig. 2). The authors selected this point because it is suitable for the variety of materials they test, especially the work-hardening materials. It is easily reproducible and gives a reference point for wear measurements.

MATERIALS

The standard workpiece size is 4 in x 15 in x ½ in. Standard test materials used are given with their condition in Table 1. Other materials may be used for special development work.

PURCHASING

To insure quality, the authors purchase drills and materials in large quantities under the following conditions:

Drills are purchased from one production run using the same heat of steel. The authors have the supplier run additional quality control inspection, giving written certification as to the quality of the drills. The points are then reground by them as previously discussed.

Table 1—STANDARD TEST MATERIALS

Materials	Condition
AISI 1045	Hot rolled and normalized
AISI 4340 AC	Cold rolled and annealed
AISI 303	Hot rolled and annealed
Inconel 718	Cold rolled or hot finished
Ti-6A1-4V	Annealed

only when the drill shows less than 0.015 in wear prior to its failure and it is reproducible. Failure of this type, or very rapid wear, can be caused by hard spots in the work materials. An experienced technician can tell by the sound and rate of wear when a hard spot is encountered. This is confirmed using a hardness tester and the hard area is marked and avoided in further testing.

RESULTS AND DISCUSSION

In early experiments, not reported here, the authors found that drill failure was not a suitable criterion for measurement of tool wear. Besides being a lengthy test, they found that the number of holes drilled before catastrophic failure was irreproducible. In these tests, it was also determined that holes drilled to a specific wear criterion of 0.015 in was more reproducible, as found by other workers *(13)*, *(12)*. The reproducibility of data obtained in this manner is shown in Table 2. The authors feel that the extreme measures taken to control the machining variables are essential to obtain reproducible results.

In this test, the performance of cutting fluids is evaluated by increasing the severity of the drilling operation and comparing the results with a test lubricant against those with a standard product. The severity is increased by increasing the drill speed on a given material and also by utilizing tougher materials.

The standard cutting fluid used for this test is a general purpose water soluble product, which consists essentially of a mineral oil and emulsifiers to form a coarse, white emulsion. The results of such a drilling test for fluids evaluated on 1045 steel in a normalized condition is shown in Fig. 5.

Product A is a high performance water soluble which contains extreme pressure lubricants and fats in the following amounts:

27 percent fat
1.7 percent active sulfur
6 percent chlorine

Product B is a synthetic product containing only completely water soluble ingredients. It forms a transparent solution which contains no mineral oil or extreme pressure lubricants such as sulfur, chlorine, or phosphorous compounds.

As the drilling speed is increased, the number of holes which can be drilled before 0.015 in wear is reduced for all fluids tested. The results show a large difference be-

Table 2—REPRODUCIBILITY OF DATA

Test No.	AISI 303 145	155	165	SFM
1	261	178	72	Holes
2	255	162	82	Holes
3	278	175	87	Holes

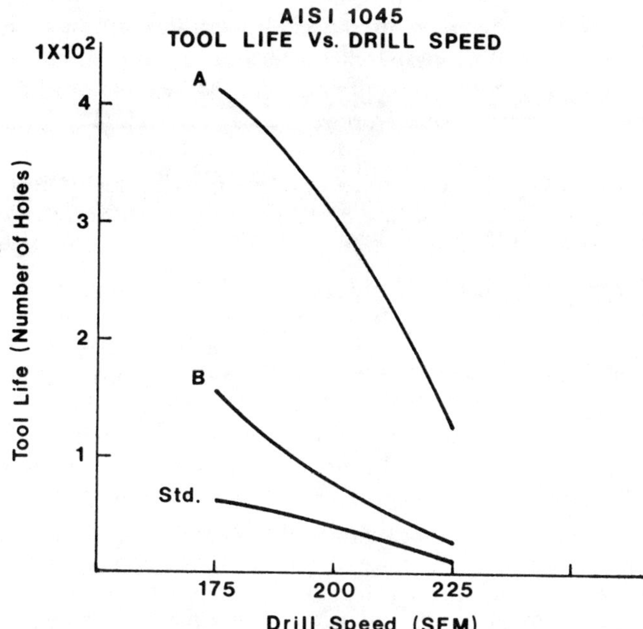

Fig. 5—AISI 1045 tool life vs drill speed

tween Product A and the standard at all speeds, indicating that the use of Product A rather than the standard could either significantly increase the tool life or speed of machining with equivalent tool life. Although the synthetic, B, shows some improvement over the standard, it is not, however, nearly as large as A.

Similar tests are shown for the same products on 4340 steel in the annealed condition as can be seen in Fig. 6, giving generally similar results to those shown in Fig. 5. Figure 7 shows the results of drilling 303 stainless steel in the annealed condition. Here, note that Product A is still better than the standard, but Product B, the synthetic, has lower performance than the standard. Also note as

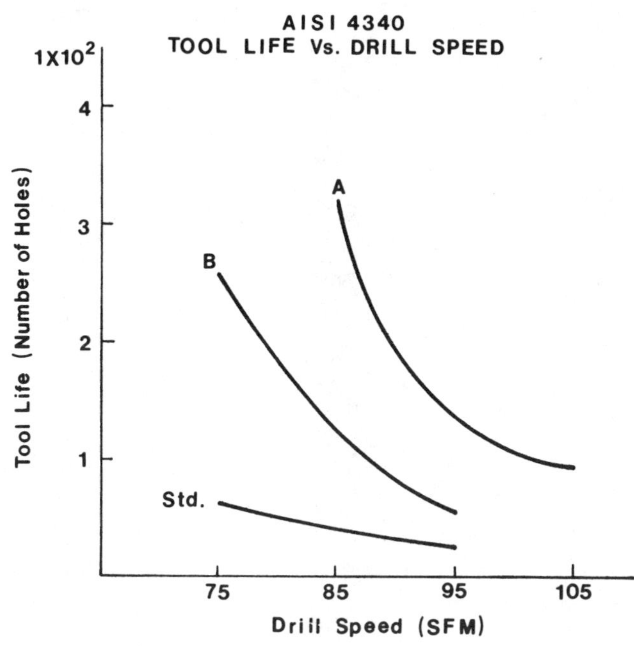

Fig. 6—AISI 4340 tool life vs drill speed

Work material is purchased with certification that it is from one heat, giving chemical and physical specifications. The authors require materials to be sawn or abrasive cut to minimize work hardening or residual stresses. They are then heat treated to their specifications, straightened if necessary, and all scale and surface defects are removed by surface grinding (Blanchard grinding).

TEST METHODS

Fluids are tested by drilling holes through a test specimen using a positive feed rate (0.005 in per revolution) until a wear of 0.015 in is reached on the drill land. At least three speeds are run on each material with a minimum of two drills at each speed. The number of holes is plotted vs the speed in surface feet per minute.

Wear is measured using a 10X measuring microscope. The back edge of the cutting facet is used as a reference. Initially, the width of the facet is measured and recorded with difference between the initial width and the worn width being the wear. A flat is ground on the heel of one land, marking the land that is used for wear measurements (Fig. 3).

In order to have an accurate and reproducible test, the authors have found that it is necessary to adopt the following procedures:

1. Maximum rigidity is maintained throughout the test by (see Fig. 4):
   a. Workpiece clamped firmly.
   b. Drill chucked to full shank.
   c. Drill to work clearance minimized.

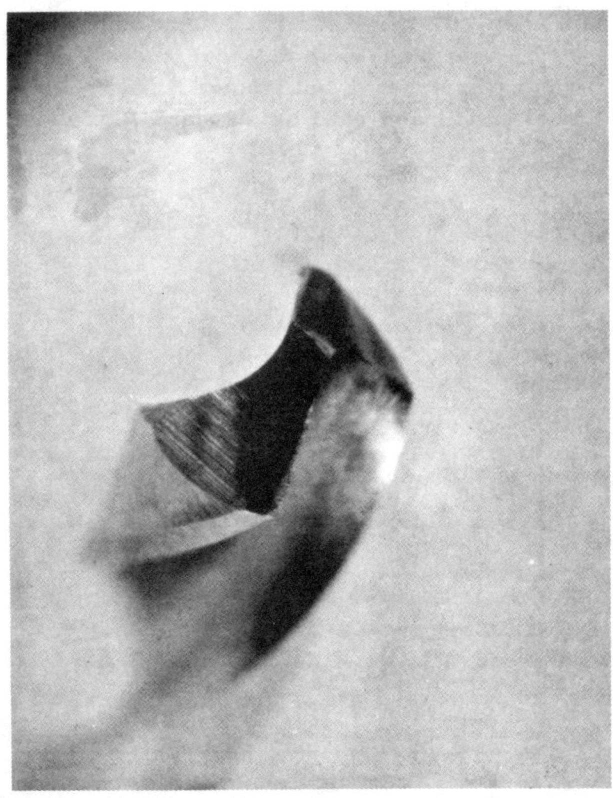

Fig. 3—Drill point with 0.015 in land wear

Fig. 4—Rigidity of work piece and drill

2. Drilling is only conducted with the machine and fluid at thermal equilibrium. This is accomplished by running machine and fluid pump for 15 min. prior to the start of drilling.
3. Holes are drilled in the work, indexing the work table, drilling a pattern of holes on $\frac{3}{8}$ in centers, maintaining a minimum border of $\frac{3}{4}$ in on the rolled edges of rolled strip.
4. Drills are only power fed into the work. (To reduce the possibility of low pressure contact causing excessive work hardening or friction wear.)
5. A depth of two drill diameters or less is used to insure fluid at the point of cut.
6. The fluid nozzle is checked frequently to insure fluid flow to point of cut. Long stringy chips may interfere with fluid flow and must be removed from drills, nozzle, and their surroundings.
7. Fluids are usually tested at a concentration of 10 parts water and 1 part concentrate. Standard procedure is to conduct tests on fresh fluids so that concentration control is not required. However, if the test is prolonged or there appears to be heavy evaporation, the concentration is determined and corrected as required.
8. The two drills run at each speed are required to have less than a 10 percent difference in the number of holes at 0.015 in wear, or further drills are run until two are within 10 percent of each other and the two are then averaged.
9. Catastrophic failure is considered a test end point,

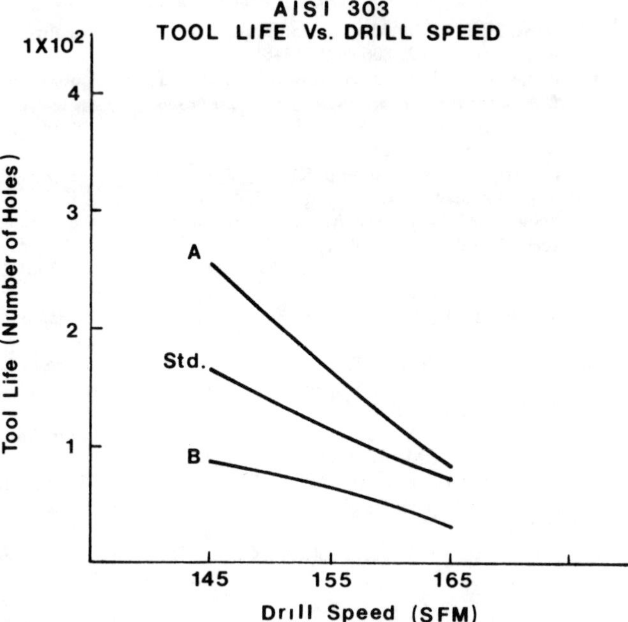

Fig. 7—AISI 303 tool life vs drill speed

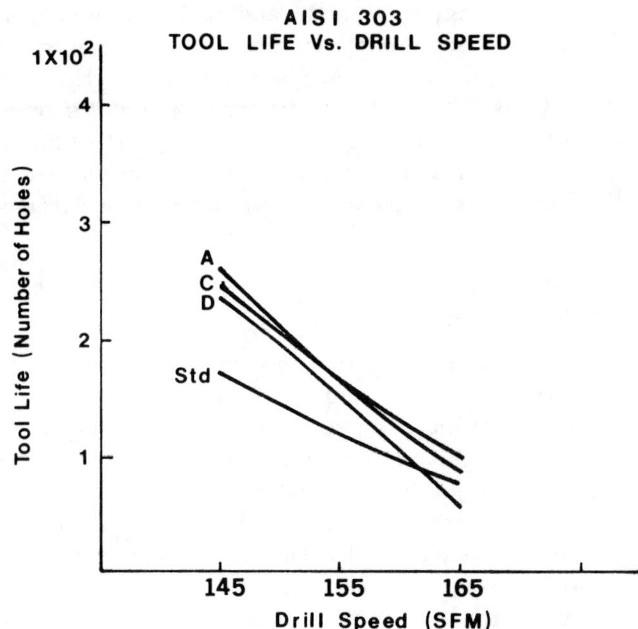

Fig. 8—Three high performance fluids tested on AISI 303 tool life vs drill speed.

the speed increases, there is no significant difference between Product A and the standard.

The reversal seen on the 303 stainless shows dramatically that results from tests on one specific work material cannot necessarily be projected to other types of materials.

It has been the authors' experience that there has been good correlation between this test and other machining operations. In virtually all operations, they found similar rating of performance and Product A (or similar products) show superior performance.

Besides the standard work materials, the authors also use other materials. In order of machining difficulty, these materials are ranked as shown in Table 3. In searching the literature, the authors found that there is no accepted standard criterion for machinability measurements and that there are many different rating schemes used (*14*), (*15*). The authors have rated products based on their experience with drilling and consequently, it should be noted that their rankings are slightly different. Obviously, this drilling test is very helpful when one is only interested in specific alloys. Thus, it has been the authors' experience that products can be developed which show outstanding performance on specific alloys. In recent years, the metalworking industry has become highly sophisticated and chemical companies have developed the capability to formulate many different types of high performance cutting fluids.

TABLE 3—MACHINABILITY OF TEST MATERIALS

MATERIAL	TEST RATING	% BASED ON B1112 = 100%	SFM FOR 30 MIN. H.S. TOOL LIFE
Inconel 718	Poor	—	39
Ti-6A1-4V	Poor	—	72
AISI 303	Fair	60	—
AISI 4340	Fair	57	98
AISI 1045	Good	72	—

The drilling results for three typical high performance products A, C, and D are compared on stainless 303 in Fig. 8. Products A, C, and D are all quite different in composition. They all contain extreme pressure lubricants and fats as follows:

	A	*C*	*D*
Fat	27%	18%	1.5%
Active Sulfur	1.7%	None	None
Chlorine	6%	9.5%	11.5%

The drilling test results show the great distinction between the high performance products (A, C, and D) and the standard product, but is not really capable of making distinctions between them. Where there are only slight changes in performance, such as these, the authors have not always been able to obtain correlation with machining operations such as broaching, sawing, tapping, threading, and form milling. In fact, generally they have found it necessary to test in the specific operation to establish which of these products is really superior.

CONCLUSION

A drilling test procedure is described, which the authors have developed and which has enabled them to accurately measure the tool wear as a function of drill speed using a four facet drill point. This procedure is shown to be very useful in evaluating the performance of cutting fluids, especially in determining whether there are significant differences in tool life performance.

The authors' field testing experience also confirms these laboratory experiments that tool life data alone will not be sufficient to separate and rank new high performance cutting fluids even though the severity of the test can be increased as they have shown here. However the authors feel that life studies may be useful if other performance parameters, such as, internal surface finish, drill size, and geometry are also taken into consideration. Further studies are planned to investigate these parameters.

USE OF THE TEST

After this paper was written, this test was used to qualify new synthetic cutting and grinding fluids. These fluids contain synthetic water soluble lubricants which aid in the machinability performance of the fluid. As shown in the following graphs, the "new" synthetics are superior to "simple" synthetic, common amine nitrite type, in the area of increased difficulty. This also has been verified in the field through extensive testing followed by years of success with synthetic products containing these lubricants.

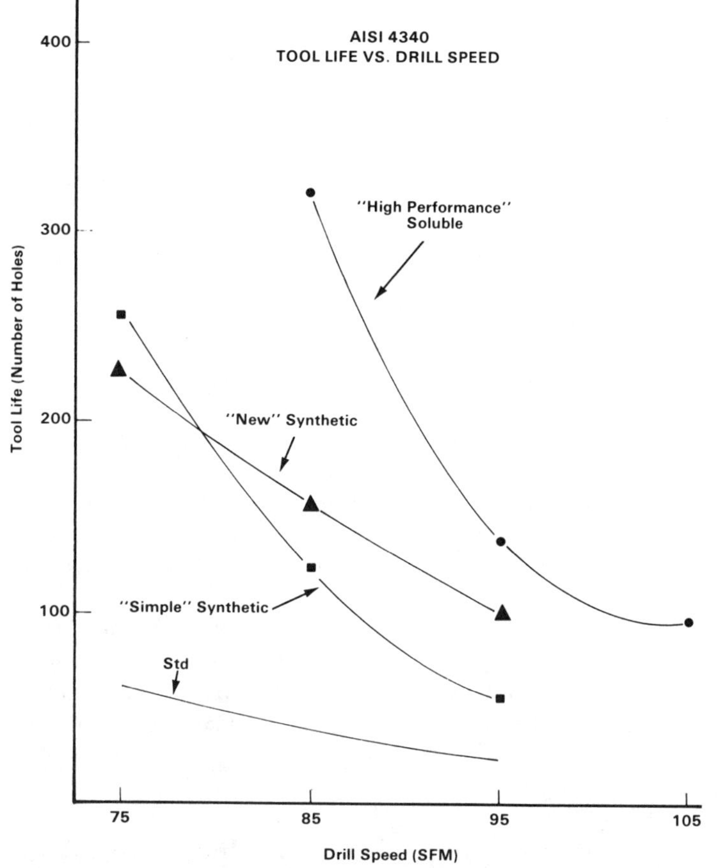

REFERENCES

(1) West, H. L. "The Mechanical Testing of E. P. Lubricants," *Jnl. Inst. Petroleum,* **32,** 206–229 (1946).

(2) Hugel, G. "Chemical Nature of Extreme Pressure Lubrication," *Lubrication Engineering,* **14,** 12, 523–526 (1958).

(3) "Measurement of Extreme Pressure Properties of Fluid Lubricants," ASTM D3233-73.

(4) Boerlage, G. D. "Four-Ball Testing Apparatus for Extreme Pressure Lubricants," *Engineering,* **136,** 46 (1933).

(5) Feng, I. M., "A New Approach in Interpreting the Four-Ball Wear Results," *Wear,* **5,** 4, 275 (1962).

(6) Feng, I. M., Gajral, A., and Shaw, M. C. "Cutting Fluid Performance," *Lubrication Engineering,* **17,** 7, 324–9 (1961).

(7) Katona, E. J., "Energetics of Metal Cutting," SME Paper MR 72-129.

(8) Greenhow, J. N. and Rubenstein, C., "The Dependence of Cutting Force on Feed and Speed in Orthogonal Cutting With Worn Tools," *Int. Jnl. Mach. Tool Design & Research,* **9,** (1969).

(9) Ladov, E. N., "A Tapping Test for Evaluating Cutting Fluids," ASLE Annual Meeting, Chicago, Ill. 1973.

(10) Clouse, R. D., and Hall, E. R., "Machinability or Cutting Fluid Evaluation by Drilling Torque Test," *Lubrication Engineering,* **14,** 11 (1958).

(11) Arzt, P. R., "Cutting Tool and Cutting Fluid Evaluation Economic Considerations for Aerospace Manufacturing," ASTME Paper MR 67-727.

(12) Holodnik, E. and Edwards, L. M., "Evaluation of Cutting Fluids With an Automatic Drilling Machine," ASLE Annual Meeting Chicago, Ill. 1973.

(13) Zlatin, N. and Christopher, J., "Evaluation of The Effectiveness of Cutting Fluids for Industry," ASLE Annual Meeting Chicago, Ill. 1973.

(14) (a) "Metals Handbook"
(b) "SAE Handbook"

(15) U.S. Air Force, Machinability Report **1, 2, 3, 4** (1950, 51, 54, 60) Final Report on Machinability of Materials ATML-TR-65-444 Contract AF33(615)-1385 (Jan. 1966) Metcut Research Inc., Cincinnati, Ohio 1966.

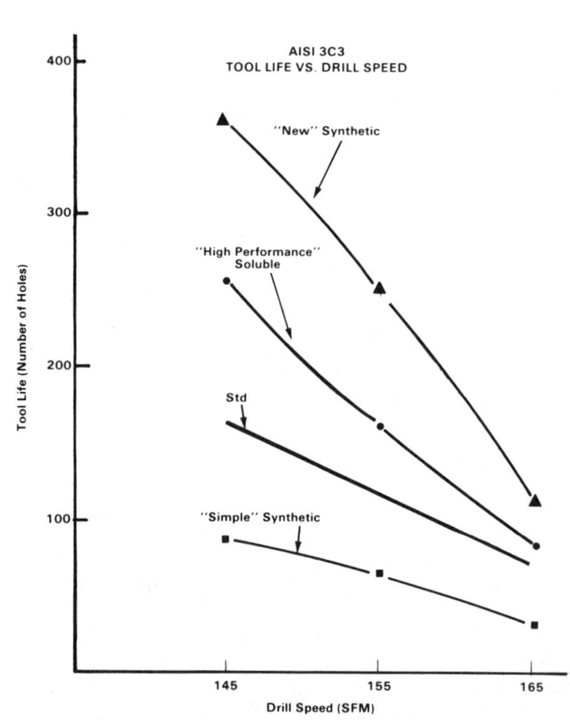

Presented at SME's Cutting Tools Program, May 1976.
Updated for Improving Production With Coolants and Lubricants.

Improving Productivity And Reducing Costs With Cutting And Grinding Fluids

By John C. Quigley
VanStraaten Chemical Company

The proper selection and application of cutting and grinding fluids is an essential element in an effective productivity improvement and cost reduction program. Despite the fact fluids are a key element in the production process, which includes the basic machine tool, perishable tooling and the parts produced, they are classified as non-productive materials by many companies. Yet, these fluids have a direct bearing on the performance of the tools as well as the finished parts. A number of new elements are having a direct affect on the economic significance of these products.

The basic cost of the fluids has increased to the point where they can no longer be used for a short time and discarded. Most companies make a significant dollar investment in processing chemicals. The cost of material used to charge a large central system will amount to several thousand dollars or more. Therefore, maximizing the life of that system provides a substantial savings.

The disposal of spent coolant may have been a relatively simple process in the past. In most cases, spent coolant must be processed before disposal or hauled away to a suitable disposal site. In either case, this is an increased burden and longer serviceable life with the fluid will help to reduce this cost.

Perishable tools have increased in cost. Therefore, increased tool life means reduced unit costs as well as a reduction in overall tool costs.

As a result, it is no longer feasible to purchase cutting and grinding fluids on a cost per gallon basis. Every plant should have formal test procedures for evaluating products on a cost/effectiveness basis. These procedures should be standardized and put in writing to assure reproducibility of procedures and results. This will help make final choices on an objective basis rather than subjective factors. Some of the factors to be included are as follows.

1. Metals Processed In Test Evaluations
2. Test Machine or Operation
3. Tools/Grinding Wheels To Be Used
4. Fluid Concentration
5. Concentration Control
6. Records and Test Reports

By adopting standardized procedures, we know that tests are being run under control conditions and we will be evaluating performance and results under the same operating conditions. It is essential that accurate records be kept so that performance can be measured on a truly objective basis.

The following is an example of a report used to analyze final test results. It has produced highly successful results when used with standardized test procedures.

CUTTING/GRINDING FLUID ANALYSIS

1. Present Fluid - Name and No.________________________________
2. Test Fluid - Name and No.________________________________
3. Test Fluids - Name and No.________________________________
4. Department - Name and No.________________________________
5. Machine - Name and No.________________________________

		#1 Present Coolant	#2 Test Coolant	#3 Test Coolant
A.	No. changes required/yr.	______	______	______
B.	Time req'd. to pump, clean & fill-Hrs.	______	______	______
C.	Labor cost/change -(B x $10.00)	______	______	______
D.	No. gals. required/change	______	______	______
E.	Make up gals. req'd/change	______	______	______
F.	Total no. gals. req'd/change (D+E)	______	______	______
G.	Coolant cost/gal.	______	______	______
H.	Coolant cost/change (F x G)	______	______	______
I.	Bactericide and conditioner cost/chg.	______	______	______
J.	Total cost/change (C + H + I)	______	______	______
K.	Pieces produced/change	______	______	______
L.	Cost/pieces (J - K)	______	______	______
M.	Total cost/yr. (A x J)	______	______	______

Recommend Fluid No. ________________ Savings/Year________________

Test requested by________________ Tested by________________

Is recommended Fluid presently authorized for use: Yes________ No________

No. of machines using #1 Fluid________________

No. of gals. of #1 Fluid used last 12 months________________

Remarks:

This information helps to make an intelligent cost/effectiveness evaluation.

An equally important part of the selection process is the development of basic Decision Criteria. This is a prioritized listing of the factors which should be considered in making a final choice. The list should be in writing to avoid any misunderstandings of the criteria to be considered or misinterpretation. Some of the factors to be included may be:

1. Acceptability of Finished Part
2. Tool Life
3. Corrosion Protection For Machine Tool and Parts
4. System/Sump Life
5. Health & Safety Considerations
6. Waste Disposal Procedures
7. Type of Fluid

It is important to note that the type of fluid ranks last. If the fluid meets all of the other requirements, the type of fluid may not be significant.

It is also important to note that these criteria may vary within a given plant from system to system as well as types of operations involved. The decision criteria for grinding fluids may be quite different than those used to determine the acceptability for cutting operations. The important factor is that, once established, they provide very effective guidelines for making intelligent selections.

The word productivity has a variety of meanings, but for purposes of this discussion it means producing results with specific benefits at effective costs. The evaluation of three fluids, under control conditions on a variety of operations, demonstrates the productivity value of proper selection and utilization of a cutting or grinding fluid.

In Figure I, we examine overall productivity of three fluids, Fluid A, Fluid B and Fluid C were run on an identical operation over a range of four concentrations. Fluid A shows the best overall performance at all concentrations. Fluid B and C offered much lower productivity rates and only slight improvements with increases in concentration. Thus, the proper choice of fluid produces the highest degree of productivity. It is also important to note the influence of concentration. Running the fluid at the proper concentration produces the highest productivity.

In Figure II, we examine the performance properties of the same three fluids on the unit cost of metal removal in a grinding operation. The close reproducibility of the results demonstrates the validity of the procedure. Fluid A provided the best overall results in terms of unit cost of metal removal as well as maximum productivity at the proper concentration.

The value of these tests can be demonstrated by examining the performance of these products on a production broaching operation. The final results corroborate with the test results. Fluid A offers the lowest cost per part produced. In this test it is extremely important to note the affect on tool life, because this is one of the essential elements in the system. By selecting and running the right fluid, the best cost/effective results are produced with respect to the cost of the tool and the unit cost of production.

By electing to use well defined test procedures and well chosen decision criteria, all of us can make a substantial improvement in productivity with substantial cost reductions with cutting and grinding fluids.

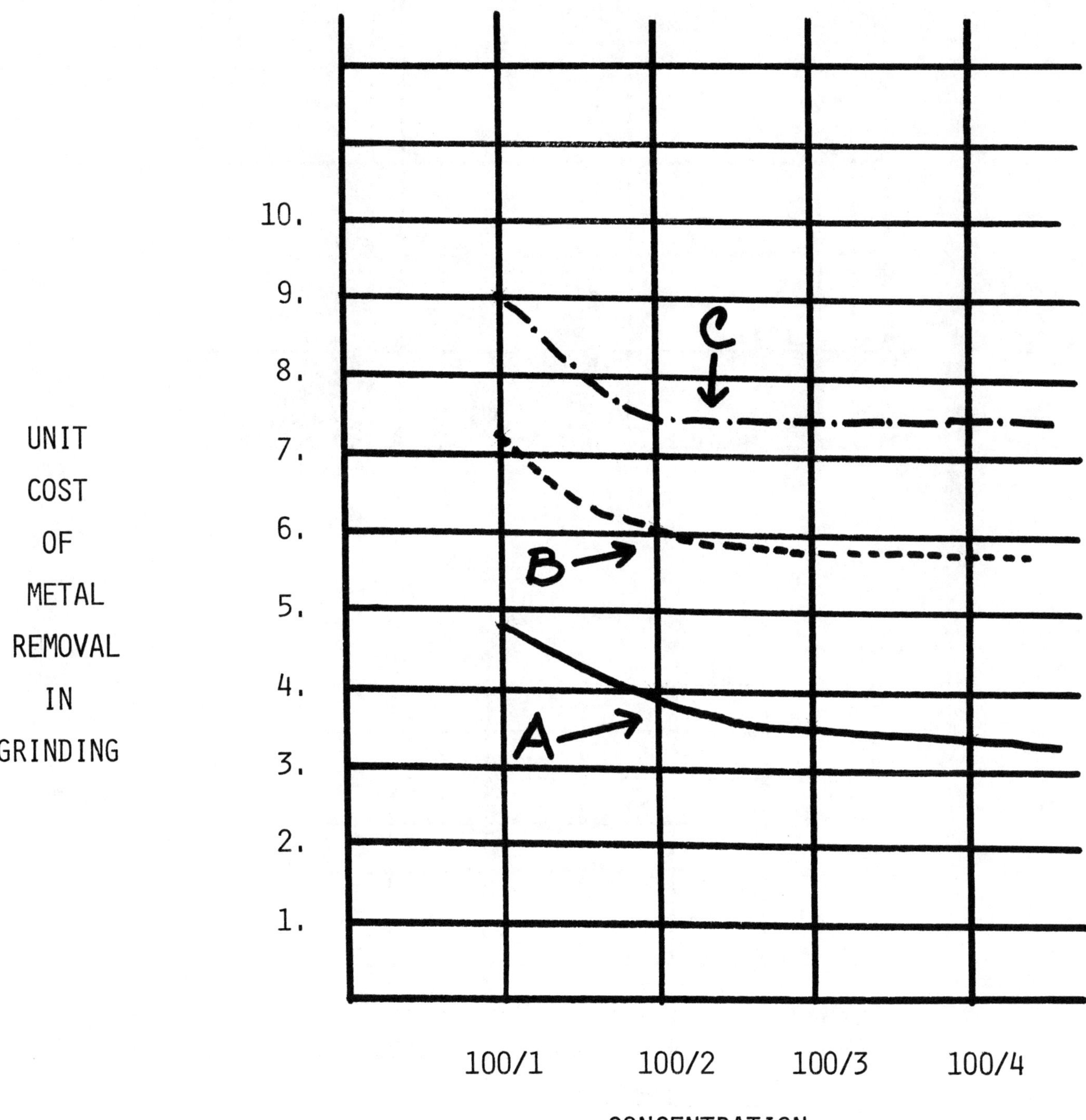
UNIT
COST
OF
METAL
REMOVAL
IN
GRINDING
10.
9.
8.
7.
6.
5.
4.
3.
2.
1.
C
B
A
100/1
100/2
100/3
100/4
CONCENTRATION

COST PER PIECE BROACHED

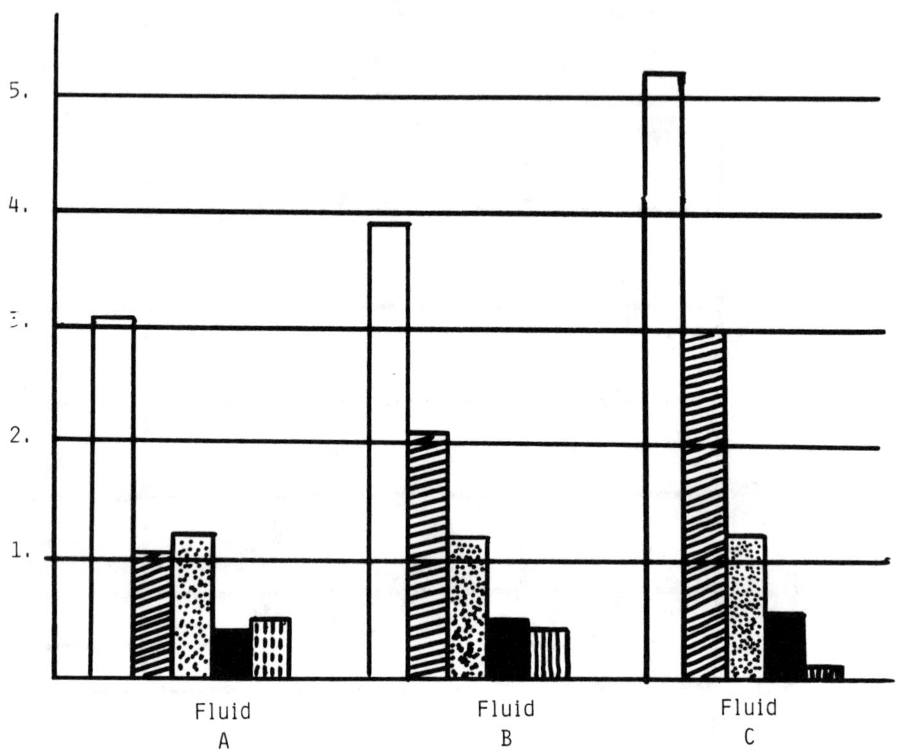

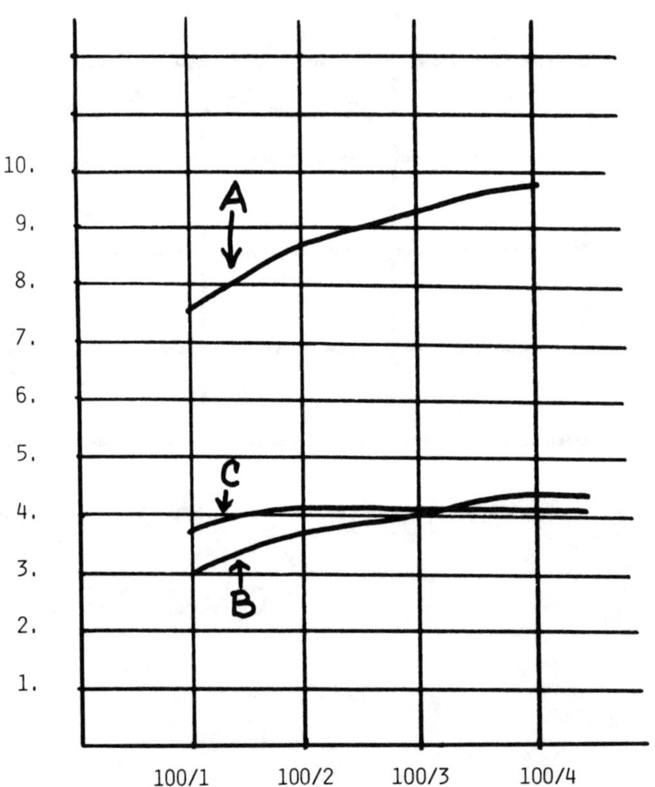

Presented at SME's International Engineering Conference, April 1970

Grinding Fluids

By Milton C. Shaw
Carnegie-Mellon Institute of Technology

In considering the performance of fluids in the grinding of metals, it is important to recognize that there are two general regimes. Those grinding operations concerned with stock removal (Stock Removal Grinding, SRG) have characteristics that are significantly different from those operations concerned with the production of fine finish and form (Finish and Form Grinding, FFG)

Typical SRG operations are abrasive cut-off and the conditioning of slabs and billets in the steel mill. Resin bonded wheels are most frequently used in this type of grinding. The stock removal rate is high and the size of individual chips is large. Coarse grain sizes are used (8 to 30) and the wheels are self dressing. Before large wear flats develop on the grains they either chip to form new sharp cutting edges or whole grains or grain clumps are pulled from the face of the wheel to provide a fresh cutting surface. Values of the grinding ratio (G)

$$G = \frac{\text{Volume of Work Removed}}{\text{Volume of Wheel Consumed}}$$

are relatively low (1 to 25) and wheels of this type are usually operated dry. Spindle vibration plays an important role in this type of operation and such wheels frequently cut intermittently, coming completely out of the cut once during each revolution.

Vitrified wheels are generally used in finish and form grinding and examples of this type of grinding are:

- Horizontal Spindle Surface Grinding
- Cylindrical Grinding both plunge and axial feed types
- Centerless grinding
- Internal grinding

The chips are smaller in size than in rough grinding and wear is usually confined to the grain tips. When the collective flat area on the grain tips reaches a value of from 3 to 5% of the apparent wheel face area the wheel must be dressed using a diamond dressing tool. This provides new sharp cutting edges. The major wheel loss frequently occurs in dressing rather than when grinding and the grinding ratio (G) is not usually a meaningful parameter for finish grinding wheels. Finish grinding wheels have a relatively fine grain size (46 to 100) and they are usually operated with a grinding fluid (water or oil base). Great care is taken to avoid spindle vibration and finish is often refined at the end of an operation by spark out (continuing to grind without further infeed).

Grinding For Stock Removal

When grinding for stock removal using a resin bonded wheel

there is little interest in surface finish. The major concerns are the rate of stock removal and the grinding ratio. For the proper wheel grade the wheel is self sharpening and there is little problem due to wheel face loading, and the formation of a built-up metal layer on grain tip flats. Thus the need for lubrication is relatively low. It is frequently met by incorporating fillers (cryolite) and grinding aids (sulfur, lead salts) when formulating the wheel bond. Stock removal wheels are mostly used dry. However, when used wet the grain should be treated with a water repellant such as silane to prevent water penetrating the grain-bond interface.

When a fluid is used in stock removal grinding it is for cooling purposes. However, the object to be cooled is the wheel rather than the work as in FFG. It is essential that the temperature of the first layer of bond which is located about half the grain diameter (g) from the cutting face of the wheel be kept below the bond degradation temperature which is about 500F.

Figure 1 shows the large grains in the face of a SRG wheel together with resin bond posts. It is essential that the temperature of the bond posts (θ_b) not exceed 500F. The mean chip temperature θ_c may be very much higher and is relatively unimportant in this type of grinding.

Temperature θ_b is reduced in the following ways:

1. By decreasing the time of cutting contact.
2. By use of large grain size (g).
3. By reducing the intensity of heating (q).
4. By removing all heat introduced into the wheel during a cut on the return portion of the revolution.
5. By having the infeed rate (d) be sufficiently high relative to the rate of heat flow into the work so that heat does not accumulate in the work.

These ways of limiting θ_b to an acceptable value may be explained in terms of Figure 2. The cutting time will be the length of cut ℓ divided by the cutting speed V, i.e. (ℓ/V). This is the time during which heat will flow from the grain tips to the first layer of bond posts. The quantity ℓ/V may be decreased by limiting ℓ to small values or by increased wheel speed (V). An increase in grain size (g) will increase the distance the heat must flow to the first row of bond posts.

The intensity of heating q will be

$$q = \frac{u(b\ell d)}{\ell d} = ud$$

where u = mean energy per unit volume required to generate chips
b = axial width of cut
ℓ = length of cut
d = infeed rate, ipm (Figure 2)

cooling and water base fluids which have a stronger cooling than lubricating capability are normally used. It is not always advantageous to employ a coolant in SRG and the best way of determining whether one should be used is by comparing specific grinding costs under optimum operating conditions (minimum specific cost) when operating wet and dry.

Details concerning the mechanics and economics of SRG grinding may be found in references 1 to 3.

Grinding for Form and Finish

In FFG, lubricants play a much more important role. Chips are relatively small and the specific grinding energy is relatively large. Surface finish and surface integrity are items of major concern.

In fine grinding, flats develop on the grain tips and there is a good deal more rubbing. Also, some of the cutting is done by the hard ceramic bond (4) which quickly develops flats. There is a tendency for metal to weld to the grain and bond flats and when this happens large weld areas result, the grains ship periodically and the time between successive dressings decreases. Active lubricants tend to prevent transfer of metal to the grain flats by contaminating the surface. This can occur by chemical or physical action. Chlorine sulfur and phosphorous additives in the fluid tend to react with the freshly ground metal surfaces to form brittle ionic inorganic compounds which tend to prevent adhesion. Fatty additives in an oil may absorb on the clean metal surfaces to screen them against welding. They may also react to form metal soaps. In any case the action of the active "lubricants" in a cutting fluid are to contaminate the freshly cut surface to prevent strong metallic bonds from being attached to the surface of the grain.

Those metals which readily form oxides (carbon steels) are less in need of contaminating action in fine grinding. The oxides form quickly upon contact with air and these oxides provide the contaminating screening layer necessary to prevent extensive metal transfer to the grain flats. The very hard friable grains work well with oxide forming steels, even dry, due to the screening action of the oxides formed on freshly cut surfaces.

Stainless steel requires a more ductile (less hard and strong grain) since it does not form protective oxides as readily. The stainless steels are also in greater need for the screening action of active ingredients in grinding fluids. The high temperature alloys are still more stable than the stainless steels and are even more in need of help from active ingredients in the grinding fluid.

The titanium alloys are the most difficult to screen against the formation of large welds between grain flats and chips. So great is the tendency for these materials to weld to Al_2O_3 in dry air, the

This is seen to depend on the product of specific energy (u), which may be decreased somewhat by grinding aids incorporated in the bond and in feed rate (d).

The main function of a grinding fluid in SRG is in connection with item 4. If a coolant is present heat may be more effectively extracted from the wheel during the time it is out of contact with the work. As a general rule, for dry grinding with a resin bonded wheel ℓ should not exceed 5% of the wheel periphery. With a copious flow of fluid all the way to the wheel face this percentage may increase. An alternative way of taking care of item 4 without a fluid is by oscillating the work relative to the wheel in a direction perpendicular to the downfeed (d). This causes the effective value of l to be less than the work width.

The wheel diameter of course also plays an important role relative to item 4. The larger the wheel diameter (D) the greater the length of cut ℓ that may be handled at a given specific grinding cost.

If d is not sufficiently high heat will penetrate more rapidly into the work than the wheel does and the temperature will rise to a high value. It is thus evident that d should not be too low from the point of view of item 5 and not too large from the point of view of item 3. We might thus expect an optimum value of d to exist and this is in fact found to be the case. When grinding ratio (G) is plotted against downfeed (d) for constant values of wheel speed (V) (d), the curve is found to have a maximum value of G at a particular value of d.

Use of a coolant may not always be advantageous in SRG. This involves the relation between θ_c and θ_b (Figure 1). It is desirable to have θ_c high in this type of grinding since this causes thermal softening of the work and a reduction in u. On the other hand, θ_b must be kept below 500F and in general θ_b rises with θ_c. The relationship between θ_c and θ_b is so complex that it is not possible to predict whether use of a fluid will give rise to an increase or decrease in the removal rate (M) and in the G value. The problem is further complicated by the fact that a machine should not be operated in SRG to provide a maximum value of (M) or a maximum value of (G) (which do not generally correspond to the same operating conditions) but under conditions of minimum specific grinding cost. The accepted procedure is to adjust the machine to operate under cost optimum conditions dry and wet and compare the corresponding specific grinding costs.

The value of q will generally increase with wheel speed since under cost optimum conditions for different speeds d/V will tend to be a constant. Thus in practice q will increase with V but the time of heating ℓ/V will decrease. Since these effects are in opposite directions it is difficult to predict what happens to specific cost with an increase in V. However, in practice it is found that specific grinding cost decreases with an increase in wheel speed (V).

In summary, the role of the grinding fluid in SRG is one of

this. Here, θ_c is the temperature of a freshly ground surface as measured by a special surface radiation technique (9) and t is the thickness of individual chips. These tests were made on a horizontal spindle rotary surface grinder using a 7 inch diameter A46-J5-V wheel at a speed of 5680 fpm. The work material was AISI 52100 steel and the work speed ranged from 20 to 260 fpm.

It is evident from Figure 4 that all fluids give about the same temperatures which are considerably lower than those for a dry cut. The temperatures with fluid are essentially the same despite the differences in specific energy (u) observed. The grinding oils gave significantly lower values of specific energy (u) but are poorer coolants. Consequently they give about the same surface temperatures as the water base fluids.

The finish and wheel life will be better for the grinding oils of Figure 4 but all fluids should give about the same surface integrity (since this depends primarily upon θ_c).

The maximum surface temperatures are seen to occur at rather small chip sizes. Values of surface temperature are seen to fall off with either an increase in chip size or with spark-out.

Water base fluids are available which contain active ingredients which cause chemical reactions to occur at the freshly cut surfaces. These water base materials are generally known as heavy duty soluble fluids.

The choice between a water or oil base material appears to hinge on the importance of finish. Grinding oils will generally give the best finish available for a given operation, particularly when the wheel wear must be maintained at a very low level for dimensional accuracy. However, if wheel wear is less critical and a softer wheel with friable grains may be used, results in FFG can be quite acceptable with a water base grinding fluid.

Water base fluids are of course to be preferred relative to ease of application. Oil base fluids often require elaborate means for pollution control due to their greater tendency to create mist and fumes in the atmosphere. Water base fluids are generally less expensive than oil base materials but this is rarely an important consideration. This is because the amount of fluid consumed per part is so low that the cost of the fluid is completely insignificant regardless of the unit cost of the fluid.

High Speed Grinding

Recently there has been a trend toward higher wheel speeds in FFG. Where the conventional speed in plunge cylindrical grinding was 8500 fpm a few years ago it is now 12,000 fpm and 16,000 fpm and higher speeds are being talked about.

The big incentive for going to higher wheel speeds is that this enables a higher stock removal rate for the same finish. The finish obtained depends primarily on the mean size of the individual chips produced and this varies as $\sqrt{\frac{v\sqrt{d}}{V}}$, where v is the work speed and V is the wheel speed and d is the feed per revolution (10). Therefore for a given chip size and finish v may be increased in direct proportion to V. The rate of stock removal in plunge cylindrical grinding is

$$M = vdb, \text{ in}^3/\text{min} \qquad (2)$$

where v = work speed
d = infeed in inches per revolution of the work
b = axial width of cut

Therefore, holding d constant, the rate of stock removal for the same finish varies as v which in turn varies directly with V.

On the negative side we have the fact that surface temperature θ_c varies as the intensity of heating (q), where for cylindrical plunge grinding

$$q = \frac{uM}{A} \qquad (3)$$

where u = energy per unit volume
A = Area being ground at any one time

The quantity A is $b\sqrt{Dd}$ (4)

where D = Wheel diameter
d = infeed, in/rev of work
b = width of cut

Therefore, from equations (2), (3), and (4)

$$q = \frac{u(vdb)}{b\sqrt{Dd}} = uv\sqrt{\frac{d}{D}} \qquad (5)$$

The intensity of heating of the workpiece (q in BTU/in^2/sec) will therefore vary directly with v. If v is increased in proportion to wheel speed V, so as to maintain finish constant then the intensity of heating q will increase approximately in proportion to V.

The main problem in high speed grinding is to prevent overheating of the workpiece. This may be done most effectively by use of a grinding fluid and making sure it gets to the wheel and work surfaces.

Since high surface temperature is the item of concern in high speed grinding we might assume that a water base fluid would be most effective. However, in practice it turns out that oil base materials give better results in this type of grinding. Figure 5 (11) shows the

G value will be as low as one in ordinary surface grinding (5). When distilled water is used as a grinding fluid the G value drops to 1/2.

Figure 3 shows a side elevation of the ions in the surfaces of $A\ell_2O_3$ and of titanium to scale. There are 2/3 as many small $[A\ell^{+3}]$ ions in the $A\ell_2O_3$ surface as there are large $[O^{-2}]$ ions. If these surfaces come close together in air a strong electrostatic bond is established which is frequently sufficient to cause rupture to occur within the $A\ell_2O_3$ when the surfaces separate. The presence of water weakens the $A\ell_2O_3$ surface (6) and causes a decrease in the G value. If, however, certain inorganic salts are dissolved in the water the G value may be greatly increased. This is due to the adsorption of positive ions (cations) on the oxide surface and negative ions (anions) on the metal surface, both of which tend to prevent the surfaces from coming close enough together to form strong bonds. The best ions in each case for this action are those which have the same size as the surface to be screened and which have the greatest charge. This turns out to be the barium ion $[Ba^{2+}]$ for screening $A\ell_2O_3$ and the phosphate ion $[PO_4^{-3}]$ for screening the metal surface. Unfortunately $Ba_3(PO_4)_2$ is not soluble in water. Therefore something close to this in behavior that is water soluble must be used ($Ba(OH)_2$ or $Ba-O_3P-(OCH_2-CH_2)_{10}-O-PO_3]-Ba$. When such solutions are used in a 0.1 M concentration the G value may be substantially increased in the surface grinding of titanium alloys.

When grinding titanium alloys or similar difficult to grind materials it is important to decrease the wheel speed (V) (7). This is because surface temperatures will be very high for such materials due to their low specific heat and thermal conductivity. The surface temperature in grinding varies as $\sqrt{V}$. Therefore, titanium alloys should be ground at wheel speeds of the order of 2000 to 3000 fpm, using a water base fluid containing effective screening ions and a wheel containing white $A\ell_2O_3$ grain.

Zorew and Klautsch (8) have studied the grinding of pure molybdenum. This material suffers from the same difficulty as titanium alloys and should be ground at low wheel speeds (2500 fpm) using an active water base fluid (6% Urotropin plus 6% $KC\ell$).

Active water base fluids were found to be superior to oil base materials when grinding both titanium alloys and molybdenum.

If low wheel speeds are not used in grinding titanium alloys high temperature alloys and molybdenum, surface cracks are apt to develop. However, at lower wheel speeds the surface roughness increases. Therefore a compromise in wheel speeds must normally be used from the points of view of surface integrity and surface finish.

In the fine grinding of steel both water base and oil base materials are effective. The water base fluids provide the best cooling action while the oil base materials generally provide the best lubrication action. Figure 4 shows some data which illustrate variation in mean grinding temperature with metal removal rate M' ($in^3/in/min$) in cylindrical plunge grinding for wheels operated at different speeds (V), both dry and with an active grinding oil. The temperature is seen to increase significantly with removal rate (M') but to a lesser extent with increase in wheel speed (V). The mean grinding temperature is very much lower when a grinding oil is used than when grinding dry. A water base fluid was found to give results intermediate between the dry and oil cases in this instance.

At high wheel speeds it is difficult to get the fluid to penetrate the boundary layer of air which clings to the wheel face. If the fluid is applied externally it must either have sufficient velocity to penetrate this boundary layer or the boundary layer must be scraped from the wheel face before the fluid is applied. To penetrate the boundary layer requires a fluid jet accelerated through a nozzle by a pressure of the order of 250 psi. The boundary layer removal technique has been applied in Germany and one of these applications developed in the Aaken Technical University (12) is shown in Figure 6. Just before the wheel reaches the work a scraper removes the air boundary layer and then fluid is immediately applied to the wheel face by means of a close fitting nozzle box. This type of system requires a large fluid flow rate at moderate pressure ($\sim$50 psi).

In high speed grinding with a fluid considerable heat goes into the fluid. The quantity of fluid in the sump must be sufficient to handle this heat without causing too high a temperature rise and normally a heat exchanger must be used to remove heat from the sump.

Practical Considerations

In the application of grinding fluids there are of course many practical considerations that may prevent a material that looks attractive in the laboratory from being used in practice. In addition to surface finish, rate of wheel wear, surface temperatures and surface integrity that would normally be looked at in the laboratory the following items must not be overlooked in application.

1. Adverse physiological action on operator.
2. Atmospheric pollution (misting and fuming tendency).
3. Adverse action to machinery (rust deposits, paint removal, reaction with nonmetallic machine elements).
4. Bactericidal growth and odor development.
5. Ease of separation from swarf and from parts.
6. Stability and life of active ingredients.
7. Ease of disposal when spent.
8. The inconvenience associated with use of a fluid of any type.

Acknowledgement

Some of the concepts presented in this paper have been supported by research grants from the Grinding Wheel Institute and the Abrasive Grain Association.

References

1. Shaw, M. C., D. A. Farmer and K. Nakayama, "Mechanics of the Abrasive Cut-Off Operation," Trans ASME, 89B, 495, (1967).

2. Farmer, D. A. and M. C. Shaw, "Economics of the Abrasive Cut-Off Operation," Trans ASME, 89B, 514, (1967).

3. Farmer, D. A. and M. C. Shaw, "Economic Factors in Abrasive Machining," Proc ASTME, Detroit 1967, MR67-595.

4. Tanaka, Y., H. Tsuwa and S. Kawamura, "Rubbing of Abrasive Grains in Grinding Process," Bull Japan Soc Precision Engrs 1 No. 3, 177 (1965).

5. Shaw, M. C. and C. T. Yang, "Inorganic Grinding Fluids for Titanium Alloys," Trans ASME 78, 861, (1956).

6. Westbrook, J. H. and P. J. Jorgensen, Trans Amer Inst Min. Metall. Engrs. 233, 425, (1965).

7. Yang, C. T. and M. C. Shaw, "The Grinding of Titanium Alloys," Trans ASME, 77, 645, (1955).

8. Zorew, N. N. and D. N. Klautsch, "Eigenheit des Schleifvorganges beim intensiven ermudungsadhasiven Verschleiss der Schleifkorner," Annals of the Intern. Instn for Production Eng Research 16, 267, (1968).

9. Mayer, J. E. and M. C. Shaw, "Grinding Temperatures," Lub Eng. 13 21, (1957).

10. Reichenbach, G. S., J. E. Mayer, S. Kalpakcioglu and M. C. Shaw, "The Role of Chip Thickness in Grinding," Trans ASME, 78 847 (1956).

11. Opitz, H. and K. Gühring, "High Speed Grinding," Annals of the International Institution for Production Eng. Research 16, 61 1968.

12. Gühring, K., "Hochleistungschleifen - Eine Methode zur Leistungssteigerung der Schleifverfahren durch hohe Schnittgeschwindigkeit," Dissertation, Aaken Technical University, 1967.

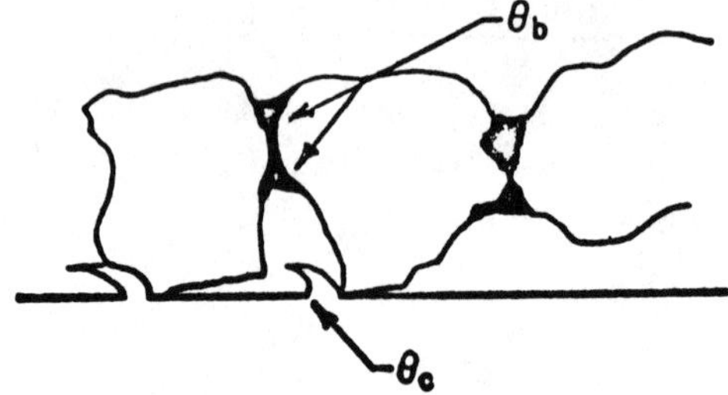

Figure 1. Schematic diagram of grain and bond. The temperature θ_b is the mean bond temperature and the temperature θ_c is the mean chip temperature.

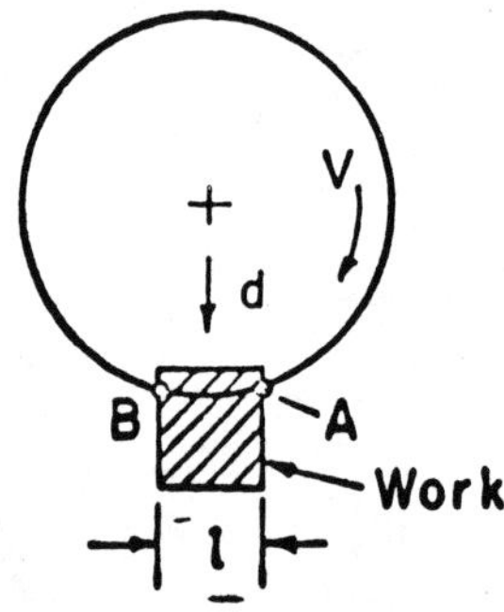

Figure 2. The abrasive cut-off operation.

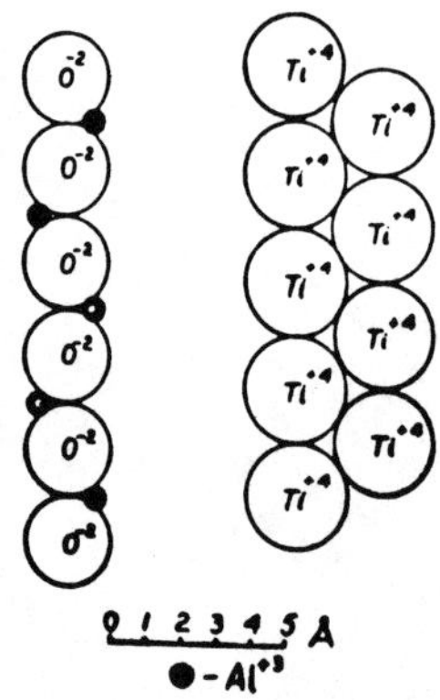

Figure 3. Side elevation of aluminum oxide and titanium surfaces shown to scale.

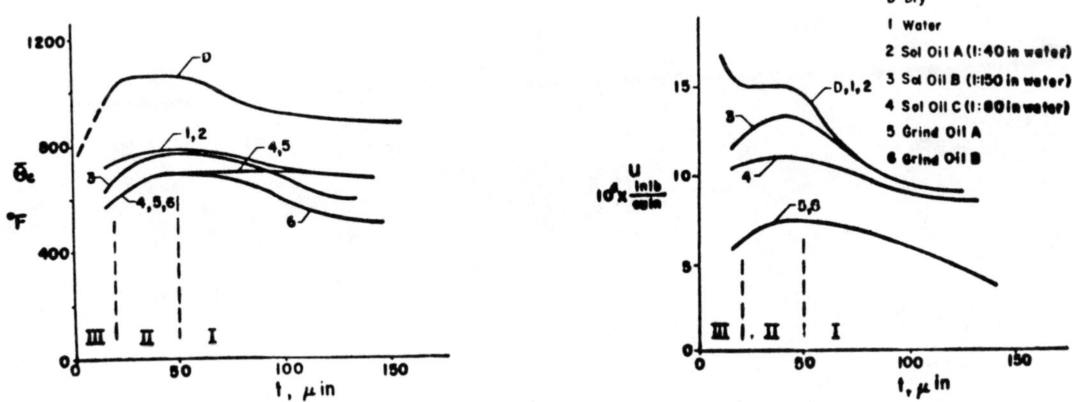

Figure 4. Variation of surface temperature (θ_c) and specific grinding energy (u) with undeformed chip thickness (t) for various grinding fluids. Work Material AISI 52100, wheel A46-J5-V, wheel speed 5680 fpm.

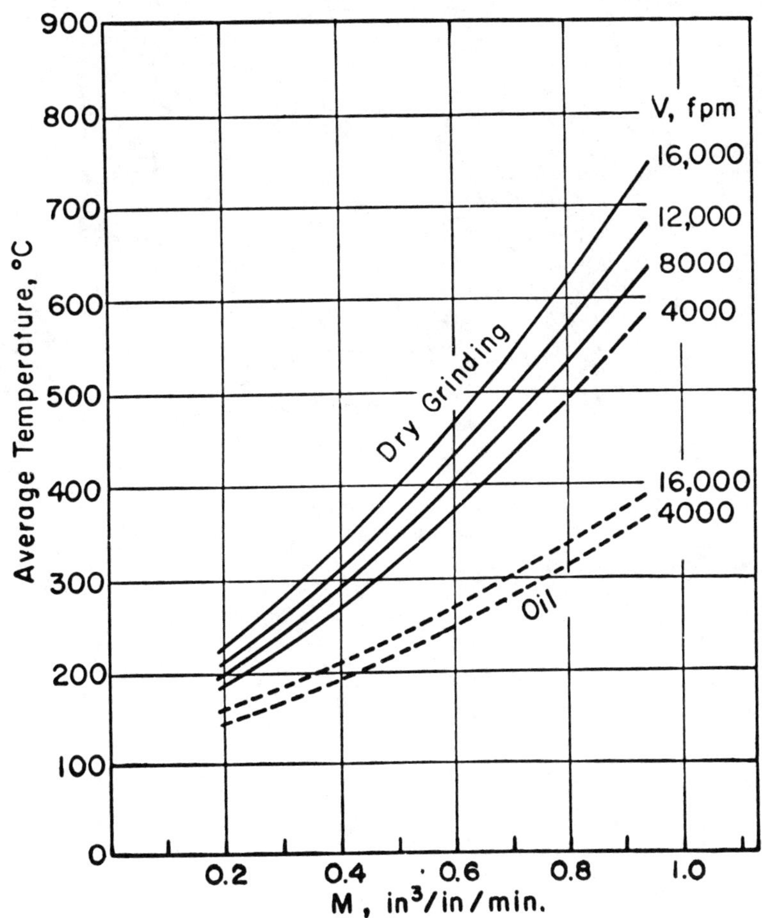

Figure 5. Variation of mean grinding temperature (θ_c, °C) with rate of metal removal (M', $in^3/in/min$) for different wheel speeds (V) and fluids. Wheel, EK80-J7-VX; work AISI 1045; work speed v = 90 fpm. (after Opitz and Gühring, 11).

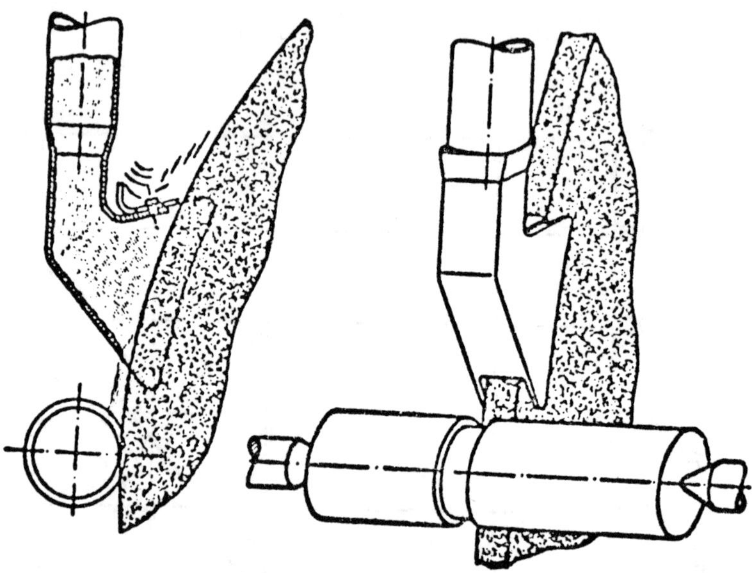

Figure 6. Special nozzle for use in supplying fluid to a high speed grinding wheel (after Gühring, 12).

Presented at SME's 1976 International Tool and Manufacturing Engineering Conference, April 1976

Machine Design Considerations For Improving Metalworking Fluid Performance

By the Metalworking Fluids Division of SME Technical Council.

This paper provides recommendations for improving machine tool design and installation practice which can significantly improve shop operations, and reduce maintenance costs. It is directed to the efficient use of water based cutting and grinding fluids, as applied to individual machine tool installations. Primary attention is given to design and location of the cutting fluid reservoir, the need for efficient removal of chips, and protecting the system from contaminants. Flow rates and optimum reservoir size are discussed. Effects of inefficient installations on maintenance costs are provided.

PURPOSE

To provide guidelines for machine tool and reservoir design which can contribute to:

* Improved machining efficiency
* Lower fluid maintenance costs
* Lower machine maintenance costs
* Improved, cleaner working conditions
* Reduced disposal requirements of fluids

SCOPE

This paper provides recommendations for improving machine tool design and installation practice which can significantly improve shop operations, and reduce maintenance costs. It is directed to the efficient use of water based cutting and grinding fluids, as applied to individual machine tool installations. Primary attention is given to design and location of the cutting fluid reservoir, the need for efficient removal of chips, and protecting the system from contaminants. Flow rates and optimum reservoir size are discussed. Effects of inefficient installations on maintenance costs are provided.

INTRODUCTION

Coolant reservoirs, piping and chip removal systems on individual machine tools are generally not adequate for modern, efficient shop operations; this is the result of overlooking cutting fluid and chip removal requirements at the machine design and installation stages. Cutting and grinding fluids are perishable items, and their change intervals significantly affect production efficiency, as machine tool downtime and associated maintenance costs are involved.

Cutting fluid contamination is a major factor which determines cutting fluid life. This contamination is in the form of chips, machine tool lubricants, and cutting fluid residue from previous charges. This paper discusses the difficulties of removing these contaminants, and provides recommendations for more efficient design and installation.

The inefficiency of some present practice is sometimes not readily apparent to management, as clean out and chip removal operations are not treated as a direct cost, but lost in a manufacturing overhead or maintenance budget. For this reason, actual cost studies of highly inefficient installations have been included in this paper.

DISCUSSION

The efficiency of a machine tool is dependent on the performance of the cutting fluid, and the ease of removing chips. The performance and life of a cutting fluid charge is affected by contaminants which deteriorate the fluid prematurely. The most harmful contaminant is the micro-organisms which enter the fluid from air, water, dirt, debris and other sources. Bactericides in the fluid are designed to inhibit normal contaminations; high levels of contaminants nullify their action. Biological deterioration is accelerated by:

a. Poor circulation in the piping (low flow rate or stagnant components), on the machine tables or in the reservoir. Stagnant fluid in contact with high levels of dirt, debris, and chips is a particularly severe condition.

b. Prolonged contact of chips and the fluid increases micro-organism activity.

c. Food sources introduced in the form of hydraulic oils and way oils.

The guidelines in the next section provide recommendations for controlling these contaminants, and for simplifying the cleaning of machine tool reservoirs.

Efficient chip removal and handling is another important consideration in machine design and installation practice. In addition to micro-organism activity, particles produced in the metalworking operation contribute to poor work finish, fluid line plugging, pump wear, machine tool wear, and other problems. It is important that the chip particles be removed from the work area, work tables, coolant lines and reservoir system, by proper design of machine beds and fluid handling systems.

The third area of mutual interest to the machine tool builder, the fluid supplier, and the cutting fluid user is machine/fluid compatibility. There is a need to consider the compatibility of fluids with machine tool components, gaging equipment, piping, gaskets and reservoir materials of construction. Attack of gasket material, galvanic corrosion of dissimilar metals, or corrosion of the piping system need to be considered in selecting materials for machine tools in contact with water based fluids. Further detail in this area is beyond the scope of this paper, and the reader can get additional information in the S.M.E. Handbook of Cutting Fluids (8).

MACHINE TOOL DESIGN AND INSTALLATION RECOMMENDATIONS

1. Machine tool builders should use external coolant tanks wherever possible, rather than reservoirs in the machine base, to allow proper sizing, facilitate cleaning, and simplify the removal of chips. Do not install machine tools in shallow pans (as a substitute for a reservoir), as the cutting fluid will be subject to many forms of harmful contamination. See Fig. I & II, with Fig. I type installation being preferred.

2. Reservoirs below floor level should have clear access for cleaning and chip removal;

   a. avoid depressed areas around the machine tool, which will accumulate dirty coolant and other debris; this will eventually contaminate the clean fluid. These depressed areas also allow drainage of floor cleaning compounds into the fluid, introducing not only dirt and micro-organism contamination, but chemicals which are not compatible with the cutting fluid.

   b. do not use floor gratings as covers; solid steel plate installed for drainage away from the reservoir is recommended.

   c. the use of concrete reservoirs is not recommended, unless the surface is properly sealed, or more preferable - covered with steel plate.

3. Where machine design or floor space requirements dictate the placement of the reservoir within the machine base:

   a. all interior surfaces should be unpainted and incorporate no ribs, channels, etc. which complicate the removal of chips, sludge etc. Compare Fig. III with Fig. IV.

   b. sufficient <u>large</u> access holes should be provided so that all interior surfaces are within arms reach of clean-out personnel.

c. avoid the use of heavy, difficult to remove coolant and chip guards, as this complicates the cleaning process, particularly on vertical turret lathes.

4. All exterior machine surfaces in contact with coolant flow, should be:

   a. smooth and free of depressions or other "dead" spots, which allow the coolant to stagnate.

   b. sufficiently sloped to promote the washing of chips or grinding swarf to the coolant reservoir.

5. All channels or troughs used to return coolant should be free of turns sharper than 45 degrees; coolant supply lines should be made from ungalvanized black iron pipe.

6. For those machines incorporating chip conveyors, installation should be designed for ease of removal, to facilitate cleaning.

7. For those machines not equipped with a conveyor, some means of readily separating and removing the chips from the coolant reservoir is desirable. Here, the fluid system should be equipped with some type of separator, trap or filter, to prevent build up of chips in the reservoir.

8. Coolant Pump Capacity:

   a. for general purpose machining and grinding operations, gal/min = machining horsepower

   b. for high production machining and grinding operations, gal/min = (2 to 4) x (machining h.p.)

9. Reservoir Capacity:

   a. for general purpose machining and grinding operations

      1. grinding tank volume = gal/min X 10
      2. machining cast iron and aluminum tank volume = gal/min X 7
      3. machining steel tank volume = gal/min X 5

   b. for high stock removal machining and grinding operations = gal/min X 10 to gal/min X 20

10. Machine gear boxes and servo mechanisms should be designed to prevent entry of fluids, and to minimize leakage of lubricating and hydraulic oils.

11. The flow of cutting fluid over machine tool ways should be avoided, or covers should be provided to minimize the

removal of way oil by the cutting fluid.

12. Electrical controls should be located to avoid exposure to high humidity conditions, or to direct splash by the cutting fluid.

13. Cutting fluid supply lines should be sized to maintain a fluid velocity of approximately 10 feet per second, to avoid build up of fines etc.

POTENTIAL SAVINGS FROM IMPROVED DESIGNS FOR MACHINE TOOLS

There are several factors which are directly affected by machine tool reservoir design. Improved designs can lead to significant cost savings to the manufacturer from:

1. Increasing productivity through reduction of machine tool down time for maintenance and clean out.

2. Lower waste disposal costs by reduction in the volume of deteriorated cutting fluid.

3. Savings in direct labor, energy and chemicals through easier maintenance.

Rossmore (1) Studied the relationship of cleaning procedures to the growth rate of bacteria on a machine tool with a 200 gallon reservoir, broaching cast iron. This study showed the following trend:

CLEANING PROCEDURE	BACTERIA GROWTH RATE
Drain, Clean, Rinse & Recharge	"X"
Drain, Rinse & Recharge	2X*
Drain & Recharge Only	20X

* assuming no significant residue of chips remaining

In a similar study, Tomko (2) showed that effective cleaning of a machine can lead to significant improvement in the fluid life and that partial cleaning lead to only slight improvement. These studies were done on vertical turning lathes.

These studies indicate that a minimum of a twofold decrease in the number of clean outs required can be obtained, with possibly much more significant number of reductions possible, due only to changes which effect the biological deterioration of the cutting fluid.

A major automotive company has analyzed in some detail the time required for cleaning automatic screw machines. This study is shown in Table 1. The average cleaning requires 9 man-hours per machine and uses four different trades at an approximate cost of $100.00 per machine tool.

Using this data, we can estimate the direct cost savings for a 100 gallon reservoir machine tool which could be obtained by improvements in design which would reduce the time required for cleaning and the number of clean outs necessary by a factor of two as:

1. Productivity - a 5% increase in annual productivity or a savings of approximately 100 - 125 machine-hours per-annum. The value of these savings will depend upon the standard overhead factor used or the value of the product produced.

2. Labor - this is estimated at 100 - 125 man-hours for machine per-annum and account costs are somewhere between $1,200 - $1,500 per machine tool per-annum.

3. Waste Disposal - a 30 - 50% decrease in the volume of waste effluent. This amounts to a savings of approximately 3,750 gallons per machine per-annum. Besides the savings in waste disposal, of course, there are the direct savings in terms of reduced use of cutting fluid and reduction in cleaners and other chemicals used in maintenance.

There are two areas in which capital costs could also be affected:

1. Maintenance Equipment - savings in decreased downtime and less frequent dumping could mean a corresponding reduction in the number of maintenance vehicles (sump wagons) required in a small-medium size operation involving 50 machines per plant or more.

2. Capital savings required for a waste disposal plant could be significantly affected; particularly if cutting fluids were one of the major components, since this volume could be reduced by as much as 50 to 60%.

The significance of these savings in productivity, direct costs, and capital costs can be seen if we examine the total metalworking industries. The average growth rate in new machine tools for the metalworking industries has been estimated to be in the range of 1.5% per year (3), or 30,000 new machines. Furthermore, the average life of a machine tool is very long well in access of 15 - 20 years.

If these guidelines for design of machine tool reservoir systems were implemented, then it is clear that the increase of productivity, savings and labor, energy and chemical will more than pay for the capital costs of the machine tool itself.

TABLE 1

COSTS TO CLEAN AUTOMATIC SCREW MACHINE

WORKER CLASSIFICATION	JOB PERFORMED	AVERAGE MAN HOURS	COST/MAN HOUR	COST
ELECTRICIAN	DISCONNECT MOTOR	0.5	11.74	5.87
TINSMITH	REMOVE SHIELDING	1.0	11.74	11.74
MILLWRIGHT	PULL CONVEYORS	1.0	11.74	11.74
BUILDING SERVICES	PUMP MACHINE TOOL CLEAN OUT CHIPS CLEAN CONVEYOR	4.0	10.19	40.76
MILLWRIGHT	REPLACE CONVEYOR	1.0	11.74	11.74
TINSMITH	REPLACE SHIELDING	1.0	11.74	11.74
ELECTRICIAN	CONNECT MOTOR	0.5	11.74	5.87
	TOTAL	9.0 man/hrs.		$99.46

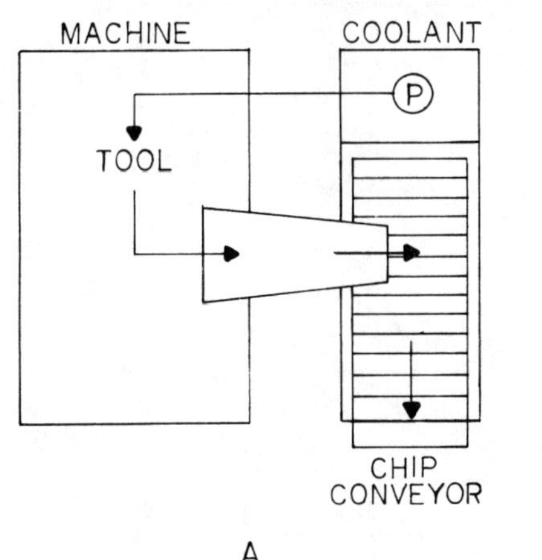

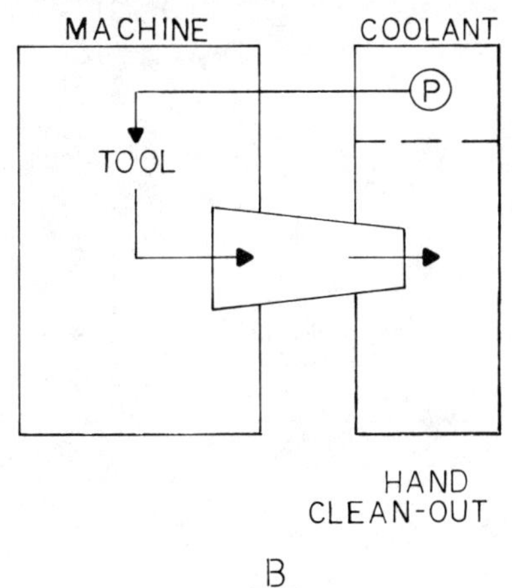

TYPICAL RESERVOIR INSTALLATION
FIGURE I

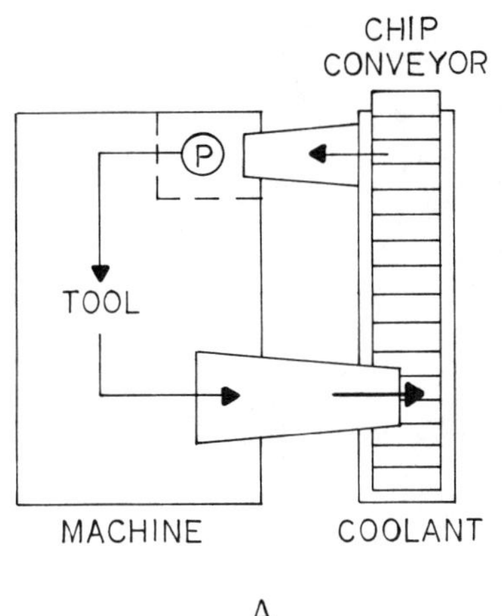

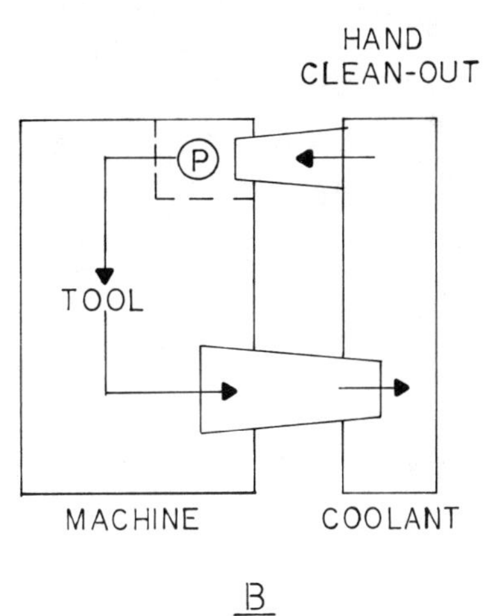

TYPICAL RESERVOIR INSTALLATION
FIGURE II

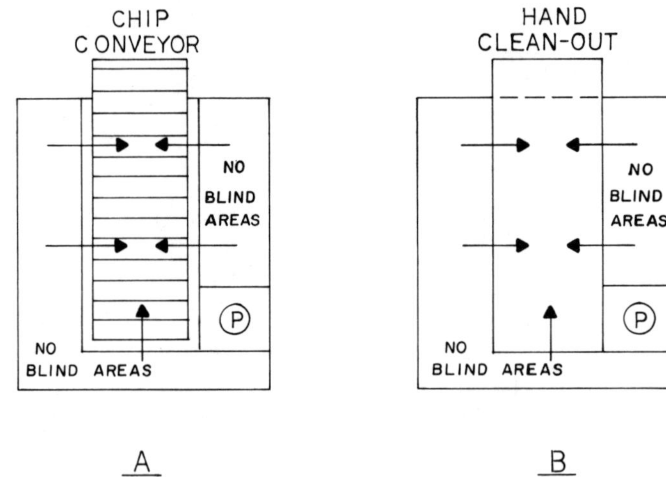

TYPICAL RESERVOIR INSTALLATION
FIGURE III

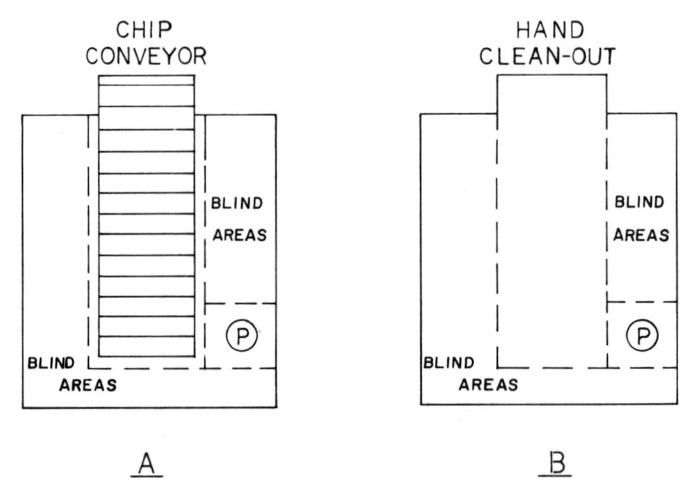

TYPICAL RESERVOIR INSTALLATION
FIGURE IV

REFERENCES

(1) H. W. Rossmore "Microbiological Causes of Cutting Fluid Deterioration" SME MR74-169.

(2) Joseph Tomko "Cutting Fluid Selection and Maintenance Factors Which Determine Product Longevity" SME MF74-170.

(3) American Machinist (1973) - 11TH American Machinist Inventory of Metalworking Equipment.

(4) Robert Brandt "Bacteria - Tramp Oil Problems in Central Coolant System - A Unit Solution" SME MR73-119.

(5) E. O. Bennett "Microbiological Aspects of Metalworking Fluids" SME MR73-826.

(6) Joseph Tomko "Cutting Fluid Selection and Maintenance Factors Which Determine Product Longevity" SME MF74-170.

(7) Mark Opachak "Central System and Cutting Fluid Control Programs" SME MR74-157.

(8) "Cutting and Grinding Fluids: Selection and Application," R. K. Springborn, Editor Manufacturing Data Series, SME.

(9) "Report on Cutting Fluid System Sizing For Machine Tools," G. F. Wilson, SME, 1972.

(10) "Swarf and Machine Tools," edited by P. J. C. Gough Hutchinson Publications, London

(11) "Cutting Fluid Filtration for Machine Tools," P. J. C. Gough Filtration and Spearation, Nov/Dec 1970, p. 706-715.

(12) E. C. Hill "Biodeter Ioration of Metal Working Fluids and Its Significance" SME MR72-214

(13) E. O. Bennett "The Effect of Water Hardness on the Deterioration of Cutting Fluids" SME MR72-226

(14) M. D. Smith and J. E. Lieser "Laboratory Evaluation and Control of Metalworking Fluids" SME MR73-120

CONTRIBUTORS

METALWORKING FLUIDS DIVISION

Joseph Tomko
S. H. Mack and Company, Inc.

Professor E. O. Bennett
University of Houston

Robert H. Brandt
Henry Manufacturing Co., Inc.

Robert Ciesko
Cincinnati Milacron, Inc.

Dr. Stephen C. Cohen
Van Straaten Chemical Co.

Leo M. Edwards
Gulf Research & Development Co.

Donald J. Edwards
Mobil Oil Corporation

William L. Fennimore
Quaker Chemical Corporation

Almiro J. Lopez
The Rams-Head Company

Dr. Peter E. McElligott
General Electric Company

Franklin J. McGee
LTV Aerospace Corporation

Todd Sutcliffe
E. F. Houghton & Company

Mark D. Opachak
Detroit Diesel-Allison Division

John S. Reeder, Jr.
Rust-Lick, Incorporated

Dr. Harold W. Rossmoore
Wayne State University

Robert E. Savoit
Texaco Incorporated

Donald Scavone
International Harvester Co.

William A. Sluhan
Master Chemical Corporation

James W. Throop
General Motors Institute

Dr. K. Trigger
University of Illinois

OTHERS

Dr. Robert Hahn
Heald Machine

Clayton Tenniswood
Hydromation Engineering

Dale De Young
General Electric, A.E.G.

Jack Meinking
General Electric, A.E.G.

Jack La Belle

COMMENT AND ACKNOWLEDGEMENT

This paper is an initial step toward improving communication between the machine tool builder, manufacturing/process engineers and maintenance engineers, with metalworking fluids research and technical service - in areas of mutual interest and concern. The metalworking fluids division of SME expects to update this paper, as additional comments and suggestions are received, and as new information is developed.

Presented at SME's Abrasive Machining Conference, September 1975

Mechanics Of Grinding Fluid Delivery

By H. Kaliszer
And G. Trmal
University of Birmingham

Efficient grinding depends to a large extent upon the method of grinding fluid delivery. The need of grinding fluid to reach the wheel-workpiece interface is seriously restricted by the pressure of an air belt formed around the wheel periphery. This requires the development of reliable and inexpensive methods of overcoming the detrimental influence of this belt of air and delivering the grinding fluids into the grinding zone. The paper analyses the existing m ethods of determining the effect of the air belt and its dependence upon such factors as the type of wheel, wheel speed, guard design etc. Existing methods of fluid delivery are also analysed. A particular attention is drawn to a new method of fluid delivery based upon the determined theoretically critical fluid velocity. The paper also describes a new more rational design of a fluid nozzle which is based upon the theoretical and experimental results discussed in the paper.

INTRODUCTION

The heat generated during grinding is very large as compared with drilling milling and other machining operations. {2} Large heat generation and high temperature of the workpiece surface which is reached during grinding requires a suitable cooling action by the cutting fluid. As a result of cutting fluid delivery the bulk temperature of the workpiece is usually reduced to no more than 12 - 15° C.

For many grinding operations where the contact area between the wheel and workpiece is increased there is also a need to reduce the temperature of the grinding wheel. Without an adequate cooling action the increased temperature of cutting edges will be responsible for lowering the hardness of the abrasive material and the abrasive grains itself will be subject to an accelerated attrition. From the above it is obvious that the function of cutting fluid is to transfer the generated heat from both the workpiece and the wheel.

As it is well known metal particles resulting from the grinding action are loaded into the wheel pores and welded to the grains. In addition the grain itself is subject to a continuous wear. All this produces large frictional forces in the wheel workpiece interface.

To reduce the friction and prevent adhesion between the grains and the metal the cutting fluid must also include lubricating and cleaning functions.

The effectiveness of cutting fluids in grinding depends to a large extent upon the method of their delivery into the cutting zone. Only a rational application of cutting fluids can result in any significant reduction of frictional forces in the wheel workpiece interface as well as improve the heat transfer and removal of debris from the cutting zone and parts of the grinding machine. All this can lead to a reduc tion in the wheel wear and wheel loading, increase the tool life and improve the quality of the ground components.

At present the best known methods of cutting fluid delivery in grinding include the following :

a) Low pressure delivery by means of a nozzle located above the wheel - workpiece interface.

b) High pressure delivery (5 to 45 kg/cm^2) by means of nozzles located above the interface or outside the frontal wheel guard opening.

c) Fluid delivery through the wheel pores.

Out of the above three methods the simplest, the cheapest and most commonly used is the free low pressure delivery. Such free flow delivery of cutting fluid in grinding however causes certain specific problems not encountered in other machining processes. The grinding wheel which rotates at relatively high speeds (30 to 80 m/s) carries on its surface a layer of air which diverts the stream of cutting fluid preventing it from reaching the cutting zone. As a result the cutting fluid can only reduce the bulk temperature of the workpiece without having any significant affect upon the grinding process itself.

FACTORS AFFECTING THE AIR LAYER

The approximate distribution of the grinding fluid and air layer in the area adjacent to the wheel-work contact zone for a conventional delivery system is shown on Fig. 1. As can be seen from the Fig. the air belt created by the rotating wheel prevents the formation of contact between the cutting fluid and wheel periphery and the access of the fluid into the cutting zone. Due to a turbulent motion of the air above the wheel-work interface only a small fraction of the cutting fluid is dragged between the wheel and the workpiece. Most of the cutting falls on the workpiece and a small amount of it is only carried into the cutting zone. For some grinding operations the amount penetrating into the zone of cutting may be adequate but higher rates of material removal and higher wheel speeds require more efficient delivery systems {1, 2} .

To determine the aerodynamic effect of the air belt formed by the rotating wheel Jashcheritsin used a wheel guard in the shape of a fan casing with a provision for a pitot tube {5} As can be seen from Fig. 2, an assumption was made that the rotating wheel will act as a conventional turbine. The dynamic pressure in the delivery part of the casing was measured by means of a pitot tube inserted in the hole (4) and micromanometric indicator. In the above method the dynamic pressure could only be measured in one fixed position. In addition the configuration of the wheel guard with a delivery opening of 30 mm in diameter replacing the frontal guard opening as well as the opening in the face of the guard could change considerably the aerodynamics of the air flow. The wheel dimensions used in the experiments were Ø 70 x 50 x Ø 20 mm.

The measuring system adopted by the authors {8} and shown on Fig 3 besides its simplicity and realistic conditions offers a convenient way to determine the distribution of dynamic pressure at various points and directions around the rotating wheel of any diameter and width. The wheel diameters used were 270 mm and 600 mm. As can be seen from the Fig a pitot tube and water column indicator were used. The dynamic pressure was measured in the tangential, normal and axial directions. Due to the fact that the pressure in the tangential direction is the highest, the air velocity which corresponds to this pressure can be calculated as follows :

$$V_A = \emptyset \sqrt{2 \times g \times \frac{\gamma_W}{\gamma_A} \times h} \qquad (1)$$

V_A = air velocity (m/s)

h = dynamic pressure (m of w.c.)

g = gravitational constant (9.81 m/s^2)

γ_W = specific mass of mater (1000 kg/m^3)

γ_A = specific mass of air (1.2 kg/m^3)

$\emptyset$ = resistance coefficient

To determine the air pressure and velocity formed by the rotating wheel a number of points were measured radially and along the wheel periphery. Some of the typical results with a wheel WA100J rotating at a peripherial speed V_s = 27 m/s (dia = 270 mm) are shown on Fig 4.

It is interesting to notice that according to Fisher{ 6} a grinding wheel rotating at a speed Vs = 33 m/s has formed on air belt with approximately the same air velocity as the wheel speed. For a wheel 150 mm wide he found that even at a distance of 12.5 mm from thr wheel periphery the air velocity V_A = 19.2 m/s. Serov{ 7}found that the distribution of air velocity in the radial direction changes exponentially decreasing with the distance from the wheel periphery (Fig. 5.) As can be seen from the Fig. the air velocity also changes along the wheel width reaching the minimum value in the middle section. Close to the wheel periphery the air velocity approaches 10 m/s.

The dynamic pressure and air velocity formed by the rotating wheel is affected by such parameters as the wheel structure abd topography, wheel speed, wheel guard etc. To establish the effect of these parameters the same measuring principle was adopted as shown in Fig. 3.

The effect of wheel structure and topography

The effect of wheel porosity is strongly emphasised by Jashcheritsin 5 As can be seen from his results (Fig. 6) obtained by the method shown on Fig. 2, the air flow (Q_A) and therefore also the air velocity for a porous wheel are greater than in case of a wheel with an average structure.

In addition it can also be seen from the same Fig. 6, that in a case when the wheel faces were covered with a thick protective paper the rate of flow and the air velocity are considerably reduced and that the re is a small difference in Q_A for the average and porous wheels when the faces for both wheels are covered. On the basis of the above he concludes that the air contained in the wheel pores is forced from its faces into the wheel periphery by the action of centrifugal forces and hence it is compared to a pumping action.

A more detailed analysis of his experimental results shows however, that only a certain proportion of air flows through the wheel body.

The main stream of air is predominantly dragged from the wheel faces around edges to the wheel periphery. This conclusion coincides with the experiments carried out in MTIRA {1}. To find the effect of wheel porosity the pressure of the air belt close to its peripherial surface was measured with a pitot tube by using a grinding wheel which was subsequently replaced by a wooden disc of the same dimensions. In both cases a similar air pressure was found to be associated with each air belt. On the basis that it is not possible for air to be sucked through the sides of a wooden wheel they concluded that the air belt is predominantly dragged by the wheel periphery. For wheels of more open structure and higher wheel speeds, however, some appreciable quantities of air may be sucked into the wheel faces and ejected at the peripheries.

Khudobin {2} has investigated the pumping action by comparing wheels on ceramic and resinoid bonding. He noticed that for wheels on resinoid bonding with almost no pores the air stream will be significantly smaller as compared to wheels on ceramic bonding.

To establish the relationship between the wheel roughness and dynamic pressure the authors were using wheels of grain size 100 and 46. This choice is based upon the fact that the higher the grain size number the smoother the wheel periphery. The measurement of dynamic pressure in the tangential direction was selected in the centre of the wheel width and as near to the wheel periphery as possible. The diameter of the pitot tube was as previously 1.6 mm. The results performed with a wheel dia = 270 mm and rotating with a peripherial speed of 27 m/s are shown on Table 2.

TABLE 2

Grain size	Dynamic pressure cm of w.c.	Corresponding air velocity m/s	Velocity ratio
100	0.9	12.0	1
46	1.65	16.3	1.36

The effect of wheel speed

With increase in the wheel speed the amount of cutting fluid delivered to the cutting zone decreases. For very high wheel speeds the cutting fluid may almost stop to penetrate into the zone and the grinding may become dry. The dynamic pressure of the air belt in wheel speed tests was measured around the wheel of 600 mm diameter rotating at 955 and 1910 RPM which corresponds to peripherial speeds of 30 m/s and 60 m/s respectively. Results of these measurements are shown on Table 3.

TABLE 3

Peripherial Speed	Max. dynamic pressure cm of w.c.	Corresponding air velocity m/s	Air velocity ratio
30 m/s	2.1	18.6	1
60	9.25	38.6	2.08

It should be pointed out that the same ratio of pressures (Table 3) was also found in all the measured points.

The results given by Jashcheritsin are plotted on Fig. 6. As can be seen from the Fig. in the region of high wheel speeds the effect of wheel speed (V_S) upon the rate of air flow (Q_A) decreases.

The significant difference in the non linear incerase of air velocity as a function of wheel speed shown in Fig. 6 and linear relationship given by the authors in Table 3 can be explained by the difference in the wheel dimensions, wheel characteristics and entirely different methods of establishing the dynamic pressure. Most investigators agree however that the velocity of the air belt increases almost proportionally with the wheel speed.

The effect of wheel guard

Besides the wheel structure wheel topography and its speed of rotation the guard in which the grinding wheel is enclosed plays also an important part. To find the effect of wheel guard the dynamic pressure was measured initially by rotating the wheel inside the guard and subsequently rotating the wheel after the guard was removed. {8} The investigated points were taken in the centre of the wheel width and with varying the distance from the wheel periphery. The average dynamic pressure when thje wheel was rotating inside the guard is approximately 50% higher as compared with the wheel rotating without the guard.

METHODS OF FLUID DELIVERY

As explained earlier the delivery of cutting fluid can be substantially improved if to reduce or eliminate the detrimental effect of the air belt formed around the wheel periphery before the fluid approaches the cutting zone. To overcome this and improve the fluid action effect various fluid nozzles have been designed in the following ways.

A. By diverting mechanically the air belt which prevent the cutting fluid from reaching the wheel.

B. By increasing the kinetic energy of the cutting fluid in the direction tangential to the wheel periphery.

C. By increasing the kinetic energy of the cutting fluid in the direction normal to the wheel periphery.

The simplest way of reducing the air belt effect is by fixing various types of scrapper plates above the nozzle delivering the fluid (Method A). A typical example of such a scrapper plate is shown on Fig. 7a. All types of scrappers are effective but require frequent adjustment to maintain a close gap between the edge of the plate and the wheel periphery.

The simplest way of increasing the kinetic energy and velocity of the cutting fluid in the tangential direction is by applying an increased pressure. The same effect however can also be achieved without using an expensive delivery pumping system. As can be seen on Fig. 7b, the fluid in shoe nozzle delivered under conventional pressure is accelerated by the wheel and leaves the nozzle with high velocity.

Fig. 7c, shows the nozzle, where the energy of the air belt is used to accelerate the cutting fluid. However such systems require a rigid control of the nozzle adjustment in reation to the wheel periphery.

In order to overcome the above disadvantages an attempt was made {8} to utilise a certain critical energy and the velocity of the fluid stream acting in normal direction to the wheel surface. As will be shown later such critical velocity is sufficient to divert the air belt and replace it by a belt of cutting fluid. On the basis of such velocity it is possible to optimise the nozzle geometry and in this way to eliminate the necessity of high pressure delivery systems.

The effect of scrapper plate

Generally with reduction in the gap between the wheel and the scrapper plate the action of the cutting fluid improves. At the same time however a small gap requires frequent adjustments due to the wheel wear and therefore may not be very practical in actual industrial conditions. The scrapper plate used in this investigation is shown on Fig. 7. The average dynamic pressure as a function of the gap which varied from $\sim$0.2mm to $\sim$5 mm is shown in Fig. 8. As can be seen from the Fig. for a relatively large gap of 4 to 5 mm the dynamic pressure was approximately 6 mm of w.c. which is comparable with the dynamic pressure when the wheel was rotating without the guard. Therefore it may be concluded that the presence of a scrapper plate even at a relatively large distance from the wheel periphery can effectively compensate the detrimental effect of the wheel guard. It is also interesting to notice that the average pressure with the guard but without the scrapper plate was 9.6 mm of w.c.

Determination of critical fluid velocity

The method C can be implemented when the cutting fluid penetrates the air layer and "sticks" to the wheel periphery. Such conditions will arise above certain critical fluid velocity {8}. Fig. 9a refers to a situation when the fluid velocity was below critical and equal to 1.3m/s As can be noticed the cutting fluid does not penetrate the air layer. Fig. 9b refers to an above critical velocity of 2 m/s. Here it can be clearly seen that the air layer was replaced by the fluid layer.

To determine the critical fluid velocity the fluid flow applied to the periphery of a stationary grinding wheel is divided into two streams - one in the upward direction, and one downwards (see Fig. 10).

The determination of a critical fluid velocity is based on the following assumption : If the momentum of flux of the air layer is higher than the momentum of flux of fluid in the upwards direction the fluid will not penetrate the air layer. If the fluid momentum is higher, the air layer will be diverted.

As can be seen on Fig. 10, the momentum of the upwards stream of the cutting fluid can be calculated by considering the sum of momentum in vertical direction,

$$M_{f1} = M_f \times \frac{1 - \sin\alpha}{2} \qquad (3)$$

where

M_{f1} = momentum of cutting fluid flow in upwards direction

M_f = momentum of the main fluid flow leaving the nozzle

α = angle of application

The momentum of the main fluid flow can be determined by considering the weight of the fluid per second striking the air belt, i.e.

$$W_f = \gamma_f \times A \times V_f \quad (Kg/s) \qquad (4)$$

where

γ_f = specific weight of the cutting fluid (Kg/m^3)

$A=B\times b$ = cross sectional area of the nozzle orifice (m^2)

V_f = velocity of the cutting fluid (m/s)

B = width of the nozzle orifice (m)

b = height of the nozzle orifice (m)

If to consider the fluid flow per unit of nozzle width then

$$A = \frac{B \times b}{B} = b \ (m^2/m) \qquad \text{or}$$

$$W'_f = \gamma_f \times b \times V_f$$

The momentum of the main fluid flow can be determined by considering the weight of the fluid W_f and its velocity V_f i.e.

$$M_f = \frac{W_f}{g} \times V_f = \frac{\gamma_f}{g} \times b \times V_f \qquad (5)$$

The velocity of the cutting fluid

$$V_f = \frac{Q}{A} = \frac{Q'}{B}$$

where

$Q' = \frac{Q}{B}$ = rate of flow per unit nozzle width

Finally therefore,

$$M_f = \frac{\gamma f}{g} \times \frac{Q'^2}{b} \qquad (6)$$

When designing a fluid nozzle (gap height b) all parameters of the delivery system (pump, filter, valves etc) must be taken into consideration, since all the parts of the delivery system will increase the hydraulic resistance and therefore affects the pressure drop in the nozzle.

For a maximum momentum

$$\frac{Q^2}{b} = \max \quad \text{where} \quad Q = f\ (b)$$

For the calculation of the momentum of an air layer the velocity profile (Fig. 10b) was substituted by an average velocity in the layer of thickness .

The momentum of an air layer can be calculated as follows :

$$M_a = \frac{\gamma A}{g} \times \ell \times V_A^2(\ell) \qquad (7)$$

where

M_a = momentum of an air layer

γ_A = specific weight of air

ℓ = layer thickness

$V_A(\ell)$ = average air velocity

The air layer velocity can be related to the dynamic pressure measured during the tests as follows :

$$V_A^2(\ell) = 2 \times g \times h_{A(\ell)} \times \frac{\gamma W}{\gamma A} \qquad (8)$$

where

$h_{A(\ell)}$ = average dynamic pressure in air layer of thickness ℓ (in w.c. height)

γ_W = specific weight of water

Substituting (7) into (8)

$$M_A = 2 \times \ell \times \gamma_W \times h_{A(\ell)} \qquad (9)$$

For the critical value when the momentum of the air layer is equal to the momentum of the upward stream of fluid ($M_{f1} = M_A$) the fluid flow rate can be calculated as follows :

$$Q_{CT} = B \times Q^1 = b \times B \times \sqrt{4 \times \frac{\gamma\ w}{\gamma\ f} \times \frac{\ell x g}{1 - \sin\alpha} \times h_{A(\ell)} \times \frac{1}{b}} \qquad (10)$$

where b x B represents the cross section of the nozzle orifice and the square root the critical fluid velocity.

The above assumptions (eq.10) were tested experimentally by means of a specially designed nozzle 8 shown on Fig. 9. The testing conditions were as follows :

nozzle gap	$b = 1$ mm
width of the nozzle opening	$B = 20$ mm
airlayer thickness	$\ell = 5$ mm (practical nozzle distance)
average dynamic pressure	$h_{A(5)} = 0.96$ cm of w.c.
angle of nozzle location	$\alpha = 15^o$

The specific weight of water and grinding fluid in eq. 10 were assumed to be equal.

On the basis of eq.10 the calculated critical rate of flow for the conditions specified above was $Q_{CT} = 1.95$ l/min, which corresponds to a fluid velocity $V_f = 1.62$ m/s.

The experimental tests also show that with the rate of flow equal to 1.6 l/min (velocity 1.3 m/s) the fluid does not penetrate the air layer (fig. 9a). With the rate of flow increased to 2.4 l/min (velocity 2 m/s) the cutting fluid replaces the layer and "sticks" to the wheel periphery.

As can be seen the agreement between theoretical and experimental results was surprisingly good. Although such good agreement may not always be expected due to a wide range of existing conditions the analysis shows that the suggested method can be satisfactorily used for an optimised nozzle design {8}. The design of such a nozzle incorporates the following features.

1. Direction of cutting fluid close to normal
2. Comparatively small orifice area for high velocity of coolant
3. Presence of a scraping plate which is built into the nozzle shape

The new and the conventional nozzles are shown on Fig. 11. The rate of flow in both cases is the same. The flow from the new nozzle "sticks" to the wheel while the flow from the conventional nozzle is diverted.

The grinding tests show pronounced improvements in the quality of the machined surface. As can be seen on Fig. 12, the surface ground with a conventional nozzle is rougher with clearly seen helix of burned marks. The surface ground with the new nozzle has a surface roughness (Ra) 4 x better than the surface ground with the conventional nozzle. The absence of visible burn marks is also apparent (in both cases the grinding conditions were the same).

CONCLUSION

The air layer rotating with the grinding wheel tends to divert the applied cutting fluid.

The velocity of air belt in the layer depends upon the wheel speed wheel surface topography (roughness and design of the wheel guard). The use of scrapper plates can improve the conditions considerably.

For any given conditions, there is a critical velocity of fluid, which can be determined theoretically. The fluid applied with a velocity above the critical value will penetrate the air layer and "stick" to the wheel.

Nozzles of a new design based upon the use of critical velocity can deliver cutting fluid more efficiently with a very pronounced effect upon the grinding results.

ACKNOWLEDGEMENT

The authors wish to thank Professor S A Tobias for providing facilities for the work and the Science Research Council for financial support.

REFERENCES

1. M T I R A : High-speed grinding with particular reference to the proper employment of grinding fluids. (1973)

2. L V Khudobin, Cutting fluids applied in grinding, Mashinostroenie, Moscow, 1971 (Russian)

3. H. Opitz : High speed grinding a way to increase the performance of the grinding process. Proc. 8th Annual Conf. of Amer. Soc. for Abrasive Methods 1969 p 48.

4. R. S. Hahn : The relation between grinding conditions and thermal damage in the workpiece. A S M E Trans. 1955.

5. P. I. Jashcheritsin : Surface quality and accuracy during grinding Minsk 1959 (Russian)

6. R. C.Fisher : Grinding dry with water. Grinding and Finishing V.11 No. 3, 1965

7. V. N. Serov : Dust free during abrasive machining, Moskow, Mashpiz 1961

8. G. Trmal and H. Kaliszer, Delivery of cutting fluid in grinding, 16th M T D R Conf. Manchester, 1975.

9. Ya. V. Polanskov, Effectiveness of cutting fluid in grinding antifriction bearing races, Vestnik Mashinostrojania No.3 1974.

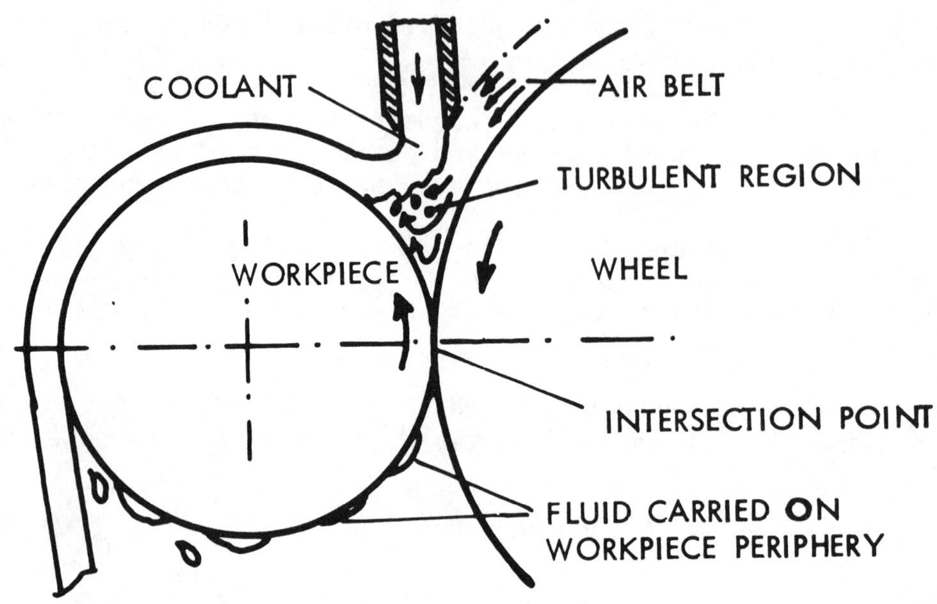

Fig 1 CUTTING FLUID AND AIR BELT AROUND THE GRINDING WHEEL

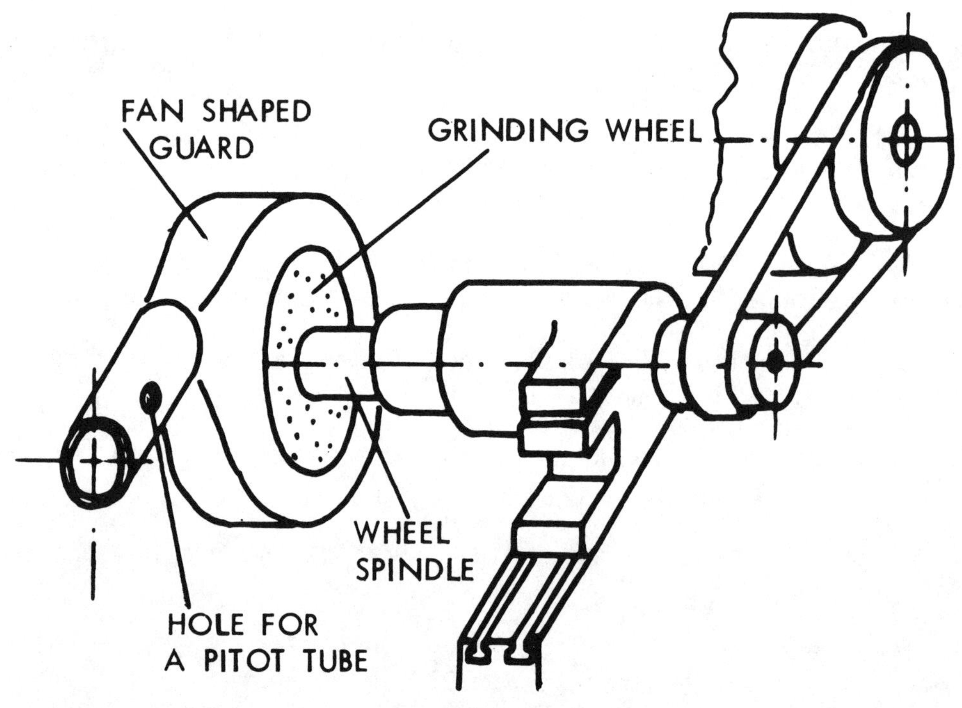

Fig 2 MEASUREMENT OF THE GRINDING WHEEL AIR PUMPING ACTION (JASHCHERITSIN)

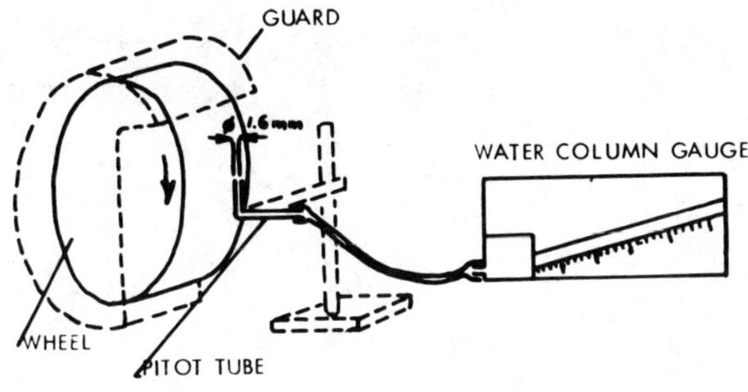

Fig 3 MEASUREMENT OF DYNAMIC PRESSURE OF THE AIR BELT AROUND THE WHEEL

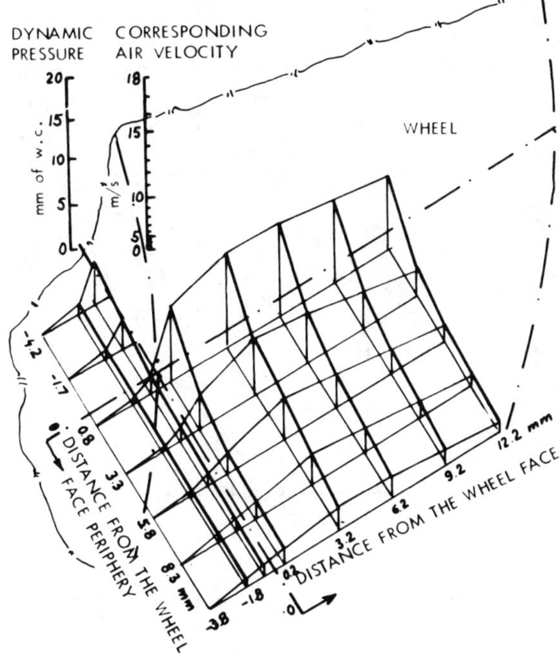

Fig. 4
DYNAMIC PRESSURE OF THE AIR LAYER AROUND THE WHEEL

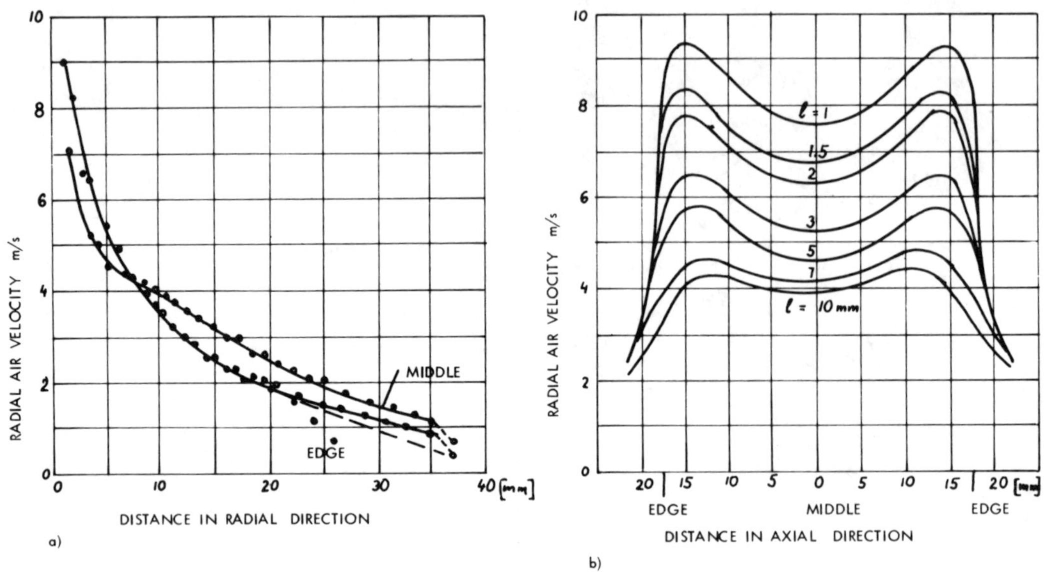

Fig 5 AIR VELOCITY IN RADIAL DIRECTION AROUND THE WHEEL (SEROV [7])

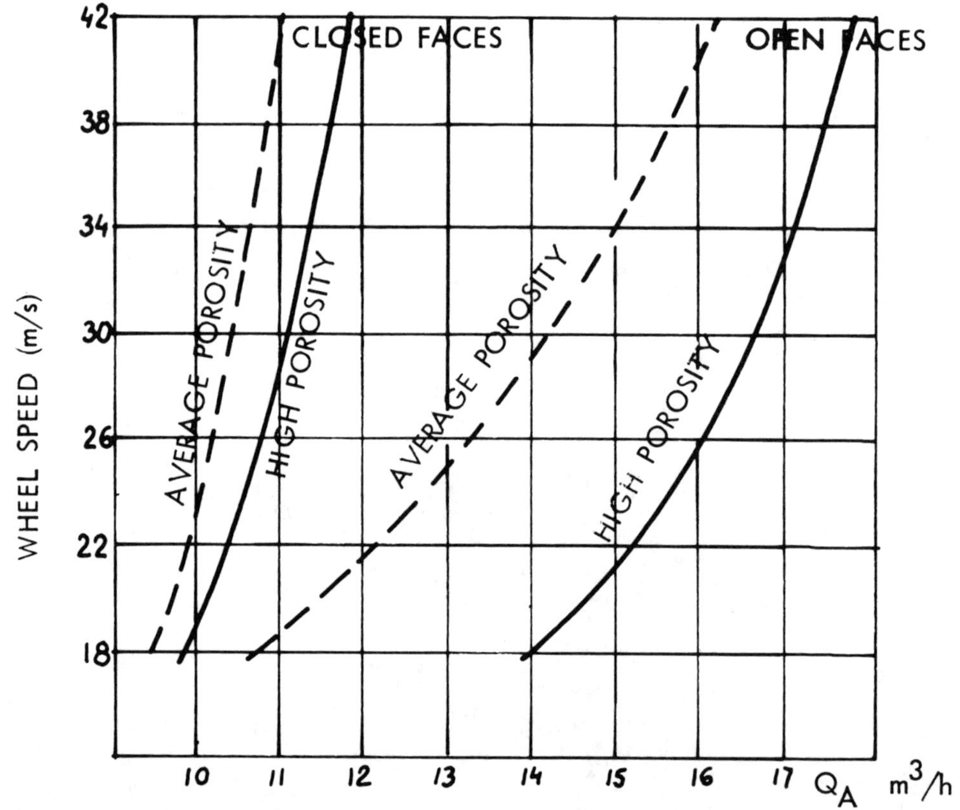

Fig 6. AIR FLOW AS A FUNCTION OF WHEEL SPEED (JASHCHERITSIN [5])

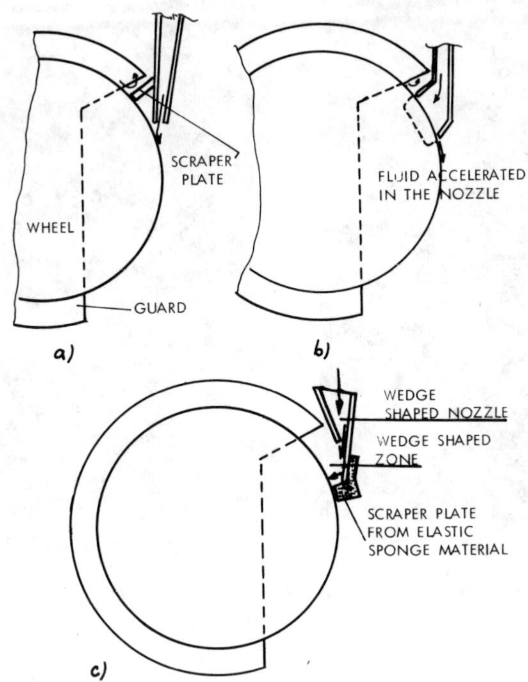

Fig. 7 VARIOUS TYPES OF NOZZLES

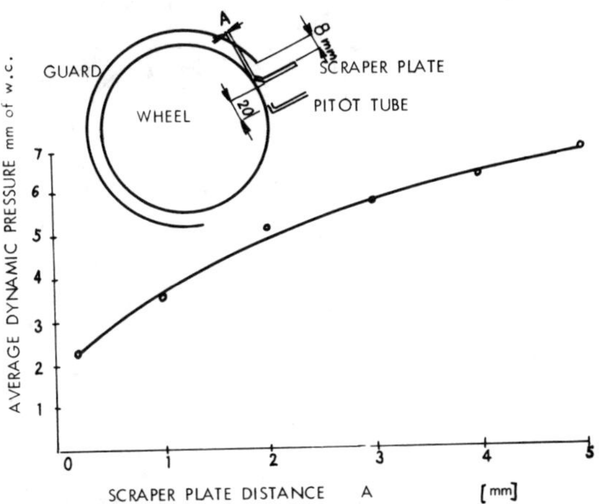

Fig. 8. AVERAGE DYNAMIC PRESSURE AS A FUNCTION OF SCRAPER PLATE DISTANCE

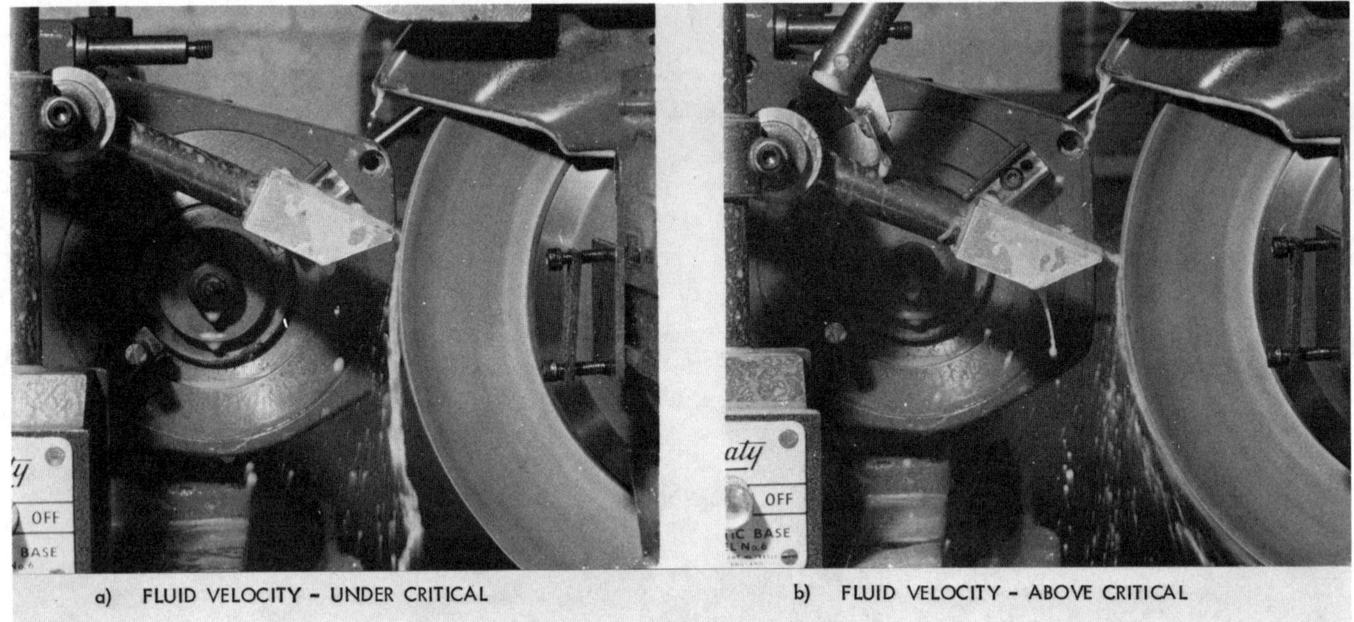

a) FLUID VELOCITY - UNDER CRITICAL b) FLUID VELOCITY - ABOVE CRITICAL

Fig 9 INTERACTION BETWEEN FLUID FLOW AND AIR LAYER

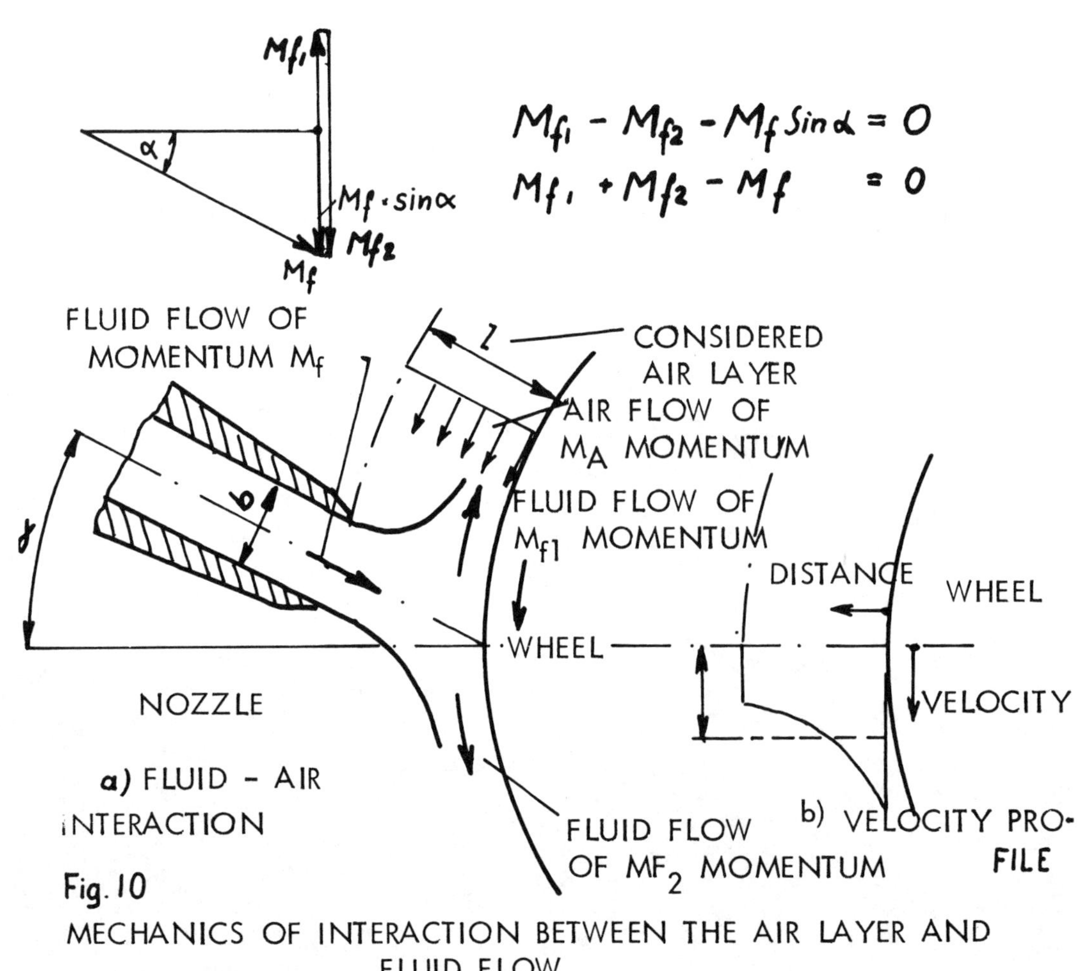

Fig. 10

MECHANICS OF INTERACTION BETWEEN THE AIR LAYER AND FLUID FLOW

a) CONVENTIONAL NOZZLE b) NEW NOZZLE

Fig 11 DELIVERY OF CUTTING FLUID TO THE WHEEL PERIPHERY

BURN MARKS GENERATED WITH THE CONVENTIONAL NOZZLE
IMPROVED SURFACE QUALITY ACHIEVED WITH THE NEW NOZZLE

Fig 12 SURFACE QUALITY OF THE GROUND WORKPIECE

Presented at SME's Cutting Tools Conference, May 1976

Advancements In Coolant Feeding Techniques

By Howard M. Whalley
The George Whalley Company

The metal working industry has been plagued with the ever increasing costs of labor, raw materials and overhead. It is almost impossible to reduce raw material costs without decreasing quality. The overhead continuously rises as inflation increases as well. The industrial worker can only produce parts as quickly as his machine will allow him. Therefore, our concentration must be on improving manufacturing methods. One of the most forgotten procedures in metal work machining is applying the coolant to the cutting tool in the proper place. In the average machine shop the coolant is either not applied at all or is misdirected. However, when coolant is directed through the cutting tool and piped from the machine's pumping system such a problem no longer exists.

There has been considerable advancements in the last ten years in the use of coolant feeding tools, adapters and hollow spindle coolant machine tools. The result of adding coolant through the tool under pressure is revolutionizing the Industry. This feature can be adapted to nearly every imaginable tool. By giving the characteristics of and suggesting guidelines for the coolant feeding process, we will demonstrate how this technique can be practically applied to the everyday machining as a cost saving innovation.

ADVANTAGES

1. Reduction of friction on the cutting edge of the tool: A thin layer of coolant forced between the cutting edge of the tool and the work piece acts as a buffer.

2. Ability to increase speed and feed: The chips forced out of the work piece by a flushing action of the coolant allows increased speeds. The coolant dissipates heat as friction of the tool's rotation occurs.

3. Elimination of recutting old chips: The pressure of the coolant separating the chips as they are being formed allow them to flow freely and independently away from cutting action.

4. Improvement of finish: The ability of the tool to run without heat, friction and galling produces finer finishes.

5. Improved accuracy: The cutting action of the tool is improved, eliminating the pressures and dulling effect on it which causes imperfect parts.

6. Reduction of horsepower: A properly lubricated tool without chips binding it requires less amperage to operate efficiently.

7. Elimination of secondary operations: As a result of improved cutting action and better finish, frequently reaming or boring operations may be eliminated.

8. Increased tool life: The dissipation of heat and reduction of friction extends tool life.

9. Reduction of air pollution: Smoke caused by heat and friction is eliminated.

10. Reduction of contamination: The constant flowing of coolant recirculating through the tool and pumping system eliminates stagnant coolant.

PROBLEM AREAS

1. Short run jobs: The higher initial cost of coolant tooling and the greater set up time involved can overcome cost saving advantages. Exhibit "A" set forth below presents an example of a solution to this problem.

Drilling Study On Cast Armor Exhibit "A"

Object: Test to prove small production runs can justify coolant fed tooling.
Material: Cast armor brinnell 217-262.
Drill Diameter: 15/16 6 1/8 flute x 10 3/4 overall length high speed taper shank.
Drill Depth: 1 1/4
Bushing: 15/16 x 1 1/2
Fixture: Circle ring plate.
Single pass no withdrawals.

Conventional Twist Drill	Versus	Oil Hole Twist Drill
15/16 diameter 2 1/8 useable flute		15/16 diameter 2 1/8 useable flute
2 parts per day		2 parts per day
36 holes per part		36 holes per part
72 holes per day		72 holes per day

Conventional Twist Drill	Versus	Oil Hole Twist Drill
Actual plant standard 2.6" I.P.M.		Oil hole drill standard 7.920 I.P.M.
6 holes per sharpening		18 holes per sharpening
Use .1562 of drill per regrind		Use .015 of drill per regrind
13.6 sharpenings before tool is consumed		158.33 sharpenings before tool is consumed
Tool drill cost $15.53 or .1903 per hole		Tool drill cost $52.33 or .01836 per hole

Based On

2 Units Per Day (72 Holes)
250 Days Per Year

Conventional drill cost only:	Oil hole drill cost only:
$3,402.00 per year	$ 330.48 per year

Gross savings: $3,071.52 or approximately 90%

This savings is based only on tool life not considering production increase, downtime or resharpening cost.

Small production runs result in substantial savings with oil hole drills.

2. Recommendations by certain machine tool builders against using coolant: The proper shielding of the machine's vital components may allow the use of coolant lubricants.

3. Inadequate horsepower: This may not permit sufficient increase in production to overcome increased tool cost. Changing the motor from AD to DC or changing the gear ratio of the machine tool may resolve the problem.

4. Loss of rigidity: The coolant holder may project the tool beyond reasonable limits for rigidity necessary to increase speeds and feeds. Several different procedures may be utilized in compensating for the tool overhang, depending upon the particular set up:

a) Collet type holders are available to grip on flute diameter which allow only the minimum length of tool necessary for cutting purposes.

b) A coolant collar may be mounted directly to the spindle.

c) A machine tool may be purchased with coolant directly through the spindle. Many machine tool builders offer conversion packages to change solid spindles to hollow ones.

d) The adaptation of a collar ground and faced to clamp onto the projected portion of the tool and pulled tight against the face of the tool holder is another possible solution.

e) The machine tool table may be raised up to the tool, allowing minimum clearance for tool removal.

f) The spindle of a machine should be as close to the column of the machine as possible when operating a radial drill press.

5. Splash problem: The use of high pressure pumping systems, combined with the rotation of the tool, may create unwanted splashing. This may be overcome by various means:

a) Completely enclose machine tool with plexiglass or lexon shielding.

SENS-A-FEED

b) Shield around individual parts.

c) Mount commercially available shields directly to coolant rotary inducer or machine tool spindle.

d) Design fixtures to enclose part to cover splash area.

e) When necessary, use misting system in place of high pressure flooding systems.

f) When necessary, use commercial pumping systems which mist upon initial engagement into work piece, then pulsate through balance of cut and return to mist upon withdrawal of cut.
Reference: Balcrank Mist Pulse Mist.

6. Leakage: Basically, two types exist--leakage through the plumbing manifold systems and accessories on the machine tool and coolant leakage when holding straight shank tools in collet type holders. Plumbing leakage can be controlled by these methods:

a) Replace low pressure hoses with high pressure rated hoses.

b) Replace low pressure fittings with high pressure fittings.

c) Use taper pipe fittings (male) inside straight pipe (female) connections.

d) Where possible, replace rubber hoses with copper, galvanized steel or hard plastic piping.

Collet type leakage can be controlled by these means:

a) Seal collet with a rubberized epoxy cement. Note: Caution should be used with this method because epoxy may extrude out of collet grooves when tightening collet holders.

b) Use a commercially available backup screw with a urethane type cap formed to concave, convex or counterbored shapes to seal the end of the tool.

COOLANT SEAL STOP SCREWS, DIFFERENT STYLES TO ADAPT TO A VARIETY OF OIL HOLES DRILLS.

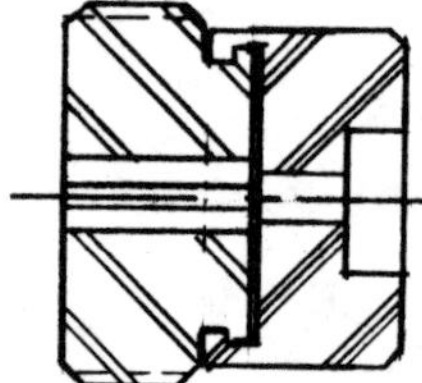

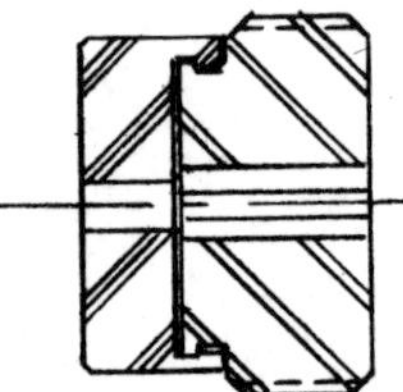

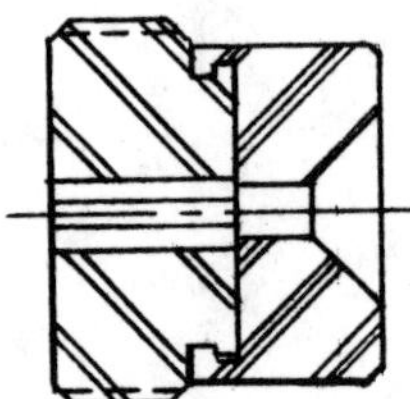

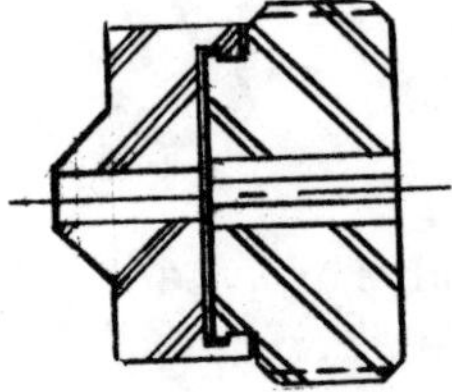

COOLANT SEAL
SPRING STOP SCREW FOR END MILL IN COLLET

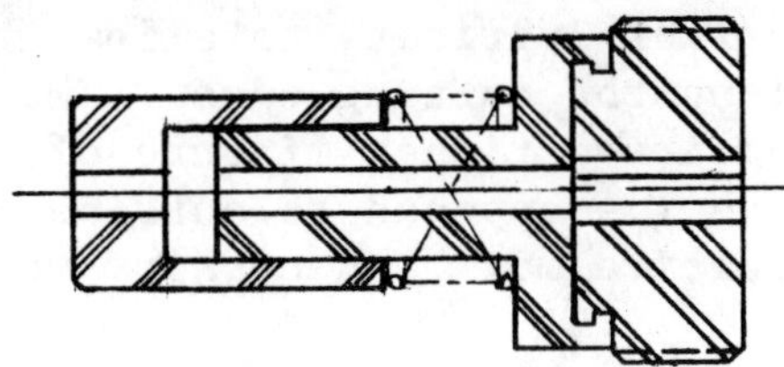

c) Use a backup screw with a pressure spring to push against end of tool.

7. <u>Blockage of coolant fed tools</u>: Poorly filtered coolant systems may cause blockage in the orifices of the coolant fed tools. This may be resolved by the following:

a) Ultra-sonic cleaning of tools will assist in dislodging the dirt particles within tool passages.

b) High pressure coolant or air may be directed through the passages to dislodge impacted orifices.

c) Wire (piano) may be forced through the orifices with pliers manually or a low rpm electric drill. Caution should be taken so as not to break wires inside coolant holes.

To prevent tool blockage a filtering device to collect small particles of contaminants should be adapted directly in the coolant pumping system at the coolant resevoir, preferably 120 mesh or finer. A cartridge type filter should be placed in the coolant supply line as close to the coolant rotary connection as possible.

CHECKLIST

Before proceeding with the installation of coolant fed tooling, a machine shop should consider these steps:

a) proper tool geometry;

b) rigid fixturing permitting ease of chip flow pattern;

c) sufficient horsepower to obtain necessary optimum results;

d) adequate rpm and feed;

e) adequate coolant pumping system. The best manufactured coolant fed cutting tool cannot perform to its utmost capability without sufficient volumn and pressure;

f) proper amount of lubricity to allow chips to flow back freely and smoothly along the cutting edges. Many commercially available coolants may require a richer mixture of coolant for through the tool applications (as opposed to outside flooding). In addition, an anti-foam additive may be necessary when using coolant under high pressures.

SPEED AND FEED CONSIDERATIONS

The following Society of Manufacturing Engineers Technical Papers include guidelines for oil hole drilling speed and feeds:

Technical Paper MR67-104
Drill Design and Application Requirements for
Optimum Coolant-Feeding Twist Drill Usage
By R. W. Brockman
And R. O. Burant, R. G. Kennedy, N. W. Marotte, R. E. Sikora

Technical Paper MR68-504
Process Report -- The Revolution in Drilling Technology
By Ray B. Donovan
And E. A. Rich

Technical Paper MR67-103
Let's Bring Drilling Into The Jet Age
By Bruce F. Woodcock

Technical Paper MR69-173
The Growing Acceptance of Coolant Fed Drilling and
Cutting Tool Systems for Drastic Cost Reduction in
Metalworking
By E. A. Rich

Technical Paper MR68-193
Drill Temperature as a Drill Performance Criterion
By Marvin F. DeVries

Additional information on oil hole twist drilling, spade drilling and oil hole reaming can be found in Machining Data Handbook by Metcut Research Associates Inc., Cincinnati, Ohio 1972.

The most valuable considerations for proper speed and feed should be determined in each individual machine shop. Here are some recommendations to follow. There are two types of standards established in the individual machine shop:

1. Original plant standard: This is the suggested operational standard usually set by the methods and engineering department for figuring job costs and efficiency.

2. Actual standard: This second standard is of even greater value. It is the true operating speed and feed as they are currently being performed in the individual shop.

A key to establishing the best starting point is working from the present actual standard. Secondly, a realistic goal should be set, pinpointing the areas of production for improvement, such as increase of rpm and feed, tool life, surface finishes, tolerances, etc.

Because most concentration in the average shop centers on the rpm, it is a good practice to start by increasing feed rates. High rpm in conjunction with low feed burns and dulls tools prematurely and often work hardens them. Increase the feed approximately 15% initially. Thereafter, as the feed is increased, the rpm should be increased proportionately less.

Observe the chip formation. There should be no discoloration when using coolant. For example, when twist drilling, an excessive rpm produces a long, stringy chip. Balance the cutting action by either increasing feed or reducing rpm. A short, tightly wound chip means the feed may be excessive. Reverse the above balancing procedure accordingly. The ideal chip formation is a short figure six shape.

As the speed and feed are continually adjusted the tool should be sharpened to obtain the best possible results. Once the best geometry, speed and feed have been found, a tool test should be run to evaluate over all performance. Care should be taken to reduce the optimum speed and feed to a level which can be obtained under actual working conditions.

CONSIDERATIONS IN DESIGNING SPECIAL COOLANT FED TOOLS

1. Coolant orifices should be located as close to leading cutting edge of the tool as possible.

2. Coolant orifices should not intersect in the center of the tool. They should be staggered for maximum cross section strength.

3. Orifice size should not exceed 1/3 of the smallest cross section or web of tool.

4. Orifices should be directed in such a manner as to force coolant freely along cutting or through tool.

5. The size of the intersecting holes combined should be no greater than the size of the main axial hole thereby maximizing strength and performance.

6. Coolant holes should not be placed in sharp corners of tools. These holes should be placed in the radius or open areas.

7. The easiesthole location may not necessarily be the most efficient. Many times cross holes must be made in tools to allow orifices to come in at proper angles to cutting edges.

PROPER CARE AND HANDLING OF COOLANT FED TOOLS

1. These tools should be stored in paper or plastic tubes to prevent dirt from entering orifices and blocking coolant flow; also to protect sharp cutting edges.

2. Tools should be coated with an oil or lubricant to avoid oxidation and rusting.

3. After sharpening tools, a good practice is to hone cutting edges to remove burrs. This practice should extend tool life and reduce chipping of cutting edge. (This is recommended practice for all tools, coolant feeding or otherwise.)

DRILLING STUDY BASED ON 3000 SERIES CAST IRON

This test is the first of a series comparing convential twist drills to oil hole drills, and also oil hole drills to gun drills. All our further testing will be based on drilling depths exceeding ten times drill diameter. The object of comparing oil hole twist drills to gun drills is to cover the area where a gun drill is not required to hold size or finish, but merely the most reliable method of drilling deep holes.

Since the first initial test is based on oil hole drilling, cast iron 5/16" diameter, 5" deep or 16 times the drill diameter, we are using as reference for comparison Metcut research data.

5/16 diameter high speed conventional twist drill, Metcut recommends 90 sfm at .005 feed (1100 rpm or 5.5 inches per minute. (Note conventional drilling would require several withdrawals to complete the 5 inch drilling depths.) 5/16 diameter gun drill, Metcut recommends 200 sfm at .0015 feed (2444 rpm) or 3.66 inches per minute. As further testing is completed, we will cooperate with the S.M.E. with all information available.

Material: cast iron	Used 1 drill
Drill diameter: 5/16 H.S.S.S. 9" Fl x 12" oil hole drill	Machine used: Sens-A-Feed
Bushing: 5/16 x 3/4 long	Bed mounted: 45 degrees
Fixture: V block bridge type	Torque sensing
RPM: 3400	Variable feed: 360 to 3600 rpm
Feed: 30" I.P.M. (.009 I.P.R.)	Pump used: 7-1 ratio pulser
SFM: 275	Finish: 125 R.M.S.
Single pass no withdrawal	Straightness: .003 per inch
150 pieces six holes five inch depth	Hole size: +.002
600" regrind .015 per regrind	

Tooling drill: $47.00

9" flute
6 1/2 minus
2 1/2 inch useable life
$47.00 = $18.80 per inch
$.285 regrind
$.000475 per inch x 5" = $.0024 per hole, Tool Cost
Machine Cost: $15.00 per hour
Per Piece:

Loading time:	20 seconds
Drilling time:	90 seconds
Index time:	20 seconds
Unload time:	20 seconds
	150 seconds gross
	24 pieces per hour gross

Based on 80% efficiency = 19 pieces per hour.
$15.00 divided by 19 pieces per hour divided by 6 holes per piece = $.1316 labor cost per hole

Conclusions of study:

Tooling cost	$.0024 per hole
Labor cost	.1316 per hole
Total drilling cost	$.1340 per hole

COOLANT FED CUTTING TOOLS AND HOLDERS

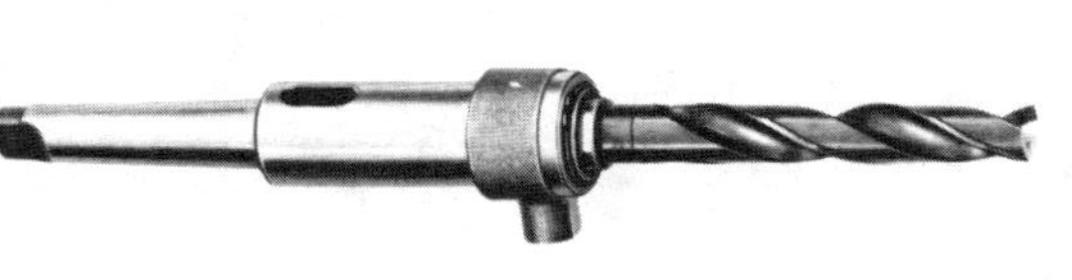

MORSE TAPER HOLDER
AND DRILL

STRAIGHT SHANK
HOLDER AND REAMER

MORSE TAPER HOLDER
STRAIGHT SHANK DRILL

FLOATING TAPER SHANK
HOLDER FOR TAPS

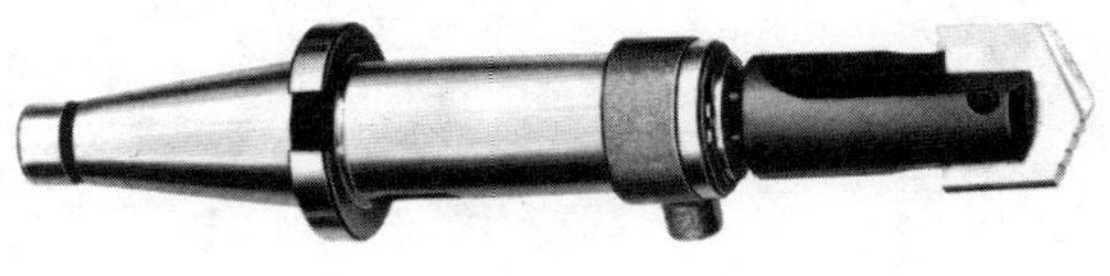

MILLING MACHINE HOLDER WITH MORSE TAPER
CROSS HOLE SPADE DRILL HOLDER

MILLING MACHINE END MILL HOLDER AND END
MILL WITH COOLANT INDEX COLLAR

MILLING MACHINE COLLET HOLDER WITH ROUGH-
ING END MILL AND COOLANT INDEX COLLAR

MILLING MACHINE COLLET HOLDER WITH
T SLOT CUTTER

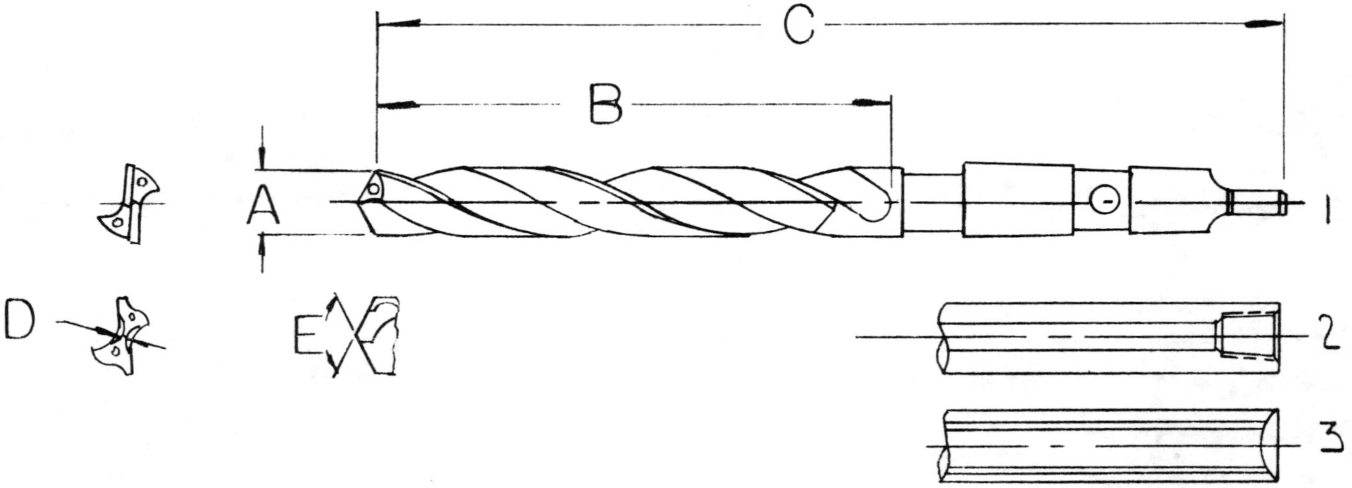

1. MORSE TAPER OIL HOLE DRILL
2. STRAIGHT THREAD PIPE THREAD ENTRY
3. STRAIGHT SHANK CUPPED END

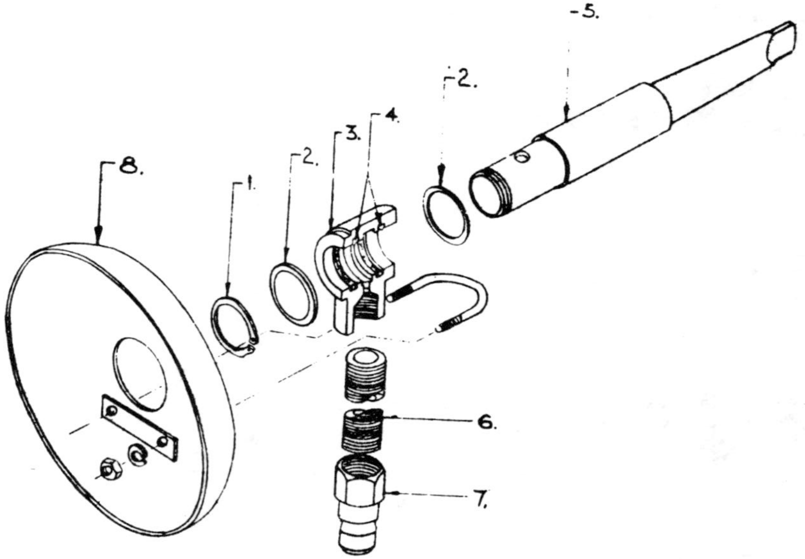

STANDARD MORSE TAPER ROTARY INDUCER

1. SNAP RING
2. THRUST WASHER
3. COOLANT INDUCER RING
4. O RINGS
5. SHANK
6. PIPE NIPPLE
7. QUICK CHANGE PIPE COUPLER
8. SHIELD

COOLANT FED SPADE DRILLS

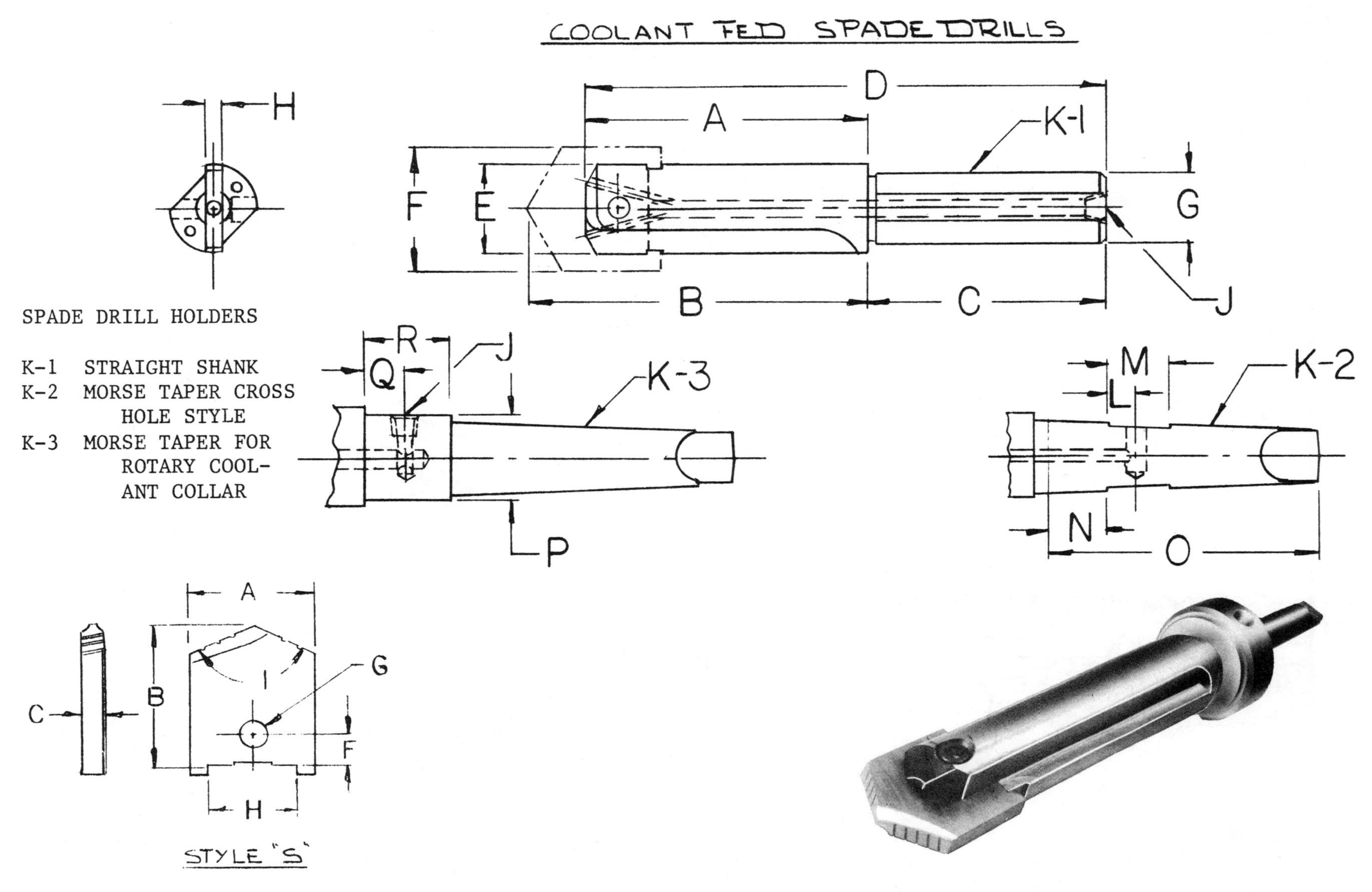

1. SPADE DRILL BLADE

2. SPADE DRILL HOLDER WITH ROTARY COOLANT COLLAR

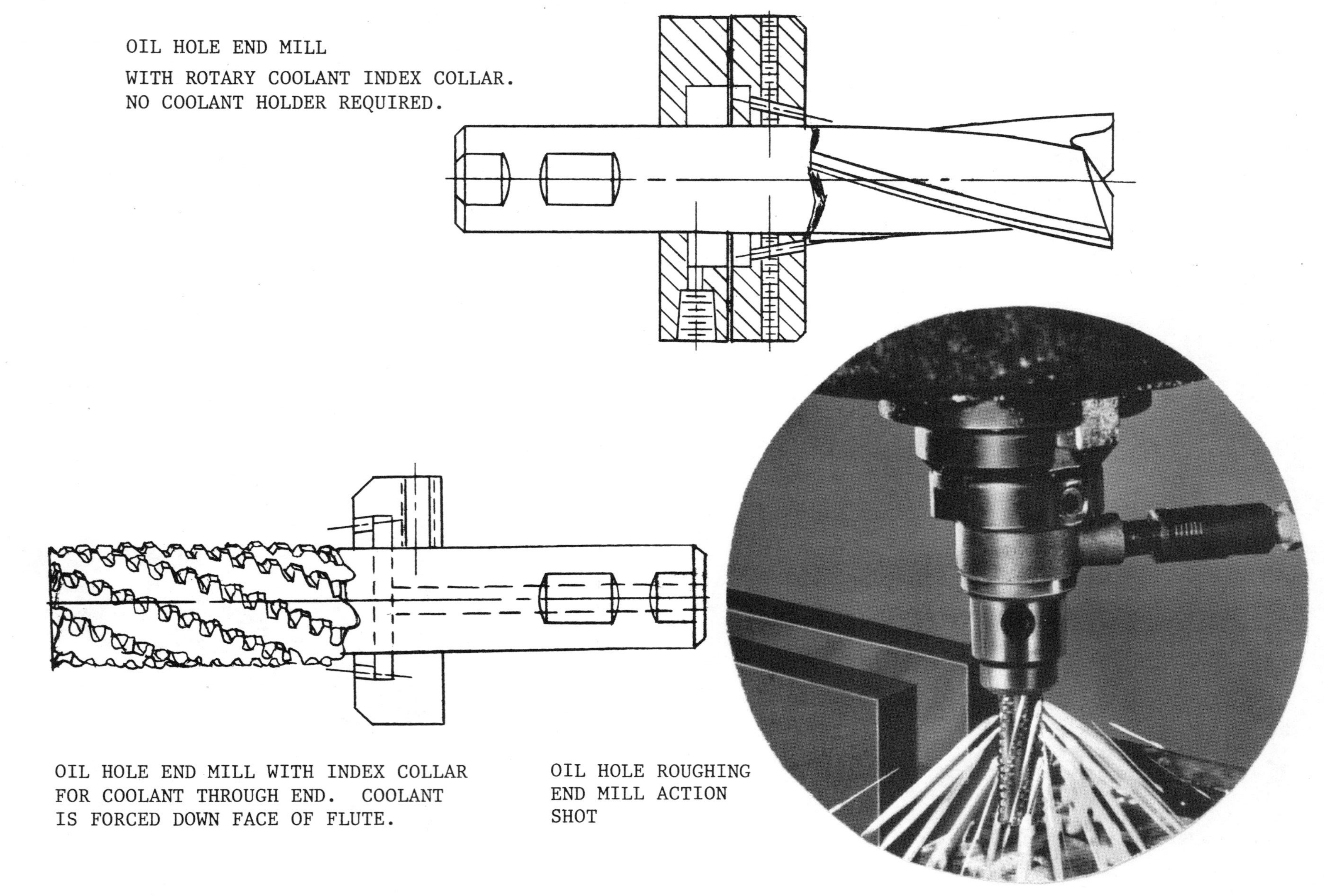

OIL HOLE END MILL
WITH ROTARY COOLANT INDEX COLLAR.
NO COOLANT HOLDER REQUIRED.

OIL HOLE END MILL WITH INDEX COLLAR
FOR COOLANT THROUGH END. COOLANT
IS FORCED DOWN FACE OF FLUTE.

OIL HOLE ROUGHING
END MILL ACTION
SHOT

COOLANT FED "T" SLOT CUTTER

1. COOLANT HOLES FORCE CHIPS OUT T SLOT AND LUBRICATE CUTTING EDGES.

2. ACTION SHOT SHOWING COOLANT PATTERN.

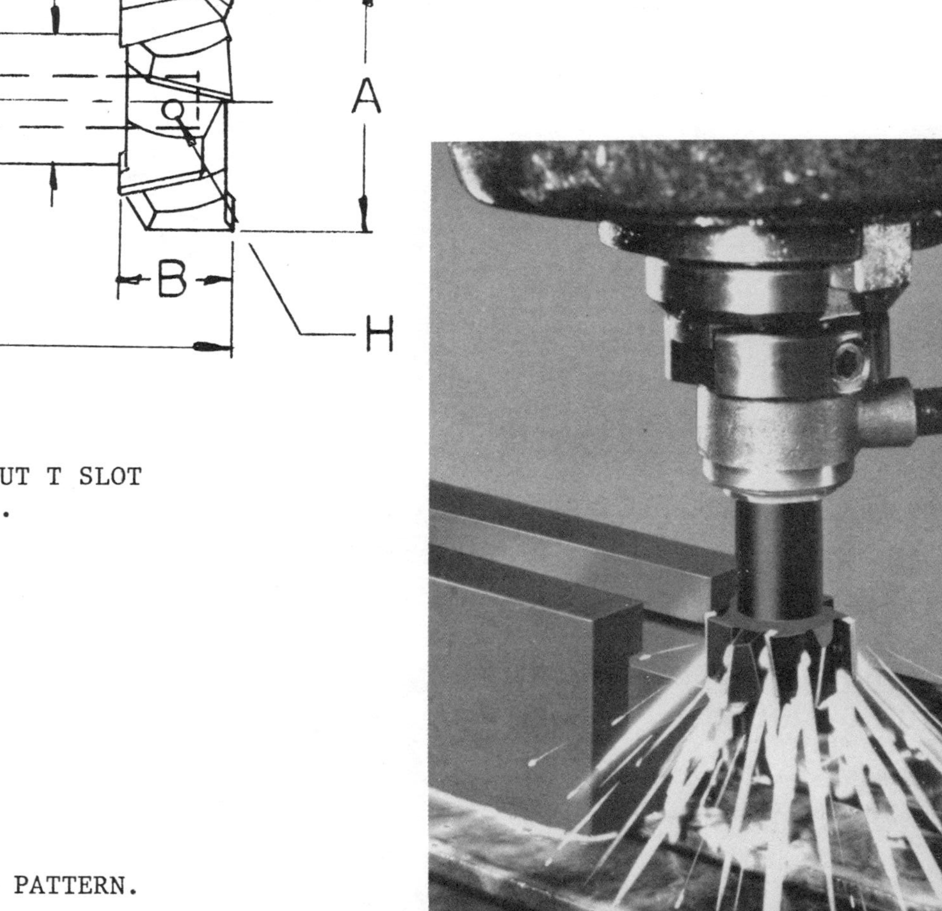

COOLANT FEED POWER MILL

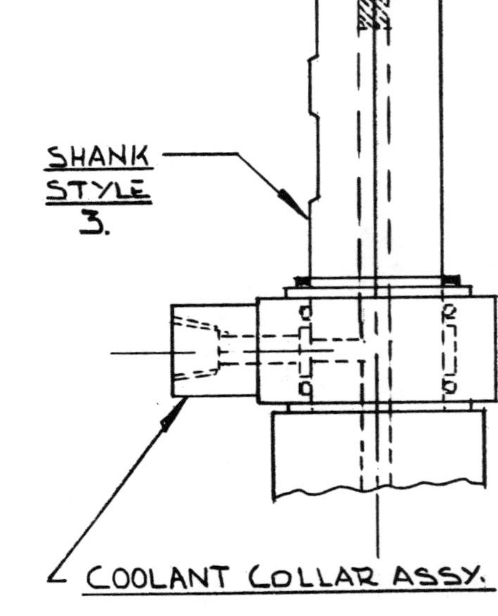
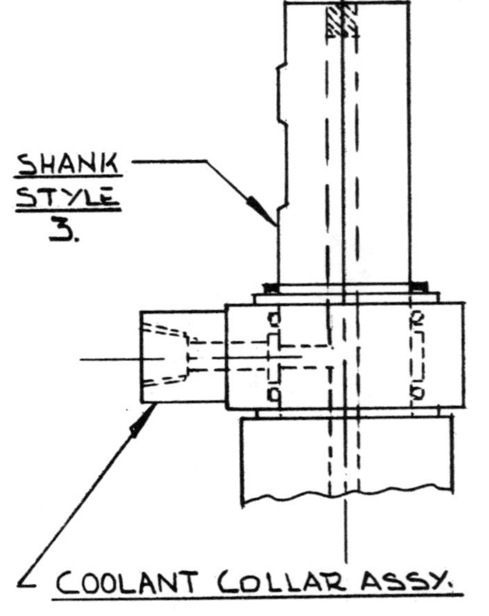

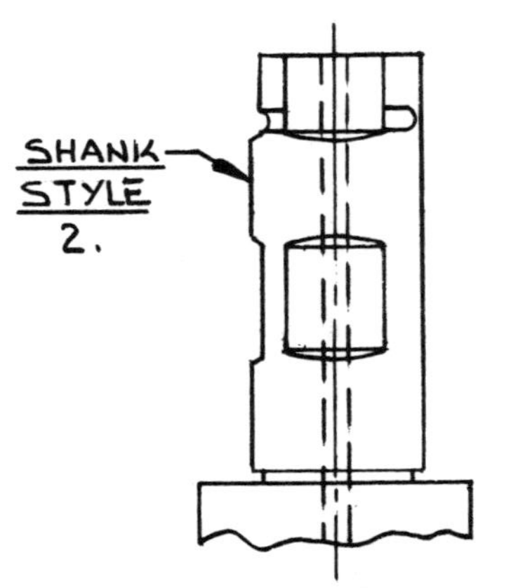

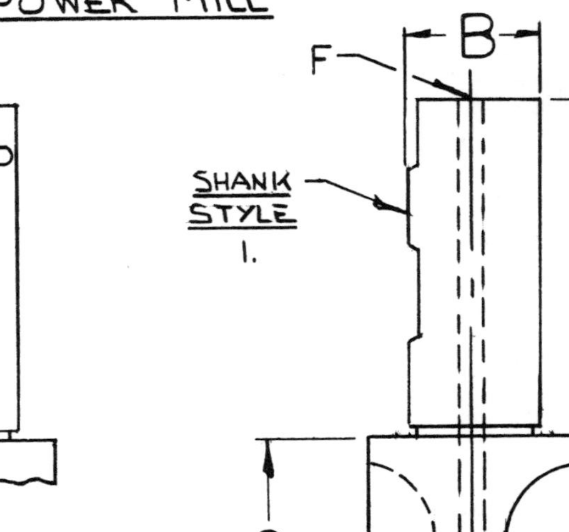

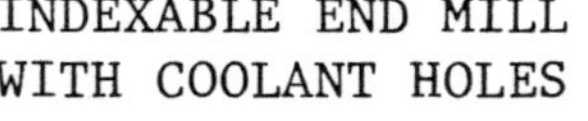

INDEXABLE END MILL
WITH COOLANT HOLES

1. COOLANT SHANK TO FIT IN ROTARY HOLDER OR DIRECTLY THROUGH SPINDLE.

2. SAME SYTLE WITH SURE LOCK SHANK.

3. ROTARY COOLANT COOLAR MOUNTED DIRECTLY TO SHANK

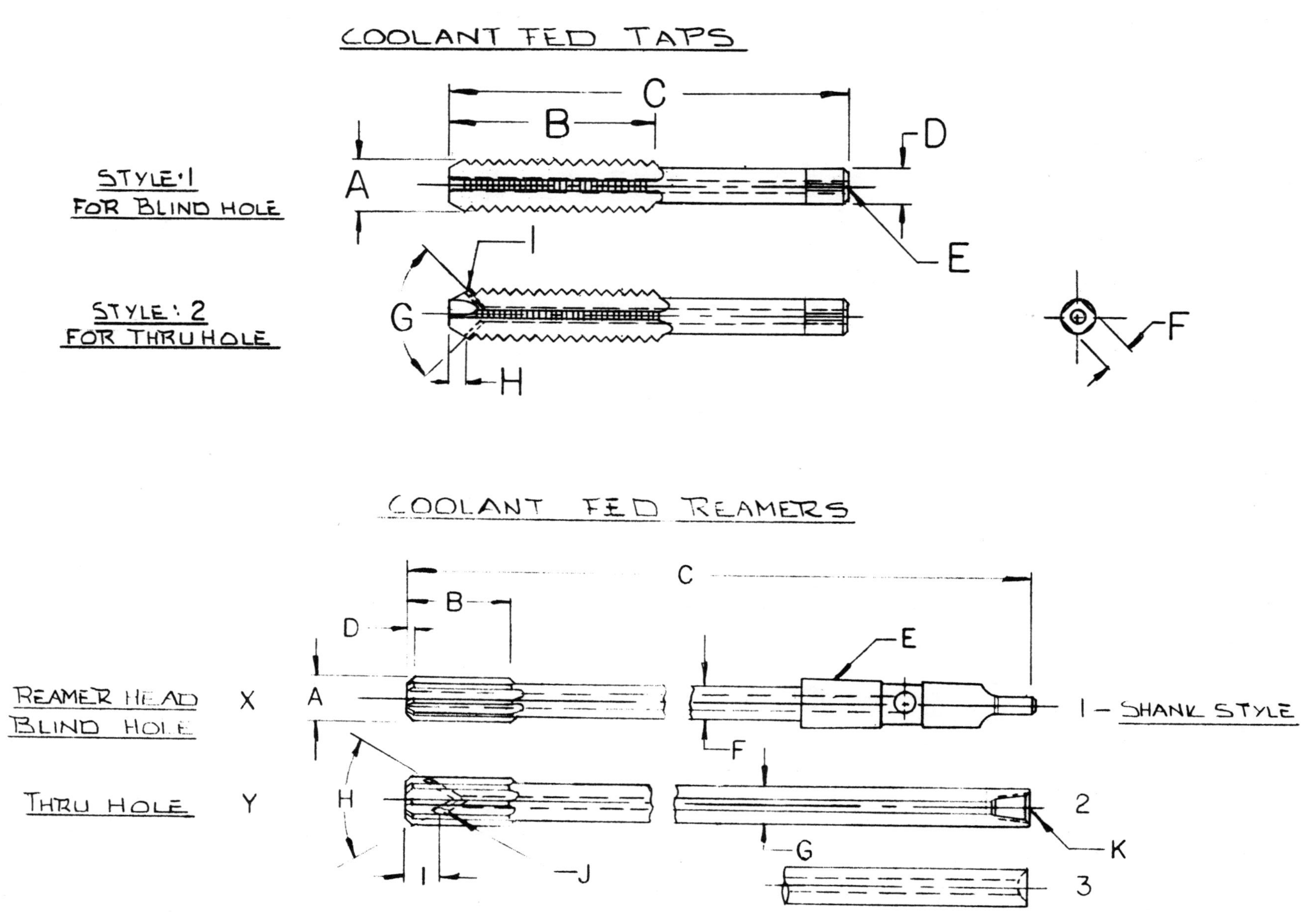
COOLANT FED TAPS
STYLE: 1
FOR BLIND HOLE
STYLE: 2
FOR THRU HOLE
A
B
C
D
E
F
G
H
I
COOLANT FED REAMERS
REAMER HEAD
BLIND HOLE
THRU HOLE
X
Y
A
B
C
D
E
F
G
H
I
J
K
1 - SHANK STYLE
2
3

Presented at SME's International Engineering Conference, April 1970

Filtration Of Coolants Pays Dividends

By Norman Williams
Eastern Filter Sales, Inc.

The cost of FILTRATION may appear to be high, but it is decidedly low compared to the cost of continuing to do nothing!!

Permit me to use this statement in as broad a sense as possible. It has been abundantly proven in recent years by a wide range of companies and over a very wide range of applications. Admitting at the same time that there are many possible applications where filtration would be "nice" but not essential to the job and may not justify any sizable expenditure.

Briefly, in what way does filtration and clean coolant pay dividends? Primarily because clean coolant week after week, eliminates a big variable from the entire grinding, turning or forming operation. The coolant is clean every day and not just on Monday because the machine sumps were mucked out over the weekend! Therefore, two, three, or more combinations of the following, depending on any particular application in conjunction with the type of equipment used, will become evident. A - Better surface finishes; B - Better quality; C - Better production rates; D - Better wheel, tool, cutter or die life; E - Extending the life of a coolant from a week or two to six months or longer since simple "old age" of a water-type coolant (especially the synthetics) would be hard to recall. And a straight oil coolant is good forever if clean. F - Generally less tendency toward heat buildup in the coolant; G - Cleanliness around the machines and improved housekeeping; H - Operator satisfaction and reduced or eliminated dermatitis if previously a problem; L - Lower maintenance labor costs through several areas, etc.

Now, what coolant to use to get best results? That is not for discussion at this time, except to say, "there is a difference in coolants." Get a couple of good coolant people to review your needs. Changes and improvements are in evidence, so your old coolant may be putting you several years behind.

The Book says: - "Filter, n. - Any device of porous substance used as a strainer for clearing or purifying liquids. Filter, is to pass through a filter. To separate solid matter from liquid by a filter." This is an easy dictionary definition that is just as easily misused.

To phrase it differently, I will quote from a recent issue of the English publication FILTRATION & SEPARATION,---"The ways in which filtration may be defined are debatable and a recent French author suggests that, "Filtration is that operation whereby particles suspended in a fluid in motion are separated from the fluid by means of semi-permeable membrane." This, at first sight, seems very logical, but of course it excludes from consideration that important class of devices which rely upon field forces for their separation, i.e., centrifuges, magnetic separators, cyclones, scrubbers, etc."

For the moment, everything that cleans or clarifies a coolant will be dropped into the category of FILTRATION. Actually, a settling tank is NOT a filter, a magnetic separator is NOT a filter, and a liquid cyclone is NOT a filter, but I shall nevertheless apply the term filtration in a very broad sense. However, I will caution against being misled by the manufacturers of equipment that use this generic definition in labeling a product a "filter" when it is not.

Filtration in the grinding and metal working applications encompasses pretty much any method of "removing solids from liquids and returning an acceptably clean liquid" back to the work area, and hopefully this is as close to the original condition of the coolant as possible.

"Acceptably Clean Liquid" may be given some boundaries in the range of; - Removal of Total Solids - 75% to 99.8% can be acceptable cleanliness!!!

Many jobs are operating today where 15% to 25% of the total solids are recirculated over the work. This may sound high but it is a fact. Not necessarily good, but nevertheless a fact.

Industrial coolants and their clarification are under discussion here so a wide variety of filter applications in waste disposal, potable water, food processing, petro-chemical, pharmaceuticals, etc., will be by-passed entirely.

Filtration, by my definition, commences where, hopefully, at least 95% of the total solids are effectively removed from industrial coolants--because this is real darn clean for most people's requirements and can be achieved in a practical and economical manner!! ·95% of total solids! Much equipment delivers a far better degree of clarification than this with no extra cost.

Total solids carried in suspension or recirculated has no direct kinship to the particle size carried back to the work area. Sometimes the total solids are much too high even though particle size is very low, as in honing or EDM work. And other times the total solids in suspension are very low but a few large particles are very detrimental to a process as in die form drawing jobs, mirror finishes and high polish roll grinding.

Again, a generally accepted range of coolant cleanliness for grinding can go all the way from 10 micron to 125 or 150 micron particles. Recognizing of course that an upper range of 40 or 50 micron particle size will keep you out of a lot of trouble and produce far better than average commercial finishes. At the same time when engineers blandly tell a filter man to supply 5 micron cleanliness they had better be prepared to pay more.

You can get anything you want as long as you are willing to pay for it. This was my reason for excluding the potable water, food processing, pharmaceuticals, etc., because they regularly specify as low as 1/4 micron absolute filtration.

Now days with requests for 5 M cleanliness, the engineer should take into account that not all coolants want to be filtered to 5 M. Simply because some of the "coarse" emulsion type coolants used for extreme pressure and high lubricity applications can have the goodies separated and filtered out due to the relatively large, up to 15 micron range, droplets in the emulsion even when properly and fully mixed in a tank.

Viscosity of an oil is always going to be a critical factor in micronic filtration because if you are to filter out micronic particles you must break up the stream of oil into micronic droplets or streams smaller than the interstices of the permeable filter cake or the filter element. Viscosity may be considered as a sort of "built-in" resistance against pulling apart and lubrication depends upon film strength, which is nearly the same property. As some additives are for increased film strength, they increase the problem of filtration even though they may not actually be increasing the rated viscosity of a given oil. Because of this resistance, you find it hard to use diatomaceous earth on oils higher in viscosity that 250 SSU @ 100 degree F. In these cases a good look at the whole problem is required. This of course applies only to filteraid filtration and not to cases where the natural chip does its own precoating. It is not known yet what the viscosity limit is in these cases, or rather that we feel only that the nature of the chip is the main factor.

Light viscosity oils cool better! Light viscosity oils as now made can take the sulfur or chlorine in true solution. Chlorine up to 25%. They used to use the higher viscosity oils to carry the sulfur in suspension as it was only partly in solution. Sulfur in suspension, or free sulfur, can give problems in filtration.

Generally speaking, the only way the oil is used (consumed) is by the carry out on parts and with the sludge. Sulfur and chlorine is spent and carried out in about this direct proportion and the normal makeup of oil may have to be supplemented by additions of the particular additives desired. Lard oil however plates out preferentially and it is extracted on the parts and sludge faster than the base oil. Also with lard oil, a bacterial growth can occur on the parts.

After much discussion with oil men and chemists from many oil companies, it is pretty much agreed that filtration (except of course with Fuller's earth) does not effect the oil or oil additives.

Some oils now have special additives to encapsulate the dirt particles to give the dirt lubricity if it is delivered back to the work area and comes between the wheel or tool and the work. This permits the dirt particle to "slide" thru without scratching. It is my feeling that this little dirt particle with its slippery overcoat may also be sliding too easily thru the filter septum or porous cake. This encapsulation also causes a lighter specific gravity of the dirt particle and it does not settle nearly as well, maybe taking up to 48 hours.

Now a moment to define a point that is all too commonly misused:

- A micron is a measurement of particulate size, in length, width and thickness and is generally viewed under a microscope as the greatest dimension - its length. A micron is 1/25,400 of an inch. .001 of an inch is 25.4 microns. Please separate entirely a "micron particle size" from a "micro inch or R M S measurement of surface finish." R M S or micro inch definition is: - A Micro Inch is not a metric term actually and is NOT related to "micron" but represents one-millionth of an inch. A 5 micron inch finish means not having depressions of a depth exceeding 5 millionths of an inch. In length it can be anything.

Now as how to get an "Acceptable Clean Liquid." This question will be limited to the commonly accepted and widely used methods of this present time. Ignoring the once used, the special applications, the wonderful new mechanical device that has been sold for a couple of jobs and then withdrawn or shelved because of limited function or mechanical faults or possibly just because industry has refused to accept the gadetry of the item. And there has been plenty of this last condition in the filter equipment field. We have all been guilty of this on a few occasions because experimentation and developmental work demands the varification of our mistakes.

A listing of some of the methods used today could be:-

1. SETTLING & SEDIMENTATION, as the oldest method. Settling is effected by specific gravity of particles, distance thru which particles must travel, and the turbulence present. So people have used tanks and pits, baffled and compartmented in a variety of ways to effect the three conditions just mentioned. Machine coolant tanks, etc., are the starting point.

Conveyorized tanks of a wide variety in design, with and without positive surface skimming of tramp oils and floating scuds. Conveyors in the main tank bottom to remove settled solids, and also by one manufacturer at least, in the clean tank section to remove fine sedimentation as this is sure to be present. In other words a mechanical janitor to put the sludge in a tote box or dumpster bucket. These are excellent for some jobs and almost the only method for straight cast iron grinding, glass edge grinding and for centerless or belt grinding of plastics (laminated types) that may have up to 40% removal of total solids by means of the surface skimmer. When properly applied - good function. - They are the least expensive to operate and maintain, with initial cost in line with other items and many times it is less.

2. FLOTATION, basically a conveyorized settling system, supplimented with an airation, foaming, skimming mechanism to gather the fine particles together and return them to the settling compartment for a second chance to drop out.

An old technique and many installations in operation. Still suited for certain applications but a number of these existing flotation systems have had some form of positive filtration added to them in recent years to upgrade the total system efficiency. Foaming coolants are not to everybody's liking anymore and flotation demands a high foaming coolant and thereby limits what coolant a shop can use.

3. MAGNETIC SEPARATORS, and most of you know this equipment. Properly applied on magnetic dirt they are okay. <u>NOT</u> on plastic or aluminum grinding as can be found in all too recent years!! Again an old time method that has performed a lot of service over the years with a reasonable initial dollar cost, low floor space required and then hogged out a lot of dirt. Not generally rated too good as to removal of total solids and many have been replaced with positive filters in past ten or fifteen years.

4. CENTRIFUGAL CLARIFIERS, Centrifuges - Suited to certain water coolants but more commonly on oils. They fit individual machine applications, are fairly simple, use limited floor space and are easy to maintain. On honing, thread rolling, wire or tube drawing and special jobs and special alloys they will often out perform other methods of clarification.

5. HYDROCYCLONES OR LIQUID CYCLONES, that belong to the class of wet mechanical separators. An old item but only in the last few years has it appeared in wider usuage on the metal working coolant jobs. It is sort of the "new thing." A great deal of debate as to their real efficiency and value. I have used them experimentally on a number of tests and not found a sufficiently good function to recommend them except on rare occasions. They're down right poor on oils. I have heard reports about "real good installations" but to my knowledge they are scarce. Their efficiency is directly dependent upon the thru put in the GPM/PSI factor and the feed to each single cyclone where they are used in multiples must be constant with respect to volume and pressure. And where they have been properly sized to begin with, a small

variance in pump efficiency or partial stoppage of the cyclone can mean a major drop of unit efficiency.

6. GRAVITY FLOW FILTRATION, thru a disposable filter media. This came into being as an industrial coolant filter in the early 1950's and it was a definite advance over other methods for many metal working jobs over other methods because it was a method of continuous, positive filtration.

POSITIVE FILTRATION - now by changes in grade of the filter media a shop could really clean up a coolant where it was required and where the expense of the expendable could be justified as to quality, production increases, coolant life increase, wheel life improvement, etc. It was a really new approach at the time and companies put them in. Many applications still are suited to this type of equipment as made by several manufacturers, with some jobs showing a superiority over any other method. Initial cost is okay, floor space requirement is fairly high and sometimes hard to find and filter material costs can be very important. Again, properly applied they are good for individual or semi-central systems.

7. SUCTION FLOW OR PRESSURIZED FLOW of coolant thru a disposable filter media is a refinement of the original gravity type filter and gets more G P M for less floor space used. This lends itself to the semi-central or large central system, especially where the swarf will build a desirable filter cake. Still the cost of the disposable filter media can be all important. Not selective generally as to whether a foaming or a non-foaming coolant is used. Several types on the market.

8. SUCTION FLOW, thru permanent or semi-permanent filter screens of stainless steel, fabric bag elements, socks or tubes. Also some have used an endless belt element of fabric. A further type uses pressure flow thru the unit. Positive filtration with good efficiency on small, medium and large flows but my preference leans toward not applying it to the really small jobs. This equipment can handle just about any type of coolant. The initial cost is in line with other large type of equipment with no expendable cost (or at least limited) and maintenance experience is generally good. A widely used type of filter and it is made by several well-known companies.

9. PRESSURE BACK-FLUSHING CARTRIDGE FILTERS, where a highly polished coolant is desired as on gun drilling oil, lapping, honing, etc. Now you are moving into the range of 2 to 15 micron demands. It is well established as to the high degree of filtration thru various types of cartridges but once loaded these cartridges are costly to throw away so the back-flushing technique has widened the range of application and lowered the cartridge costs.

10. PRESSURE BACK-FLUSHING, of filters using tubular nylon or woven wire mesh or screen elements. At least one design that fits this category is made by nearly all the industrial coolant filtration companies. Some companies have several basic designs. These filters are widely

applied as self-coating or as a pre-coat type unit. Self-coating where the particle size and nature of the dirt in the liquid will build its own permeable cake and continue to deliver a reasonable GPM capacity per square foot ratio. Or else pre-coated where the dirt is too fine or slimy or where a very high degree of filtration is required. They can be designed and sized to deliver 1/2 micron clarity on honing oils, grinding oils, E C M, E D M, solvents and water type coolants. Much can be said about this type of equipment, so much in fact that it has enough meat in it for a complete session, but I will leave this to the interest you may develop or to the man you will call in to discuss the application.

They are very efficient, expensive (sometimes very expensive) and can be a fairly elaborate system with conveyorized tanks, pumps, control panels and automatic features that must be properly evaluated as to future maintenance. This general type of filter has a wide range of uses beyond the ordinary industrial coolants.

With ten various types of clarification equipment listed above and slides of each general type to be shown later, I ask your attention for the most important minute of this session. Particularly, when you consider a system, small or large, please view carefully:-

A. FUNCTION - It must meet the demand put upon it! It must perform! Missapplications of filter equipment over the years has been extremely well documented.

B. OPERATIONAL COSTS - Maintenance, expendables or throw away material, power costs as on pumps, etc., as this is more important than some take into account.

C. SERVICE - From the salesman, field reps, and factory people. If you don't buy proper service as part of a filter system, you are really being short-changed.

D. INITIAL COST OF EQUIPMENT & INSTALLATION COSTS.

And this order of preference is partly my own, but is is largely accepted by companies that have used filtration over the years.

Having given at least some of your time and thought to the above mentioned considerations, a further aspect of how you can get, or fail to get, dividends from coolant filtration will be touched on. If you have filtration equipment now or have future plans in mind involving some phase of coolant clarification, do not ignore the fact that many things can effect the operation of a filter system and introduce a very marked change, leaving you with something other than what you planned for or what you have experienced in the past.

1. Missapplication of equipment.

2. Poorly mixed or blended coolants, assuming you have a good one at first.

a. Greatly affected by water hardness ($CaCO_3$) and iron content. Water from certain wells and even the Mississippi or Missouri rivers are so high in iron that a large percentage of the rust inhibitors may be used up immediately upon mixing the coolant. Wells in North Jersey have tested at 1200 PPM hardness and three coolant suppliers gave up because the coolants would not mix and I finally removed the filter equipment because it would not function.

b. Water temperature when mixing and the pH of water.

c. Bacteria from fresh water, - streams, ponds or wells. To quote from: - "Ground Water and Wells", by Edward E. Johnson, Inc. (a well-known manufacturer of well screens and filter screens)"---several types of iron bacteria which change dissolved iron to insoluble iron and produce slime that plugs screen formation in the vicinity of a pumping well." This also occurs in coolants and plugs filters.

d. A major change in coolant used. Also when a customer decides to blend his own coolant by mixing a couple different coolants that very well may be incompatible and he winds up with nothing, plus filter trouble.

3. Bacteria from shop sources: - spitting in tanks, old dirt on and in the machines, dirty lines, dirty return trenches with special regard for the concrete trenches that harbor bugs. The presence of bacteria will reinfest a system that has just been cleaned and pretty soon a new "bug farm" is in full production. A run up in bacteria count means filter efficiency drops way off, with higher cost of expendables or much shorter cycles with other types due to plugged elements.

4. Oil in water, or water in oil. Filters can handle either one or the other <u>but not both</u> at once. With water coolants tramp oil, lub oil, or hydraulic oils cause a variety of problems not the least of which is the start of, or the increase of bacterial growth in a system. With oil coolants, water (often condensate) throws a system completely out of balance. Also when hydraulic oil is added by error to a grinding oil system, then look out.

5. Wheels of some special characteristic such as high rubber bond and the dolomite or magnesite compositions can greatly effect filter function and cost of operation.

6. Residual contaminants brought to the coolant, on the stock, from previous operations such as oil from a stamping operation, lime on steel rods or wire, or heavy rust and scale from exposed storage facilities.

7. Special conditions, metals, alloys or material being worked. Mixing of jobs where hard rubber or a plastic are occasionally run; where cast iron is run for a day or two in a system designed to handle steel; where a leaded steel is worked; where graphite comes from the metal or is introduced as an external ingredient; where a small amount of aluminum is worked in a steel or cast iron system and an electrolytic action can form gaseous bubbles on the aluminum chips that will float unbelievably large chips of 1/2" or 3/4" length; and many other factors that have been observed thru the years.

The Analysis of Particulate in "Filtered" Coolant

Presented at the 27th Annual Meeting of the American Society of Lubrication Engineers, May 1972

ROBERT H. BRANDT
Henry Manufacturing Company, Inc.
Bowling Green, Ohio 43402

This paper presents a gravimetric procedure to investigate industrial filtration equipment. This procedure is simple and inexpensive to run. It provides basic information about the filtration and coolant system and particulate contaminants in industrially "filtered" coolant. General information is given from various in-plant observations on current coolant and filtration practices.

INTRODUCTION

Industry has been concerned with production, tooling, machines, and, more recently, the total aspects of inside and outside plant ecology. However, it has, for the most part, overlooked the coolant and "filtration" system which has been with them for some time. In addition to its initial cost, it is a continuing expense and has a direct bearing on overall production operation.

Recognition of the importance of coolant and its interaction with the metalworking process has now come of age (*2, 7, 10, 11*). Many companies are using coolant and have gone to the further expense of putting in "filtration" equipment. This is a big step toward better efficiency and reduction of production costs. There is, however, very little knowledge or correlation between varying degrees of filtration, tool life, and manufacturing quality. However, the act of installing a coolant system is only the beginning in realizing its full potential..

Many plants have initiated coolant control procedures which vary in thoroughness. Performance of the filtration system, once installed, is one area that has been somewhat overlooked. The users of filtration systems see a big improvement over the days before filtration systems were used, and, in general, feel they have good operating systems, even though the improvement amounts only to reduced coolant line-plugging problems. But are they really getting what they paid for; do they have to change coolant often, every week or month or four times a year, because their coolant goes sour? Are they experiencing difficulty in maintaining tolerances, tool life or production schedules? Admittedly, a good coolant and good coolant control procedures have a great deal to do with the performance of a filter system, but the performance of the filter also plays an important part in the whole system (*4, 12*). The claims made by the filtration experts sound encouraging until trouble occurs, and the customer is left to work out his own difficulties. The coolant must be kept clean enough to do the job satisfactorily, while the maintenance costs for expendables and upkeep are kept within tolerable limits.

The problem presented is solvable if a method were available to evaluate the performance of the filter equipment, determining the amount of particulate left after filtration has taken place. This paper deals with a method based on gravimetric technique; it is easy to perform, economical in equipment costs, and is a composite of standard and accepted procedures (*1*). It is called the gravimetric method.

With the development of the membrane filters and pore sizes which are fairly exact and are in the micron or nanometer range, the gravimetric method can be applied with a degree of precision. The use of membrane filters also allows for microscopic examination of the remaining residue and for easy particle-sizing of the particulate.

BACKGROUND INFORMATION

The term "filtration" applied very generally in industry includes the individual sump or circulation tank; the gravity separator; centrifugal and magnetic devices; the media filter including paper, cloth, screen, permanent membrane; and also the precoat or diatomaceous earth filters. This wide range of filtration equipment is geared for various functions and various cleanlinesses. With such a wide range of filtration equipment available, the proper application of the filtration equipment is important. Manufacturers are responsible for applying the filters properly and for giving the customer the necessary particular application information and service. In this competitive area, manufacturers of filtration equipment tend to indicate a cleanliness or clarity of the coolant that will be found after the filter or filtration step. This cleanliness is usually given in very generous terms such as "filtration to 40 micron," "absolute filtration to 5 micron," etc. Confused by these claims of absolute filtration to certain particle sizes, the industrial purchaser has been at the mercy of the filtration equipment supplier and has been unable to get proof that these industrial filters will perform as expected. The manufacturer of industrial filtration equipment not only should be able to take prospective customers to similar installations that are functioning in the field, but he also should be able to present data as to the performance of the filter at these various other locations. The customer who now has industrial filtration equipment should be able to run tests on his filtration systems currently in operation—not only to check out their operation, but also to allow him to gather data and criteria for future installations of filtration equipment.

It has become very apparent that a method is needed that can be applied to industrial coolants and industrial filtration equipment. A literature search revealed that certain procedures had been established for various industrial fluids, particularly those concerned with jet aviation fluid and hydraulic systems in the aerospace industry (*3, 8, 9*). Using these established procedures, application to industrial coolant systems was begun.

SAMPLING

In any type of analytical procedure, it is imperative that a representative sample is being analyzed. This, therefore, involves the determination of the best sampling procedure for industrial filtration systems.

In order to determine the particulate concentration that is observed by the tool work-piece interface, samples should be obtained from the distribution lines going to the machine. At this point the liquid should have gone through the filter, through the distribution line, and back to the work. As the actual procedure calls for a 100-ml sample, that should be the minimum sample size. The sample should be procured from a line which has been free flowing for a period of time. If a line is chosen which has not been flowing liquid, the line should be allowed to flow liquid for a period of time to insure any entrapped particulate matter on the wall of the line is removed or flushed out. A collection device should be used such that it captures the total flow from the line in the sampling bottles with none allowed to overflow. Care should be taken that the lip of the bottle does not touch the end of the pipe or that the pipe is not jarred so any particulate adhering to the outside of the pipe is not jarred loose, contaminating the sample. Once the sample has been taken, the test procedure should be run in a short period of time. The reason for this is two-fold: 1.) so that the coolant does not separate into possible tramp oils or other floatable debris; and 2.) so that the particulate does not settle to the bottom of the container or adhere to the sample container wall (so that it cannot be put back into suspension for running the tests).

PARTICULATE DETERMINATION

Once the sample has been taken, the particulate content of that sample must be determined. As this is based on gravimetric technique, it is first necessary to weigh a piece of membrane filter paper. This procedure uses a 47-mm diameter, 8 μ-size polycarbonate membrane filter paper (referred to as "membrane filter" hereafter). The membrane filter should be weighed to the nearest tenth of a milligram or a hundredth of a milligram. Care should be taken at this point as some of the membrane filters do have an electrostatic charge and inaccurate weighings can occur. An electrostatic dissipator can be used in the balance case if this presents a problem. Once the membrane filter has been weighed, it can be placed in the appropriate holding apparatus and connected to a vacuum pump which will allow a vacuum of 15 inches of mercury to be placed on the membrane filter. The sample should be shaken vigorously for a period of time to ensure it is back in a homogeneous state. A 100-ml sample should be placed in a holding container and 15 inches of mercury vacuum placed on the membrane filter, and the sample filtered. Then the surface of the membrane filter, still under vacuum, should be washed with petroleum ether or some other liquid which will dissolve extraneous oils and other hexane-soluble materials. Using a material which has a low vapor pressure can also help in the drying process because of rapid evaporation.

With the vacuum released, the membrane filter is removed from the holding apparatus and allowed to air-dry. At this time the sample residue and membrane filter should be weighed again. Knowing the initial weight of the membrane filter and the weight of the membrane filter with the sample's particulate on the surface, the difference will give you the quantity in grams of debris present in the sample. As a 100-ml sample was used, it is easy to determine the parts per million this represents in the total system by the following equation:

$$\frac{\begin{pmatrix}\text{Wt of membrane}\\ \text{filter and residue}\end{pmatrix} - \begin{pmatrix}\text{Wt of membrane}\\ \text{filter}\end{pmatrix} \times 10^6}{\text{Wt of 100-ml sample}} = \text{PPM}$$

With this simple procedure, the parts per million of the quantity of debris remaining in the liquid after the industrial filtration can be determined.

In the procedure two separate membranes were not requested to be weighed, one as a control which would be placed underneath the top membrane filter upon which the sample would be poured. In several procedures, this is done so that any changes of either loss or gain in weight of the membrane filter itself can be taken into account in the final calculation. There are currently three suppliers of membrane filters; this procedure has been tested with two of the three suppliers' products.

One manufacturer has membrane filters in various types and pore sizes, i.e., a cellulose acetate membrane which is a general membrane, and other special-material membrane filters.

It was found that there was an interaction using the membrane filters of the cellulose acetate type between the various industrial coolants from the chemical fluids to the soluble emulsions with the membrane filter itself; so much so that it was hard to be consistent in the determinations even using a control filter.

Another manufacturer produces a polycarbonated plastic membrane. Using the 8 μ-size polycarbonate membrane filter paper, consistent results were obtained. It was found that it is not necessary to use a control membrane filter unless very precise work needs to be done. So with the selection of the proper membrane filter, the parts per million particulate can be run with the aforementioned procedure.

After this determination is made, additional determinations can be made using optical microscopic examinations at a 100 $\times$ magnification of the surface of the membrane filter. Two things were looked for in this microscopic examination; 1.) the largest individual particle left on the membrane filter; and 2.) modified particle count or distribution so that the average size particle could be determined. In all of the examinations of the membrane filters from actual industrial filtration installations that have been made to date, even when a positive media is used that is rated at a 40 μ or 20 μ size, larger particles are observed than this nominal size or cut-off size of the positive media. This is consistent with the work done by P. A. Knowles (*6*). So, in actuality, the purchaser of industrial filtration equipment buying only on the guarantee that a certain absolute filtration will be achieved is being misled. It might be that the average size of the particles will be a certain nominal size, but all particles will not be removed to that size..

Knowing the total particulate content or residue from the filter sample, microscopic examination for largest particle, and the average size of particles, the purchaser of industrial filtration equipment now has a yardstick with which to compare the operation of the industrial filtration systems he has installed. If the operation of his particular industrial filter is such that it is "doing the job" it was purchased to do, the cleanliness criteria required for future systems can be changed accordingly. This becomes the basis for acceptable performance for both purchaser and filter manufacturer. One such criteria was included in a recent request for quotation for filtration equipment (reproduced in part in Appendix A).

ADDITIONAL INFORMATION

Other observations can be made with this simple test. Although it is a harsh test, the rate or the time for a 100-ml sample to be passed through the membrane filter can be taken when the initial charge of coolant goes into the industrial filter system. If this same test is performed on a weekly basis, at some point the filtration rate may begin to slow down. This test can be an early indication that something is changing within the industrial coolant and corrective procedures might be necessary.

It was found in the author's investigations that several samples of industrial coolant could not be filtered readily through the membrane filter. The following three factors influenced the filtration of the fluid through the membrane filters.

1. **Particulate size**
   If the particulate size is very similar to that of the pore size of the paper, plugging of holes in the membrane filter can occur.
2. **Precipitates**
   Water precipitates of insoluble hydroxides; soap scum precipitates that occur with hard water conditions; and certain oil-soluble emulsions and concentrations of tramp oils acted upon by other contaminants to become gelatinous or slimy.
3. **Microbiological contamination** (*5*)
   Either aerobic, anaerobic or fungus.

When poor filtration is observed through the membrane filter, one helpful step is to use a slight backpressure on the membrane filter to clear the surface of the paper; then going back to a vacuum, alternating pressure and vacuum, a sample can be brought through the membrane filter. This is another reason for the selection of the 8 μ polycarbonate membrane filter paper—its tensile strength is such that it can stand a slight back pressure without rupture.

CONCLUSION

At this time no specific guidelines can be given those who would like to test their filtration systems as to what results to expect from all systems. No standards have been established. However, over the last two years the experience of doing many samples has shown what can be expected in various applications, using equipment manufactured by one specific manufacturer.

Because each plant functions somewhat differently, with varying care and maintenance of coolant and machines, each plant's parts per million will be variable. This simple test can be performed by almost any metallurgical laboratory or in-plant personnel. The cost involved is less than $250.00 for the equipment and membrane filter paper.

Interpretation of the results may be misleading until experience and familiarity with the operation of this procedure are gained. Interferences that might cause inaccurate results must be learned.

This total procedure will give the industrial purchaser and the filtration manufacturer the data to present basic information and general guidelines for the operation of current installations as well as possible criteria for future installations.

REFERENCES

(1) Brandt, R. H., "Particulate Contaminant in Machine Coolants," Henry Technical Bulletin 2-01 (March 1970).
(2) Dwyer, Jr., J. J., "The Production Man's Guide to Cutting Fluids," *American Machinist*, 105–120 (1964).
(3) "A Dynamic Test Method For Determining The Degree Of Cleanliness of Downstream Side of Filter Elements," Soc. of Automotive Eng., ARP599 (1962).
(4) Gough, P. J. C., "Cutting Fluid Filtration For Machine Tools," *Filtration and Separation*, 706–715 (1970).
(5) Hostetler, H. F., and Powers, E. J., "Bugs, Surfactants and Woes," Fuels Session, Div. of Refining—Am. Pet. Inst., 28th midyear meeting, 1963.
(6) Knowles, P. A., "Classification of Filter Media by Particle Counting," Second Fluid Power Symposium, Jan. 1971.
(7) Mason, Jr., J. A., "Cost Savings Through Cutting Fluid Selection," Technical Paper MR69–259, Am. Soc. of Tool and Manf. Engr. (1969).
(8) "Particulate Contaminants in Aviation Turbine Fuels," ASTM D2276-65T, 839–853 (1965).
(9) "Procedure for the Determination of Particulate Contamination in Hydraulic Fluids by the Control Filter Gravimetric Procedure," Soc. of Automotive Engr., ARP 785 (1963).
(10) Sluhan, C. A., "Economics of Metalworking Fluids," Tech. Paper MR70-549, Soc. of Manf. Eng. (1970).
(11) Sluhan, C. A., "Solving the Coolant Problem," Am. Soc. for Abrasive Methods, Nat. Tech. Conf., 91–108 (1968).
(12) Wilson, A., "Coolant Filtration Pays Off In Better Grinding," *Machinery*, 94–102 (June, 1963).

APPENDIX A

DESIGN DATA

E. The coolant clarity specification defines the size and quantity of solid particles remaining in the coolant when it reaches the cutting tools. This specification is composed of three (3) parts. The mean size of particles in micron (μ), the maximum permissible particle concentration in weight of solids per fixed weight of coolant (Parts Per Million PPM) The degree of clarity requested has been determined by studies made at this location using the attached "Coolant Clarity Testing Procedure" on similar machining operations. After the system has been installed, the successful bidder shall obtain independent laboratory test using the "Coolant Clarity Testing Procedure" to determine system capability and make necessary adjustments to meet the clarity specifications.

SPECIFICATION DATA SHEET

CENTRAL COOLANT AND CHIP COLLECTION SYSTEM

C. Type of Chip:

Nodular iron chips—small ½-in long × ⅛-in wide
Some borings ½-inch diameter × 3 in long

D. Coolant Clarity:

1. Mean size of particles	25 microns
2. Maximum permissible particle size	60 microns
3. Maximum particle concentration in weight of solids per fixed weight of coolant	80 PPM

COOLANT CLARITY TESTING PROCEDURE

1. The purpose of this procedure is to quantitatively analyze the solid contaminants in machine coolant. This procedure should not be considered as the only or best way to obtain data, but rather as the standard adopted for uniform testing and reporting.
2. The following test description does not detail every step of the procedure. It is understood that good laboratory techniques will be used at all times.

A. Coolant samples shall be taken from a machine cutting tool nozzle when the coolant system is running at or near design capacity. The date, time, and number of machines running at the time of sampling must be recorded. The minimum sample size is 150 ml, and testing should be performed within 24 hours of the time of sample collection.

B. A 100-ml sample shall be filtered through an 8 μ polycarbonate plastic filter media, 47 mm in diameter, with the aid of a vacuum pump. The solids of the media shall be washed with petroleum ether to remove any oily residue. The concentration of solids shall be determined using the weight of solids deposited and the specific gravity of the coolant. The concentration shall be expressed in parts per million (PPM).

C. The filter media shall then be placed under a 100-power microscope with a calibrated eyepiece. The largest particle found shall be measured and the size recorded. The representative area of the media shall be scanned and the approximate percentages of each of the following determined:

16–30 micron particle
31–45 micron particle
46–60 micron particle
61–75 micron particle
76–90 micron particle
91–105 micron particle
105 and over micron particle

The mean particle size can then be determined using the cell midpoint X, cell frequency f, number of measurements of object N, and the relation

$$\bar{X} = \frac{\Sigma X f}{N}$$

3. Test Reporting:

The test report shall contain the following items:

Name of the coolant system
Date and time of sampling
Number of machines running at time of sampling
Date and time of testing
Size in microns of the largest particle
Coolant specific gravity
Solids contamination in PPM
Particle-size distribution data
Calculated mean particle size.

Presented at SME's 1974 International Engineering Conference, April 1974

Effect Of Metal Working Fluids On The Skin

By M. H. Samitz, M.D.
University of Pennsylvania

Cutting and grinding fluids (cutting oils) are used extensively in those industries requiring various machine-cutting and drawing operations. It has been estimated that slightly over one million machine tool operators, machinists, layout men, instrument makers, tool and die makers and setup men are exposed to these compounds.[1] The engineering industry is one of the largest involved and a majority of its employees are exposed significantly to contact with these compounds. They represent an important and frequent cause of occupational dermatoses. Up to one percent of workers using these fluids have skin disorders as a result and they are the prime cause of time lost due to occupational disease.[2]

These compounds are called cutting oils, cutting fluids, grinding fluids, cutting compounds, lubricants or coolants. Currently, the term cutting and grinding fluids has gained wide acceptance. These fluids fall into two principal categories [3]: (1) straight oils, and (2) water-miscible fluids. The composion of these compounds is well described in other reports [1-6].

The straight cutting oils (insoluble oils) are generally composed of petroleum lubricating oil fractions and small amounts of animal or vegetable oil with additives such as sulfurized fats and chlorinated paraffins.

The water-miscible fluids fall into three general categories: (1) soluble or emulsion oils, basically suspensions of oil in water containing an emulsifier, a rust inhibitor, anti-foaming agents and a dye (usually fluorescein); (2) synthetic fluids, which are aqueous solutions generally composed of triethanolamine and sodium nitrite with surface active agents and other additives (but containing no petroleum oil); and (3) semi-synthetic fluids that usually contain 15 to 40% petroleum oils, emulsifiers and rust inhibitors (sometimes including nitrites and other additives). Germicides are added to all water-miscible coolants to preserve them in use through inhibition of microorganisms. At the present time the water-miscible fluids are the most commonly used. They now comprise from one-half to two-thirds the cutting fluid market.[7]

The cutting oils have two primary functions: to cool the cutting tool, thus prolonging tool life, and to lubricate, which minimizes heat due to friction. Secondary functions include their use as anti-rust agents and to flush away metal chips. Their actions on the skin cause mechanical blockage of the follicular orifices, clinically manifest as oil acne, and degreasing of the skin, presenting as eczematous dermatitis (Figure 1).

CUTTING OILS: ACTIONS ON METAL AND SKIN		
	Straight Oils	Water-Miscible
Types of Cutting Oils	Insoluble oils	a) soluble oils b) synthetic fluids c) semi-synthetic fluids
Function	1. Cool the cutting tool, prolonging tool life 2. Lubricate, which minimizes heat due to friction 3. Anti-rust action 4. Flush away metal chips	
Action on Skin	Mechanical blockage of the follicular orifices	Solvent and alkaline actions cause degreasing of skin
Effect on Skin	Oil acne: comedones and folliculitis	Eczematous contact dermatitis a) irritant b) allergic

Figure 1

Skin disorders resulting from cutting oils depend upon the summation of effects of the physical forms of the lubricant for use, the individual compounding ingredients (including additives and diluents), the contaminents acquired during use, and conditions of the working environment. A schematic representation of skin problems arising from the use of cutting oils is shown in Figure 2.

SKIN PROBLEMS ASSOCIATED WITH THE USE OF CUTTING OILS

1. Mechanical injuries from metal slivers
2. Eczematous contact dermatitis (irritant)
   Solvent action from emulsifiers, soaps, detergents and high alkalinity
3. Eczematous contact dermatitis (allergic)
   Germicides, rust inhibitors, chromates and nickel salts leached from metals
4. Oil acne and folliculitis
   Predisposing factor - usually hairy individuals
   Forearms, thighs and buttocks because of oil impregnation of work clothes
5. Pigmentary changes:
   Hyperpigmentation: (a) direct from irritation of oil
   (b) secondary to folliculitis
   (c) sunlight as a contributory factor
6. Neoplastic lesions: papillomas, keratoses and skin cancers
   Infrequent in U.S.A. In England from contact with shale oil

Figure 2

The insoluble oils are well known for causing a follicular acneiform eruption which occurs on exposed hairy surfaces or on areas in contact with work clothing that becomes soaked with oil. Common sites are the backs of the hands, forearms, thighs, face, and back of the neck. These characteristic lesions are the result of mechanical blockage of the follicular openings by oil producing comedones and perifollicular inflammation. According to Key et al.[1], follicular occlusion is believed to be primarily mechanical, similar to blockage by grease, carbon black and various dusts. Inflammation around the blocked follicles may be the result of irritation by the oil, rupture into the tissues of retained sebum, or the entrapment of surface-resident bacteria. Insoluble cutting oils are relatively free of bacteria, and the microorganisms cultured from the oils are not responsible for this condition. This reaction pattern has been demonstrated in laboratory animals and in experiments on humans: Kligman[8] showed that the application of straight cutting oils on rabbit ears is strongly comedogenic. In volunteer human subjects, he applied the compounds under continuous occlusion for four weeks to the back of twelve young adult males with a previous history of acne. Biopsies showed a purely comedonal reaction, not inflammatory; the effect was comparable to 20% crude coal tar.

The severity of the folliculitis depends upon the type of job and certain environmental factors. In a study conducted at an engineering facility by Finnie [9], capstan-lathe and automatic-lathe operators suffered a significantly higher degree of moderate and marked oil folliculitis than workers in other types of jobs. The duration of exposure to oil was not an important factor. Insoluble oils were incriminated in these high-risk occupations.

Personal cleanliness is an important factor. For the worker exposed to cutting oils, skin damage is frequently aggravated by methods used to remove the oils from the skin. Watch how the worker removes the day's grime: the machinist and mechanic resort to gasoline, kerosene or other solvents to remove grease adhering to their skin, or they will speed up cleaning by using abrasive cleaners or soap powders of high alkalinity. These contribute to the basic problem and the worker becomes the victim of a triad of skin hazards.

Because of decreased use of insoluble oils, folliculitis is now less common. In the past, when chloracnegenic hydrocarbons were used in oils, chloracne would develop, but this effect is now rare.

Primary irritant contact dermatitis is now the major skin disorder of workers using cutting oils. The soluble cutting oils and synthetic coolants can act as low-grade irritants causing defatting of the skin. This is attributed to repeated exposures to the soap-like action and/or to the alkalinity inherent in these preparations. These agents cause denaturation of keratin and remove water holding substances and surface lipids from the skin. Unprotected skin immersed in or saturated with a detergent or solvent solution several hours a day often develops damage of the horny layer and inflammatory changes. The reactions are characterized by dryness and fissuring, subsequently progressing to frank eczematization.

The response of different individuals to the same substances, and between the response to different irritant substances on the same individual, varies depending on age, the type of skin, previous exposure, the concentration of the irritating compounds and the duration of contact. The onset is ordinarily not abrupt, but insidious and slow. It is frequently difficult to determine whether the cutting oil or the harsh cleaning methods used after work are chiefly responsible: frequently they are cumulating in their effects. The fingers, hands and forearms are the sites commonly involved. Instances of true allergic contact sensitivity are uncommon, and when reported are related to bactericides and breakdown products, rust inhibitors, nickel salts and chromates.

Prolonged exposure indicates clearly that these compounds have a keratogenic action, in the form of keratotic lesions or squamous cell cancers. They have been reported in Great Britain from contact with shale oils, but are rare in the United States (although definitive data are wanting).

Hyperpigmentation of exposed areas may result from direct irritation of oil or may be secondary to folliculitis. Sunlight may be a contributing factor. This is an uncommon finding.

Workers handling cutting fluids may also develop mechanical or abrasive injuries when rags or waste contaminated with metal slivers are used for wiping the skin or cleaning a machine. These wounds are sometimes complicated by secondary infection. Occasionally allergic contact dermatitis results from nickel or chromates leached from the machined metal.

In essence, the formerly common problem of oil folliculitis has decreased significantly because the insoluble oils are being replaced with water-miscible fluids. In this changed environment, irritant contact dermatitis has become the major type of occupational skin disease in workers using the water-miscible compounds.

The incidence of dermatitis depends on several factors: the type of cutting oil used, the cleanliness of the work environment and the personal cleanliness of the worker. Often little attention is paid to hygiene. In conducting a study in several plants in the Philadelphia industrial area we found that in the factory where good housekeeping and good personal hygiene were maintained the incidence of skin disorders was much lower when compared to those plants where they were not maintained.

Control of the Cutting Fluid Hazard.

As in all other aspects of preventive medicine, the recognition of the mechanisms of reactions to cutting oils and the measures to prevent them are far more effective than the treatment of such reactions after they occur. For this, the cooperation of management, labor and the health professions is necessary.

Management. Two areas are important: (1) Can you provide a dermatitis-free cutting oil? (2) Can you establish a clean working environment, install safeguards pertaining to the machinery and to the worker, and provide adequate and proper facilities for the workers' personal hygiene.

Though it may seem logical to develop a dermatitis-free cutting oil, such a possibility does not appear likely. Furthermore, as Gellin cautions, in perfecting a non-dermatitis-producing cutting fluid, the needs of the industrial process cannot be overlooked. For example, to remove the extreme pressure additives only because they might produce dermatitis would prevent the deployment of techniques requiring greater lubrication than could otherwise be accomplished.[5]

The following procedure, however, can be helpful: The cutting fluids should be monitored at regular intervals for cleanliness, acidity, bacterial and fungal concentration, and deterioration.

The safeguarding of machinery in order to decrease the risk of skin contact with the cutting oils is essential. This is a highly developed industrial art which includes devices such as splash guards on automatic screw machines and adequate exhausts. During warm weather, the maintenance of a more equable environmental temperature in the workshop will result in a lower incidence of skin reaction because irritating materials are less likely to adhere to the skin. The planning by industrial engineers of devices to enclose manufacturing processes can contribute significantly to the prevention of industrial dermatoses.

The maintenance of scrupulous cleanliness in the environment is of the greatest importance, both in removing irritants and sensitizers from it, and in serving as a constant reminder to the worker of the importance of personal cleanliness. A dirty, unkempt shop inevitably breeds carelessness and disease or injury. Adequate washing facilities should be reasonably close at hand; if they are distant, they simply will not be used. The dressing rooms should include adequate locker facilities. Impervious protective clothing, such as aprons and sleeves, the provision of clean work clothes and of adequate laundry facilities are necessary. Under some conditions, it may be highly advisable to provide showers for the use of workers at the end of each working day.

Cleansing agents should be available in adequate amounts at each working station. A wide variety of detergents other than soap have been devised and have achieved varying effectiveness. Abrasives are often necessary to facilitate removal of grease and dirt and, of these, finely granulated cornmeal is reasonably effective and rarely irritant. We advise its use in combination with a bland liquid soap.[10] Waterless hand cleaners are currently in vogue. They are especially useful for removal of oil, although sometimes they can cause irritation/ or sensitization.

Protective ointments have been developed to protect against gross contamination by irritants. Is is the opinion of most observers that the value of such creams is limited; probably their use gives added assurance that the worker will wash the skin thoroughly at the end of the work period.

Management has the responsibility to keep the worker informed, to post warning signs, and to enforce safety measures.

Labor. It is imperative that the worker develops a better understanding of the skin hazards associated with cutting fluids. Safety training and the motivation to develop safe work attitudes should be a part of job instruction. Labor-management health and safety committees should describe specifically the health hazards and the techniques for prevention. Any evidence of a skin reaction should be brought immediately to the attention of the plant supervisor, nurse or doctor.

Poor personal hygiene is an important predisposing factor in cutting oil dermatoses. The degree of personal cleanliness often depends on the standards set for the cleanliness of the work environment. The education of the worker in the correct way of cleansing and of using the proper type of cleaner is essential.

Health Professions. The dermatologist, with his knowledge and information obtained by on-the-job examination and by laboratory research, should work in consultation with management and labor. His function is to establish the cause of the eruption, to differentiate irritant from allergic reactions (by patch testing), to institute treatment for the presenting disorder and to advise preventive regimens. For the latter he can provide the direction, but management and labor must assume responsibility in carrying out the program.

He must also consider the problems of rehabilitation and the evaluation of impairment and disability. Rehabilitation is an essential part of dermatologic therapy, and arrangements should be made for proper job placement during and after recovery from the dermatosis caused by the cutting oil. At times the dermatosis may result in varying degrees of impairment and disability. To evaluate these problems it is advisable to refer to the report of the AMA Subcommittee on Skin which provides a workable guide for evaluating impairment and disability.[11]

Measures used for treating dermatoses caused by cutting oils have been reported previously.[10] Unfortunately, there is little awareness among industrial physicians of treatment and control of irritant dermatitis. Home care for the prevention of irritant contact dermatitis is advisable. We have suggested the following set of instructions (Figure 3) on posters in the washrooms, in the plant newspapers and in letters to the worker. They provide an inexpensive and convenient way to hold the interest of the worker and to motivate him to practice safety on and off the job.

ATTENTION

1. A special soap has been provided for your cleansing (pHisoderm and cornmeal). Avoid harsh soaps.

2. Because of your contact with cutting oils and solvents and the degreasing action of these agents, the following routine is recommended to help retain the normal suppleness of the skin. This routine can be carried out at home.

   a. Soak your hands in lukewarm water for 20 minutes.

   b. Do not dry them. While they are wet anoint your hands with white petroleum jelly or vegetable fat (such as Crisco or Spry).

   c. Allow the ointment to remain on your hands for 20 minutes. Remove with a towel.

   d. This may be repeated once or twice daily.

   e. This routine will help to prevent dryness, cracking, and fissuring of your skin.

Figure 3

References

1. Key MM, Ritter EJ and Arndt KA: Cutting and grinding fluids and their effects on the skin. Amer Ind Hyg Assn J 27:423, 1966

2. Gellin GA: Cutting fluids and skin disorders. Industr Med 39:65, 1970

3. Barker GE (Chairman Subcommittee on Metalworking Fluids, Society of Manufacturing Engineers). Personal communication December 27, 1973

4. Arndt KA: Cutting fluids and the skin. Cutis 5:143, 1969

5. Gellin GA: Is a dermatitis-free cutting oil possible? J Occ Med 11:128, 1969

6. Hodgson G: Cutaneous hazards of lubricants. Industr Med 39:68, 1970

7. Sluhan C: Cited by Gellin GA (Reference #5)

8. Kligman A: Personal communication, January 16, 1974

9. Finnie JS: Oil folliculitis. A study of 200 men employed in an engineering factory. Brit J Industr Med 17:130, 1960

10. Samitz MH: Occupational Dermatoses in Current Therapy, 20th Ed p 591. Edited by Howard F Conn, Philadelphia, WB Saunders Co 1968

11. Birmingham DJ, Buckley WR, Kenney JA Jr, Robinson HM Jr, Samitz MH and Suskind RR: The Skin - Guides to the Evaluation of Permanent Impairment. JAMA 211:106-112, 1970

Reprinted from Materials Performance, June 1979

Corrosion Inhibitors as Preservatives for Metal Working Fluids — 2. Morpholine Compounds*

E. O. BENNETT, I. U. ONYEKWELU, and J. E. GANNON
Department of Biology, University of Houston, Houston Texas

Test results of the biocidal and inhibitive properties of 33 morpholine compounds in metal working fluids are reported. Laboratory tests showed, among other results, morpholine compounds are more effective against moulds than against bacteria, and formaldehyde is incompatible with the compounds tested. Another preservative morpholine compound was studied also because of its antimicrobial value. While the most active compound was methylene dimorpholine, several others had significant antimicrobial properties. Morpholine compounds should not be used with nitrites because of their carcinogenic properties. Several suggestions are made concerning the usefulness of further investigations of compounds which may have both corrosion inhibitive and bactericidal properties. The investigation considered information from 83 sources, listed as references.

SPECIALIZED LUBRICANTS which contain oil-in-water or water-in-oil systems are commonly used by industry. Two problems may be encountered with these types of products. First, the emulsions are subject to biodeterioration while in use. For this reason, they usually require the addition of some type of antimicrobial agent at periodic intervals. Second, due to the presence of water, they must contain an inhibitor in order to prevent corrosion of the metals they come in contact with.

In the past, the most common corrosion inhibitor employed in these products has been sodium nitrite; however, in recent months it has been detected that the presence of this chemical may result in the production of carcinogenic nitrosoamines in these lubricants.[1] Since this information has become available, there has been considerable interest in developing substitutes for sodium nitrite.

Unfortunately, the substitution of other corrosion inhibitors has resulted in a major cost increase, since no comparable corrosion inhibitor can compete favorably with sodium nitrite from the cost standpoint. In order to minimize costs as much as possible, it would be of considerable value if these lubricants could be formulated with ingredients which have several different functions within the products. Antimicrobial agents and corrosion inhibitors constitute two important components of these lubricants. It would be of value then to search for chemicals that can function both as corrosion inhibitors as well as antimicrobial agents.

The concept of employing a compound that functions both as a corrosion inhibitor and as an antimicrobial agent in a lubricant is not new. More than 20 years ago, the use of 2,5-dimercapto-1,3,4-thiadiazole was suggested for use as a corrosion inhibitor and preservative for cutting fluids.[2] Several years later, other workers studied the antimicrobial properties of several corrosion inhibitors with negative results.[3] Potassium tetraborate has been suggested for use in cutting fluids because it exhibits both properties.[4] More recently, amine salts of boric acid-polyol complexes have been found to inhibit pseudomonads and sulfate reducing bacteria while at the same time producing corrosion protection of mild steel plates;[5] chromates have been employed to control both corrosion and microorganism;[6] and several piperazine compounds have been produced which exhibit both properties as well as cooling, lubricating, and cleaning properties.[7]

The purpose of this investigation was to determine if morpholine compounds that might be employed in these lubricants exhibit antimicrobial properties. This is the second of a series of reports pertaining to this subject.[8]

Experimental Procedure

All of the compounds employed in this investigation were 95 to 99% pure and were obtained from a number of chemical specialty supply houses.

The test units consisted of quart jars placed in rows. Above each row, a metal framework was constructed to support an aeration system which consisted of aquarium valves connected together with plastic tubing. The amount of aeration of each unit was controlled by adjusting the valves. Capillary pipettes were employed as aerators to produce a fine stream of bubbles in the diluted coolants.

Five hundred mL of tap water (Table 1) was added to each jar. Each compound was used as obtained from the manufacturer, and the required concentration (Wt/Vol or Vol/Vol) was added to each unit along with 15 mL of coolant concentrate to produce the desired oil-water ratio. Each unit was then made up to a total volume of 600 mL by adding additional tap water.

Each test unit was inoculated with a mixture of bacteria and moulds which were obtained and maintained as described previously.[9] Each unit was inoculated once each week with 1 mL of a 50-50 mixture of both inocula.

Each unit was examined once each week for its microbial content for as long as the count remained below 100,000 organisms/mL. Two consecutive counts in excess of this figure at weekly intervals was considered to constitute a failure, and the test was discontinued at that time. All units which contained less than 100,000 organisms/mL were studied for a 105 day test period and discarded at that time.

Two different types of control experiments were included with all tests. Each shipment of fresh coolant was tested upon arrival in this laboratory to determine if the product exhibited any inhibitory properties as defined in the previous paragraph. All the coolants used in the investigation were especially prepared by coolant manufacturers for this work and did not contain a preservative. None of the products employed in this investigation exhibited any inhibitory properties, and they failed in the first week of testing.

Since this communication constitutes a continuation of past work by this laboratory in regards to cutting fluid deterioration, an additional set of controls (4 test units) were included. It may be noted that the inocula employed was a mixture of microorganisms and constitutes a dynamic system susceptible to changes in sensitivity of the organisms to antimicrobial agents. The second set

*Voluntary manuscript submitted for publication April, 1978.

TABLE 1 – Composition of Tap Water Used

	mg/L	Other
pH		7.90
Total Dissolved Solids	118	
Total Hardness		
HCO_3^-	34	
$CO_3^=$	<1	
Ca^{++}	9	
Mg^{++}	5	
Chloride	32	
Resistivity		5200 ohm cm

of controls consisted of a particular cutting fluid preserved with a commonly used cutting fluid preservative. Normally, these control units fail in 21 to 28 days due to mould growth and have done so for the past several years. These controls functioned normally during the test period.

Since the test units were under constant aeration, there was considerable evaporation from each unit. The units were calibrated at the 600 mL mark, and once or twice each week, depending upon environmental conditions, distilled water was added to each unit to bring the liquid level back to this mark. Distilled water was used in order to avoid a buildup of inorganic salts in the test units.

Results

The results pertaining to the antimicrobial properties of morpholine compounds are presented in Table 2. The most active compound tested was methylene dimorpholine. This material was obtained from a small chemical specialty house outside the United States, and no domestic source of the chemical could be found so that the accuracy of the data could be confirmed. While repeated experiments with this material confirmed the results reported, confirmation with a second sample from a different source is the usual practice when outstanding activity as shown here is encountered.

Other compounds showing significant antimicrobial activities included N-aminomorpholine, N-(2-chloroethyl) morpholine HCl, N-(3-aminopropyl) morpholine, 2-nitrobutyl morpholine, and morpholine borane. While some of these compounds exhibited very poor activities in certain products, they could have important dual functions in others.

It appears that morpholine compounds exhibit better antimicrobial properties against moulds than against bacteria, as indicated hy the large number of failures due to bacteria. It is of interest to note that mould problems did occur in several instances, and which were, with the exception of one compound, associated with aminomorpholine compounds. Similar results have been noted in this laboratory in regards to studies with other groups of chemicals containing amino groups.

A preservative consisting of a mixture of 4-(2-nitrobutyl) morpholine and 4,4-(2-ethyl-nitrotrimethylene) dimorpholine is available (Bioban P-1487) for use in industrial biodeterioration problems such as metal working fluids. The material can be formulated into synthetic coolants; however, its greatest advantage is due to the fact that it is oil-soluble and can readily be incorporated into petroleum base products. Since the product is well known to have good antimicrobial properties, it was studied in a concentration of one-half that employed in the studies with the other chemicals. The results of this study may be observed in Table 3. It may be noted that the mixture exhibited significant antimicrobial properties in the coolants employed in this investigation.

Possibly the best known and most commonly used compound included in this group is morpholine. Undoubtedly this chemical will be of greatest interest to those interested in information in this area because of its relative cost, toxicity, and practical experience that is available pertaining to its use in lubricants. It may be noted that while a concentration of 2500 mg/L of the compound showed some antimicrobial properties, this activity was not outstanding. Several higher concentrations were studied in order to develop information as to the concentration range required to produce significant antimicrobial activity. Table 4 presents data showing that the concentration required to produce antimicrobial properties in metal working fluids is in the range of about 5000 mg/L.

There are reports in the literature noting that morpholine has the capacity to react with formaldehyde[10,11] and to neutralize the damaging effects of the compound upon microorganisms.[12] Many of the commonly used cutting fluid preservatives are known to be formaldehyde releasing compounds; therefore, it is of practical value to know if morpholine compounds can interfere with these preservatives. Tables 5, 6, and 7 exhibit data showing the effects of mixing the morpholine preservative shown in Table 3 with several formaldehyde releasing inhibitors. It may be noted that with mixtures of preservatives, it is in most cases incompatible. It would appear that formaldehyde preservatives should not be employed in lubricants containing morpholine compounds.

Discussion

It is well established that morpholine compounds[13-19] and their reaction products[20] are corrosion inhibitors. In addition, some compounds in this group also exhibit antiwear properties[21] and are emulsifying agents.[22] Lubricant detergents and viscosity index improvers can be produced by reacting morpholine compounds such as aminopropyl morpholine with styrene and mallic anhydride.[23]

For these reasons and others, these chemicals are already employed in a number of lubricants such as hydraulic fluids,[24,25] rolling oils[26-29] and cutting fluids,[30-33] as well as in metal treatment solutions.[34] It is also known that morpholine exhibits antimicrobial properties against a wide variety of organisms.[35-41]

A number of investigators have shown that other morpholine compounds such as 4-(6-hydroxycaproyl) morpholine,[42,43] 4-petroselinoylmorpholine,[42,43] 4-decanoyl-2,6-dimethylmorpholine,[43,44] 4-oleoyl-2,6-dimethylmorpholine,[43] 4-(6(7)hydroxystearoyl)morpholine,[42] 4-(12-hydroxystearoyl)morpholine,[42] 4-(12-propionoxyoleoyl)morpholine,[42] N-decanoylmorpholine,[44] N-nonanoylmorpholine,[44] N-lauroyl-morpholine,[44] N-(11-undecenoyl)morpholine,[44] 1-tallow dimethyl-3-methylmorpholino propane diammonium chloride,[45] coco-morpholine-acetate,[45] 1-tallow-3-morpholino-propylene diamine diacetate,[45] morpholino-10(9)hydroxystearyl amine,[45] 4-tridecyl-2,6-dimethylmorpholine,[46] and N-cyclododecyl-2,6-dimethylmorpholinium acetate,[47] all appear to exhibit antimicrobial properties against a number of different moulds, yeasts, and bacteria. Others, including N-palmitoylmorpholine, N-oleoylmorpholine, N-oleoyl-2,6-dimethylmorpholine, N-linoleoylmorpholine, and N-stearoylmorpholine exhibited low grade activity against microorganisms.[44] It appears that morpholine and 4-ricinoleoyl-morpholine are the most active compounds noted in the literature.[42]

More recently, several other derivatives of morpholine[48] have been studied for their antimicrobial properties. It was found that N-9(10)-actyl-thiostearoylmorpholine and N-9(10),12(13)-diacetylthiostearoylmorpholine had the greatest activities, while N-11-acetylthioundecanoylmorpholine and N-11-acetylthioundecanoyl-2,6-dimethylmorpholine also exhibited some antimicrobial properties. The compounds appeared to be more active against fungi than against bacteria.

A considerable number of mixtures of antimicrobial agents containing morpholine compounds have been patented as effective products for controlling microbial growth in a number of industrial situations. Many of these combinations involve the use of N-(2-nitrobutyl)morpholine mixed with other compounds such as N-alkyl dimethyl benzyl ammonium chloride,[49] hexachlorodimethyl sulfone,[50] 5-chloro-2-methyl-4-isothiazolin-3-one calcium chloride,[51] 2-methyl-4-isothiazolin-3-one,[51] 3,4,5-tribromosalicylanilide,[52] 3,5-dimethyltetrahydro-1,3,5-2H-thiadiazine-2-thione,[53] 2,4,5-trichlorophenol,[54] penta-chlorophenol,[55] 2,2-dibromo-3-nitrilopropionamide,[56] cyanoacetamide,[57] and 1,3-dichloroacetone oxime acetate.[58]

Other combinations reported to be effective involve the use of a mixture of N-(2-nitrobutyl)morpholine and 2-nitro-2-ethyl-1,3-

TABLE 2 – Antimicrobial Activities of Morpholine Compounds in Cutting Fluids

Coolants/Compounds:	Days of Inhibition 2500 mg/L Concentration																
	1	2	3	4	5	6	7	8	9	10	11	12	13	14	15	16	17
A	0	28	0	0	0	0	0	0	0	105[(1)]	0	0	35	0	28	21	
B	0	21	0	0	0	0	0	0	0	105[(1)]	0	0	7	0	21	21	
C	0	21	0	0	0	0	0	0	0	105[(1)]	0	0	21	0	42	28	
D	0	21	0	14	0	0	0	0	7	42	0	0	21	0	63	28	
E	0	21	0	0	BE	0	0	0	0	BE	0	0	14	0	105[(1)]	21	
F	21	21	0	0	0	0	0	0	0	0	14	0	14	0	105[(1)]	28	
G	14	42	0	0	0	0	0	0	0	0	21	0	21	14	49	28	
H	35	21	0	0	49	28	0	0	0	14	0	0	21	28	105[(1)]	28	
I	14	21	21	14	0	0	0	0	21	BE	0	0	35	0	28	28	
J	7	21	0	0	0	0	0	0	14	105[(1)]	0	0	0	0	35	28	
K	21	42	0	0	0	0	0	14	21	21	0	0	21	35	105[(1)]	21	
L	14	14	0	0	0	0	0	0	0	0	0	0	21	0	28	21	
M	14	21	0	0	0	0	0	0	0	BE	0	0	0	0	63	28	

Coolant/Compounds:	Days of Inhibition 2500 mg/L Concentration															
	18	19	20	21	22	23	24	25	26	27	28	29	30	31	32	33
A	56	0	0	0	0	0	0	35	0	0	0	0	0	0	0	0
B	105[(1)]	0	0	0	0	0	0	0	0	0	0	0	0	0	0	0
C	105[(1)]	0	0	0	0	0	0	0	0	0	0	21	0	0	0	0
D	49	0	0	0	0	0	0	21	0	0	21	105[(1)]	0	0	0	0
E	105[(1)]	0	0	0	0	0	0	0	0	0	0	21	0	0	0	0
F	105[(1)]	0	35	7	0	14	0	21	0	0	0	0	0	0	0	0
G	105[(1)]	0	0	28	0	0	0	105[(1)]	0	21	21	14	0	14	21	0
H	105[(1)]	0	0	14	0	0	0	105[(1)]	0	42	28	14	0	21	28	0
I	105[(1)]	0	0	0	0	0	0	0	0	0	0	0	0	14	0	0
J	105[(1)]	0	0	0	0	0	0	0	0	0	21	0	0	0	0	0
K	105[(1)]	0	0	42	0	0	0	63	0	0	21	0	0	35	47	0
L	14	0	0	0	0	0	0	0	0	0	14	21	0	0	0	0
M	105[(1)]	0	0	0	0	0	0	105[(1)]	0	0	0	0	0	14	0	0

Notes: 1. 1-40 oil to water ratio.
2. Underlined number indicates failure due to moulds.
[(1)]Still inhibitory when taken off test.

Compounds:

1. Morpholine
2. N-Aminomorpholine
3. N-Methylmorpholine
4. N-Methylmorpholine-4-oxide
5. 2,6-Dimethylmorpholine
6. N-Ethylmorpholine
7. N-Hydroxy ethylmorpholine
8. N-Beta-hydroxyethylmorpholine
9. N-(2-aminoethyl)morpholine
10. N-(2-Chloroethyl)morpholine HCl
11. 2-(Morpholino)ethanolamine
12. N-(Beta-hydroxypropyl)morpholine
13. N-(3-aminopropyl)morpholine
14. N-Butylmorpholine
15. 2-Nitrobutyl morpholine
16. N-(2-aminoisobutyl)morpholine
17. N-(2-nitroisobutyl)morpholine
18. Methylene dimorpholine
19. 4,4-Ethylene dimorpholine
20. 4,4-Dithiodimorpholine
21. Morpholineethane thiol
22. Methyl-beta-N-(2,6-dimethylmorpholino)isobutyrate
23. 4-Hexadecylmorpholine
24. Morpholine oleate
25. Morpholine borane
26. Morpholine propionitrile
27. 4-Phenylmorpholine
28. N-(2-hydroxy-2-phenylethyl)morpholine
29. p-Diazo-morpholine-2,4-diethoxybenzene fluoroborate
30. 2-Amino-4-morpholine-s-triazine
31. N-Morpholino-1-cyclopentene
32. N-Morpholino-1-cyclohexene
33. 1-Cyclohexyl-3-(2-morpholinoethyl)thiourea

dimorpholinopropane in combination with bis(trichloromethyl)sulfone,[59] and 3,5-dimethyltetrahydro-2H-1,3,5-thiadiazine-2-thione.[60]

Morpholine compounds have been reported to exhibit antimicrobial properties in metal working fluids. It has been noted that N-(2-nitropropyl)morpholine and N-(2-nitrobutyl)morpholine[61] and a mixture of N-(2-nitropropyl)morpholine and 2-ethyl1-1,3-dimorpholine-2-nitropropane inhibited microorganisms in these lubricants. The results of the present investigation confirm previous work that N-(2-nitrobutyl)morpholine does indeed exhibit antimicrobial properties in cutting fluids. Unfortunately, samples of the other compounds could not be obtained in order to confirm the other previous reports.

It is conceded that many of the compounds included in this investigation are available only through specialty chemical supply companies. In addition, the literature pertaining to these chemicals is very sparse and in many instances includes information concerning only the synthesis of the chemical. On the other hand, it has been reported that N-methyl morpholine,[63,64] 2,6-dimethyl morpholine,[65,66] N-aminoethyl morpholine,[67] N-aminopropyl morpholine,[67] N-butyl morpholine,[68] and 4-phenyl morpholine,[69] all exhibit anticorrosion properties.

TABLE 3 — Effectiveness of a Morpholine Mixture(1) as a Preservative for Cutting Fluids

Coolants	Days of Inhibition 1250 mg/L Concentration
A	35
B	63
C	63
D	42
E	36
F	105(2)
G	56
H	77
I	35
J	35
K	105(2)
L	35
M	35

Notes:
1. 1-40 oil to water ratio.
2. Underlined number indicates failure due to moulds.

(1)Preservative composed of 4-(2-nitrobutyl) morpholine (70%) and 4,4-(2-ethyl-2-nitrotrimethylene)dimorpholine (20%).

TABLE 4 — Inhibitory Activity of Morpholine in Cutting Fluids

Coolants	Days of Inhibition 5000 mg/L Concentration
A	35
B	21
C	56
D	24
E	21
F	105(1)
G	105(1)
H	105(1)
I	70
J	49
K	105(1)
L	49
M	42

Notes:
1. 1-40 oil to water ratio.

(1)Still inhibitory when taken off test.

TABLE 5 — Effectiveness of a Morpholine Preservative and a Formaldehyde Releasing Preservative in Cutting Fluids

Coolants	Days of Inhibition		
	1	2	3
A	42	21	28
B	63	42	21
C	77	48	28
D	14	21	14
E	77	BE	21
G	42	14	21
I	42	14	35

Notes:
1. 1-40 oil to water ratio.
2. Underlined number indicates failure due to moulds.

Compounds:
1. Bioban P-1487, 1000 mg/L.
2. Tris(hydroxymethyl)nitromethane, 1000 mg/L.
3. Combination of both 50-50 mixture, 1000 mg/L.

TABLE 6 — Effectiveness of a Morpholine Preservative and a Formaldehyde Releasing Preservative in Cutting Fluids

Coolants	Days of Inhibition		
	1	2	3
A	42	98	56
B	63	91	35
C	77	112	35
D	14	112	63
E	77	14	35
G	42	49	49
I	42	119	49

Notes:
1. 1-40 oil to water ratio.
2. Underlined number indicates failure due to moulds.

(1)Still inhibitory when taken off test.

Compounds:
1. Bioban P-1487, 1000 mg/L.
2. Grotan, 1000 mg/L.
3. Combination of both 50-50 mixture, 1000 mg/L.

The most extensive study pertaining to the corrosion inhibition properties of morpholines presents work on 19 chemicals using the Static Water Drop Test.[70] These workers found no anticorrosion properties exhibited by N-methyl morpholine, N-butyl morpholine, N-phenyl morpholine, N-3-phenylpropyl morpholine, dithiomorpholine, 4,5-dimorpholinocatechol, 2,4,6-trimorpholinomethyl phenol, and N-2-(2-ethylhexamido)ethyl morpholine. Other compounds in this group such as morpholine, N-aminoethyl morpholine, N-dimethyl-aminoethyl morpholine, N-diethylaminoethyl morpholine, N-amino-propyl morpholine, N-hydroxyethyl morpholine, N-2-methoxyethyl morpholine, N-2-ethoxyethyl morpholine, N-2-hydroxybutyl morpholine, and 1,2-diN,N-dimorpholionethane all exhibited anticorrosion properties.

It is no longer possible to formulate products without consideration of their health hazards or their effects upon the environment. For this reason, any communication pertaining to the potential use of chemicals must give consideration to what is known about their toxicity, undesirable effects or their impact upon the environment.

The toxicity of a number of morpholine compounds has been studied, and the data are available in the literature. The LD_{50} dosage for morpholine in rats has been established to be between 1.05 and 1.6 g/kg,[71,72] 1.78 g/kg for 4-ethylmorpholine, 6.13 g/kg for 4-acetylmorpholine,[71,72] 5.66 g/kg for 4-aminopropylmorpholine,[73] 2.72 g/kg for 4-methylmorpholine,[74] and 6.7 g/kg for

TABLE 7 – Effectiveness of a Morpholine Preservative and a Formaldehyde Releasing Preservative in Cutting Fluids

Coolants	Days of Inhibition 1	2	3
A	42	77	35
B	63	77	35
C	77	91	14
D	14	140(1)	21
E	77	140(1)	49
G	42	42	35
I	42	35	35

Notes:

1. 1-40 oil to water ratio.
2. Underlined number indicates failure due to moulds.

(1) Still inhibitory when taken off test.

Compounds:

1. Bioban P-1487, 1000 mg/L.
2. Milidin TI-10, 1000 mg/L.
3. Combination of both 50-50 mixture, 1000 mg/L.

dithiodimorpholine.[75] Dithiodimorpholine was slightly irritating to rabbit skin in a 1% ointment but was tolerated by human subjects.[76]

When morpholine was fed to rats at the rate of 0.17 mg/kg (LD_{50} 0.1) per day for 30 days, it was found that the neuromuscular excitability was decreased significantly, the arterial pressure increased, reticulocytosis and leukopenia were noted and hippuric acid excretion in the urine was decreased.[77] Exposure of rats or guinea pigs to morpholine at the rate of 0.008 to 0.07 mg/L of air for 4 hours daily for 4 months was found to alter the electrical activity of the brain and lowered the leukocyte count in the blood as well as producing dystrophic changes in the liver, kidneys, and myocardium.[78] Application of morpholine to the skin damaged the excreting tubules of the kidneys and produced lung damage similar to that encountered with ammonia inhalation.

Morpholine compounds should not be employed in conjunction with nitrites in products where human contact can be expected. When 500 mg/L sodium nitrite and 250 mg/L morpholine was fed to rats in drinking water for 75 weeks, tumors were noted in the animals.[79] Other workers found that diets containing 1000 mg/L sodium nitrite and 1000 mg/L morpholine produced hepatocellular carcinomas in almost 100% of the test animals.[80] Only 3 of 160 rats developed liver carcinoma when the morpholine concentration was reduced to 5 mg/L, even though the sodium nitrite concentration was maintained at 1000 mg/L. The tumors produced were identical to those encountered when N-nitrosomorpholine was fed to the animals. Similar results were noted in studies by other workers.[81,82]

The environmental effects of morpholine compounds are also of considerable interest; however, only one study could be found in the literature pertaining to the biodegradation of these chemicals. One report notes that while morpholine has practically no BOD value, it does not inhibit the organisms commonly involved in sewage stabilization in reasonable concentrations.[83]

A major contribution of this study is that it is the most extensive investigation of this type that has been reported. It was done under uniform conditions, during the same period and in the same laboratory. It provides corrosion engineers and lubrication engineers comparative data that can be used to formulate new products.

The results show that certain morpholine compounds have significant inhibitory properties in metal working fluids. The employment of such compounds in conjunction with surfactants, glycols, and other lubricant ingredients with similar properties could result in the formulation of products that offer rancidity control without the requirement of using preservatives in order to accomplish this objective.

Other potentially profitable investigations for the future may be noted as a result of this investigation. It may be possible to combine morpholine compounds with other antimicrobial agents other than formaledhyde releasers to produce new products that have anticorrosive, surfactant, antimicrobial, and perhaps other beneficial properties in regard to their use in lubricants. In addition, it would be of considerable interest to develop information pertaining to the anticorrosion and antimicrobial properties of mixtures of chemicals such as morpholines and alcoholamines.

There are some areas that require extensive research by lubricant formulators before these compounds may be employed in these products. More information must be developed pertaining to the effects of these chemicals on corrosion inhibition, tool life, cutting temperatures, cooling qualities, lubricity, and others. Such information must be developed by evaluating new products containing these components under laboratory and industrial conditions.

Acknowledgments

The following companies are acknowledged for their cooperation and financial support which has made this report possible: Abbott Laboratories, North Chicago, Illinois; Applied Chemicals, Pty. Ltd., Melbourne, Australia; Anderson Oil & Chemical Company, Inc., Portland, Connecticut; Bellucco & Co., S.A.S., Torino, Italy; Blaser + Co AG, Switzerland; D. A. Stuart Oil Company, Chicago, Illinois; De Mille Chemical Corporation, Jersey City, New Jersey; Dow Chemical Company, Midland, Michigan; G. W. Smith & Sons, Inc., Dayton, Ohio; Glyco Chemicals, Inc., Williamsport, Pennsylvania; Henry E. Sanson & Sons, Inc., Britol, Pennsylvania; International Chemical Company, Philadelphia, Pennsylvania; International Minerals & Chemicals Corporation, Hillside, Illinois; International Refining & Manufacturing Co., Evanston, Illinois; Lehn & Fink Products Co., Montvale, New Jersey; Monroe Chemical Co., Inc., Hilton, New York; Norton Co., Worcester, Massachusetts; Olin Corp., New Haven, Connecticut; Onyx Chemical Co., Jersey City, New Jersey; Polar Chip, Inc., Irvine, California; Quaker Chemical Corp., Conshohocken, Pennsylvania; Reynolds Metals Co., Richmond, Virginia; Rohm and Haas Co., Philadelphia, Pennsylvania; S. H. Mack and Company, Inc., Aurora, Illinois; Shell Oil Co., Houston, Texas; Sun Oil Co., Marcus Hook, Pennsylvania; Tapmatic Corp., Irvine, California; The Rams-Head Co., Des Plaines, Illinois, The Sherwin-Williams Co., Chicago, Illinois; Tower Oil & Technology Co., Chicago, Illinois; Union Carbide Corp., Tarrytown, New York; Union Oil Co. of California, Brea, California; and Van Straaten Chemical Co., Chicago, Illinois.

References

1. Lijinsky, W., Epstein, S. S. Nature, Vol. 225, p. 21-23 (1970).
2. Roberts, E. N., Fields, E. K. U.S. Patent 2,703,785, March, 1955.
3. Lundgren, D. G., Krikszens, A. Appl. Microbiol., Vol. 7, p. 292 (1959).
4. Sluhan, C. A. U.S. Patent 2,999,064, September, 1961.
5. Jones, L. W. U.S. Patent No. 2,272,170, 1968.
6. Iverson, W. P. Corros. Prev. Contr., Vol. 16, p. 15-19 (1969).
7. Schuster, D. Ger. Patent 1,620,447, August 14, 1975.
8. Bennett, E. O. Lub. Eng., in press (1978).
9. Bennett, E. O. Prog. Indust. Microbiol., Vol. 16, p. 119-147 (1974).
10. Henaff, P. L. Bull. Soc. Chim. France, Vol. 1965, p. 3113-3130 (1965).
11. Fernandez, J. E., Butler, G. B. J. Org. Chem., Vol. 28, p. 3258-3259 (1963).
12. Nash, T., Hirch, A. J. Appl. Chem., Vol. 4, p. 458-463 (1954).
13. Rosenfeld, I. L., Persiantsiva, J. P., Terentev, P. B. Zhur. Priklad. Krima, Vol. 34, p. 2049-2057 (1961).
14. Farbwerke Hoechst, A. G. French Patent No. 1,540,150, 1968.

15. Subramanyan, N. Ramakrishnaish, K. Indian J. Technol., Vol. 8, p. 369-372 (1970).
16. Watanabe, T., Tomura, S. Japan Patent No. 70,00,974, 1970.
17. Bouillot, M. J., Baumgartner, P. French Patent No. 2,126,564, 1973.
18. Humphrey, E. L., Morse, W. B. U.S. Patent No. 3,785,975, 1974.
19. Patel, N. K., Patel, K. C. J. Electrochem. Soc., India, Vol. 24, p. 187-188 (1975).
20. Kuhn, K. Ger. (East) Patent No. 77,236, 1970.
21. Magne, F. C., Mod, R. R., Sumrell, G., Parker, W. E. U.S. Patent No. 3,746-644, July 17, 1973.
22. Shea, T. J. Ind. Hyg. Toxicol., Vol. 21, p. 236-245 (1939).
23. Coleman, L. E. Ger. Offen. 1,963,761, 1970.
24. Banko, S., Ban, M., Bruckner, E. Hung. Teljes 2977, Nov. 22, 1971.
25. Ueda, A., Tanaka, K., Yamane, Y. Japan Patent No. 74 34,167, 1974.
26. Humbert, J. R. U.S. Patent No. 3,195,332, 1965.
27. Dickey, J. R., Pattison, D. A., Pawson, B. A. U.S. Patent No. 3,268,447.
28. Berger, J. E. U.S. Patent No. 3,444,080, May 13, 1969.
29. Hunter, D. P., Wells, J. E., Rendall, H. L. Brit. Pat. No. 1,354,480, 1974.
30. Meyers, R. K., Pier, S. M., U.S. Patent No. 3,089,854, 1963.
31. Brunel, H. U.S. Patent 3,320,164, May, 1967.
32. Moll, L., Moll, C. British Patent 1,272,100, April, 1969.
33. Marx, J. Ger. Offen, 2,043,855, Sept., 1962.
34. Dell, G.W., Goodspeed, E. W. U.S. Patent No. 3,380,859, 1968.
35. Siebenmann, C., Schnitzer, R. J. Can. Pub. Health J., Vol. 32, p. 74-75 (1941).
36. VanAssche, C. J., Vanachter, A. Parasitica, Vol. 26, p. 117-126 (1970).
37. Khorshavin, A. N. Uch. Zap., Perm. Gos. Univ., No. 277, p. 123-127 (1971).
38. Khoroshavin, A. N. Kataeva, I. V., Zaitseva, T. A. Nauch. Tr., Perm. Nauch. Issled, Ugol. Inst. No. 13, p. 255-261 (1971).
39. Strehlke, P. Ger. Offen. No. 2,131,888, Dec. 28, 1972.
40. Jaeger, G., Buechel, K. H., Grewe, F., Frohberger, P. Ger. Offen. No. 2,128,700, Jan. 4, 1973.
41. Baker, J. W., Schumacher, I. U.S. Patent No. 3,719,675, 1970.
42. Novak, A. F., Fisher, M. J., Fore, S. P., Dupuy, H. P. J. Amer. Oil Chem. Soc., Vol. 41, p. 503-506 (1964).
43. Mod, R. R., Skau, E. L., Fore, S. P., Magne, F. C., Novak, A. F., Dupuy, H. P., Ortego, J. R., Fisher, M. J. U.S. Patent No. 3,285,812, Nov., 1966.
44. Novak, A. F., Solar, J. M., Mod, R. R., Magne, F. C., Skau, E. L. Appl. Microbiol., Vol. 18, p. 1050-1056 (1969).
45. Hueck, H. J., Adema, D. M., Wiegmann, J. R. Appl. Microbiol., Vol. 14, p. 308-319 (1966).
46. Kradel, J., Pommer, E. H. Meded. Fac. Landbouwwetensch., Rijksuniv. Gent., Vol. 36, p. 343-347 (1971).
47. Marchand. D. Def. Veg., Vol. 27, p. 144-152 (1973).
48. Mod, R. R., Harris, J. A., Arthur, J. C., Mange, F. C., Sumrell, M. G. J. Oil Chem., Vol. 53, p. 641-643 (1976).
49. Shema, B. F. Brink, R. H., Swered, P. U.S. Patent No. 3,881,008, 1975.
50. Shema, B. F., Brink, R. H., Swered, P. U.S. Patent No. 3,917,834, 1975.
51. Shema, B. F., Brink, R. H. U.S. Patent No. 3,929,563, 1975.
52. Shema, B. F., Brink, R. H. U.S. Patent No. 3,862,034, 1975.
53. Shema, B. F., Brink, R. H., Swered, P. U.S. Patent No. 3,860,713, 1975.
54. Shema, B. F., Brink, R. H., Sewere, P. U.S. Patent No. 3,860,516, 1975.
55. Shema, B. F., Brink, R. H. U.S. Patent No. 3,861,422, 1975.
56. Shema, B. F., Brink, R. H., Swered, P. U.S. Patent No. 3,897,554, 1975.
57. Konagai, Y., Konya, K., Muto, M., Takita, K. Japan Patent No. 75,117,933, 1975.
58. Swered, P., Girard, M. A. U.S. Patent No. 3,903,279, 1975.
59. Shema, B. F., Brink, R. H. U.S. Patent No. 3,664,821, 1972.
60. Shema, B. F., Brink, R. H., Swered, P. U.S. Patent No. 3,647,703, March 7, 1972.
61. Hodge, E. B. U.S. Patent No. 3,183,189, May, 1965.
62. Harrison, H.R. U.S. Patent No. 3,759,828, Sept., 1973.
63. Kamlet, J. U.S. Patent No. 2,475,186, 1949.
64. Weichert, K., Hegemann, H. Ger. Offen, 1,201,121, 1965.
65. Nankee, R. J., Carswell, R. U.S. Patent No. 3,753,912, 1973.
66. Shell International Research. Neth. Appl. 6,607,803, 1965.
67. Schwoegler, E. J., Berman, L. U. Lub. Eng., Vol. 11, p. 381-388 (1955).
68. Wachter, A., Stillman, N. U.S. Patent No. 2,432,840, 1947.
69. Sawyer, A. W., Csejka, D. A. Brit. Patent No. 1,214,171, 1970.
70. Schwoegler, E. J., Berman, L. U. Corrosion, Vol. 15, p. 128t-130t (1959).
71. Carpenter, C. P., Weil, C. S., Pozzani, U. C. Arch. Ind. Hyg. Occupat. Med., Vol. 10, p. 61-68 (1954).
72. Smyth, H. F., Carpenter, C. P., Weil, C. S., Pozzani, U. C. Arch. Ind. Hyg. Occupat. Med., Vol. 10, p. 61-68 (1954).
73. Smyth, H. F., Carpenter, C. P., Weil, C. S. Arch. Indust. Hyg. Occupat. Med., Vol. 4, p. 119-122 (1951).
74. Smyth, H. F., Carpenter, C. P., Weil, C. S. J. Ind. Hyg. Toxicol., Vol. 31, p. 60-62 (1949).
75. Belova, G. B., Shurupova, E. A., Stankevich, U. V. Sin. Issled. Eff. Khim. Polim. Mater., Vol. 1969, p. 466-475 (1970).
76. Krieger, H. Arzneimittel Forsch., Vol. 11, p. 798-801 (1961).
77. Migukima, N. V. Toksikol. Gig. Prod. Neftekhim. Neftekhim. Proizvod., Vol. 1968, p. 53-55 (1969).
78. Migukina, N. V. Toksikol. Nov. Prom. Khim. Veshchestv., Vol. 1973, p. 92-1000 (1973).
79. Garcia, J., LiJinsky, W. Z. Krebsforsch. Klin. Onkol., Vol. 79, p. 141-144 (1973).
80. Newberne, P. M., Shank, R. C. Food Cosmet. Toxicol., Vol. 11, p. 819-825 (1973).
81. Sander, J. Arzneim-Forsch., Vol. 21, p. 1572-1580 (1971).
82. Sander, J., Buerkle, G., Schweinsberg, F. Top. Chem. Carcinog., Proc. Int. Symp., Vol. 1972, p. 297-312 (1972).
83. Swope, H. G., Kema, M. Sew. Indust. Waste Eng., Vol. 21, p. 467-468 (1950).

Reprinted from Cutting Tool Engineering's Diamond/Superabrasive Directory 1981-82

Diamond Honing Automotive Engine Cylinder Bores with Water-based Cutting Fluids

A DIAMOND ABRASIVE honing process which uses water-based cutting fluids has been developed by Barnes Drill Company in Rockford, Illinois. The process was developed as a method to use water based coolants, but may also be able to replace certain boring operations in the production of cylinder bores in automotive engine blocks, which is clearly a high production operation. Much of the data presented here should be considered "preliminary" to the adoption of the process in production, but most of the problems have been successfully solved, and only a few details need proving out before the process can be considered "on-stream."

Some of the benefits which accrue from the use of honing and water-based cutting fluids are obvious. First, mineral oils and kerosenes are rising in price continually. These cost additions to the production process are becoming more significant, daily. Water, while not free, is certainly much cheaper than mineral oils and kerosene. And, you can *get* water, while the mineral oil and kerosene products are often on allocation from major suppliers. (Oil as a fuel for diesel engines, or kerosene for jet fuel, apparently is a more profitable proposition than cutting fluids.) Water is also not subject to supplies from foreign nations and the price and availability fluctuations suffered from those sources of supply.

Second, water is easier to deal with from an environmental standpoint, whether the environment be the in-plant industrial environment . . . or the environment outside the plant where the waste fluids must eventually be dealt with.

Without going into the complex chemistry of petroleum and water-based fluids, we can say that water-based formulas can, if properly devised, be far safer as products in the industrial environment. Water-based coolants and lubricants don't flash or ignite and burn. They can be safer on contact or in breathing. They are easier to maintain effective at the machine . . . easier to filter and treat to keep them useful . . . than mineral oils or kerosene. Depending on the production processes, clean-up of the piece parts can be a lot easier, cheaper, etc. And water soluble coolant/lubricants are much easier to dispose of when spent. Sometimes only dilution is required for safe dumping. Other times, you need further treatment . . . but at least you *can* get rid of it fairly easily. Clearly not the case with alternative fluids.

Honing has always been a part of the cylinder conditioning process in the automotive engine business, since honing provides the fine finish required for minimum wear on the piston skirt and rings while still providing a suitable surface for rapid (in fact, almost immediate) "seating in" of the sealing and rubbing surfaces. Further, the cross-hatch pattern of the honing process is critical for the retention of lubrication on the cylinder wall.

This honing has always been a final finishing operation, following boring operations. Traditional honing is a "clean-up" using vitrified abrasives with petroleum-type cutting fluids. While the hone was expected to eliminate *some* of the out-of-round, or taper, or other undesirable results from previous machining processes, the basic configuration was determined by a rigid, precision boring operation, done inside of a cylinder shape rigidly fixtured by the block, itself, as well as the fixtures retaining the block to the machine. And here, we can assume, we find part of the rea-

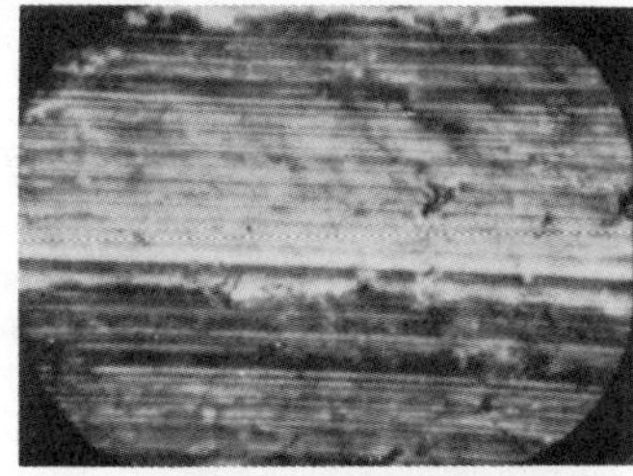

100X Magnification

Figure 1. Photomicrographs of cast iron 4-cylinder block after it was bored, but prior to honing.

50X Magnification

Figure 2. Different results achieved with diamond honing (bores 1, 2 and 4) and abrasive honing (bore 3) show the capabilities of diamond abrasive. Bore number 4 is close to target set for surface finish and cross-hatch angle.

BORE #1 honed with D11-D1 diamonds. Surface finish 46 A.A. Cross hatch angle 8°. 100X magnification.

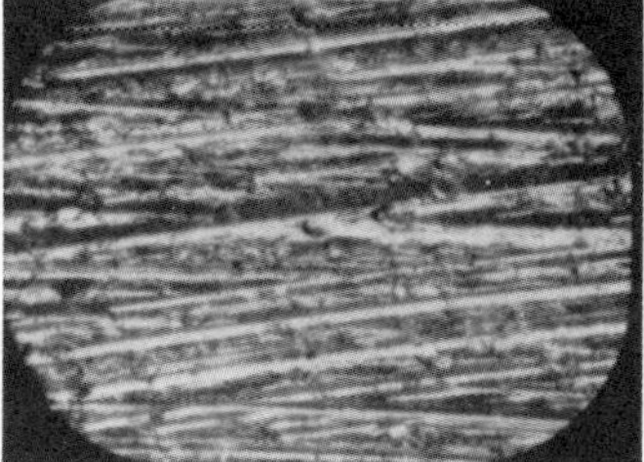

BORE #2 honed with D11-D1 diamonds. Surface finish 54 A.A. Cross hatch angle 11°. 100X magnification.

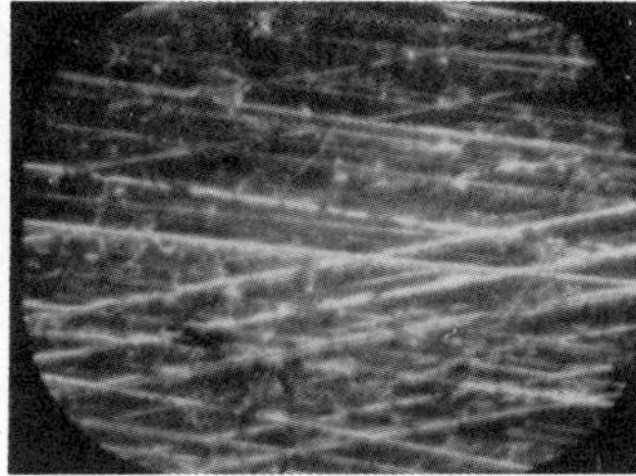

BORE #3 honed with 220 grit abrasive. Surface finish 7 A.A. Cross hatch angle 16°. 100X magnification.

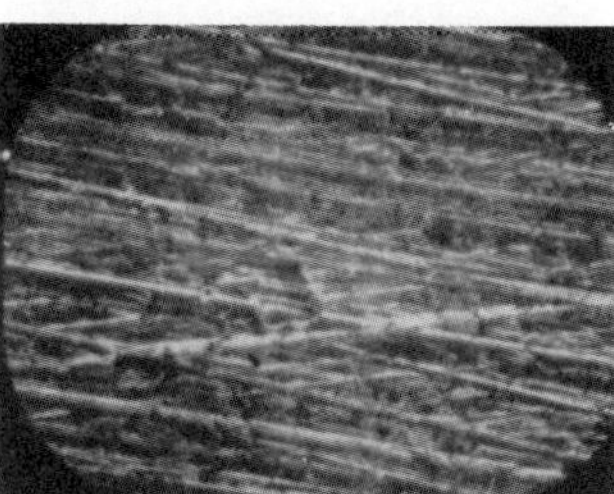

BORE #4 honed with D14-D2 diamonds. Surface finish 16 A.A. Cross hatch angle 16°. 100X magnification.

son for the very real potential of diamond honing as a high production metal removal process.

In years past, the automotive engine block was a rather "coarse" casting. To accommodate potential core shifting, extra material was allowed in the cylinder wall area. To allow for future increases in engine sizes (long before high-priced gasoline, you'll remember) even more material was allowed in the cylinder wall. Naturally, to maintain an even wall thickness for casting purposes, other walls had to "come up to" this thickness. Also, to provide additional strength for high horsepower levels, it was common to "beef up" the block with even more thickness and additional webbing. In fact, the

Figure 5. Laboratory trace of the 40 micro-inch finish in the cylinder, described in Figure 4, without the finishing pass. This is, in many cases, a wholly acceptable automotive gasoline powerplant cylinder finish.

Figure 4. Microphotograph (100X) taken from the cylinder bore of a 6-cylinder engine. The honing process, as outlined here, is actually a three-step process, with two roughing passes and a finishing pass. This 40 A.A. surface finish, produced on a roughing pass, is actually acceptable by most engine designers. In fact, low A.A. values are often not wanted.

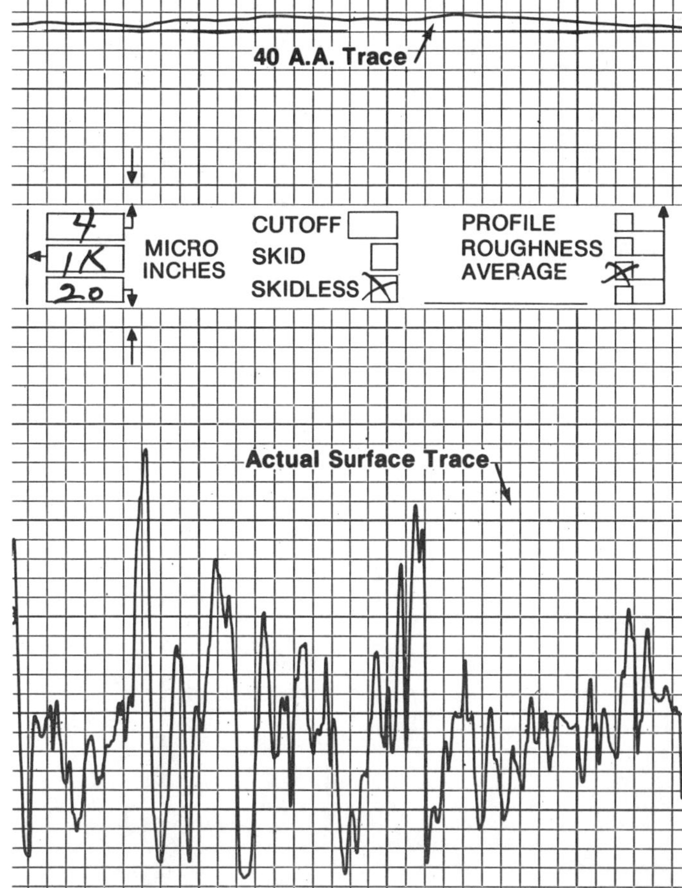

Figure 3. Microphotographs were taken from this sectioned four-cylinder block. On this block, material has been removed from the cylinder wall during the honing process, which means this is not the full wall thickness as received by the honing machine. Nevertheless, the extreme thinness of the cylinder wall (not to mention the bearing webs and other sections) is apparent. A truly artful casting job, and a difficult machining job, too.

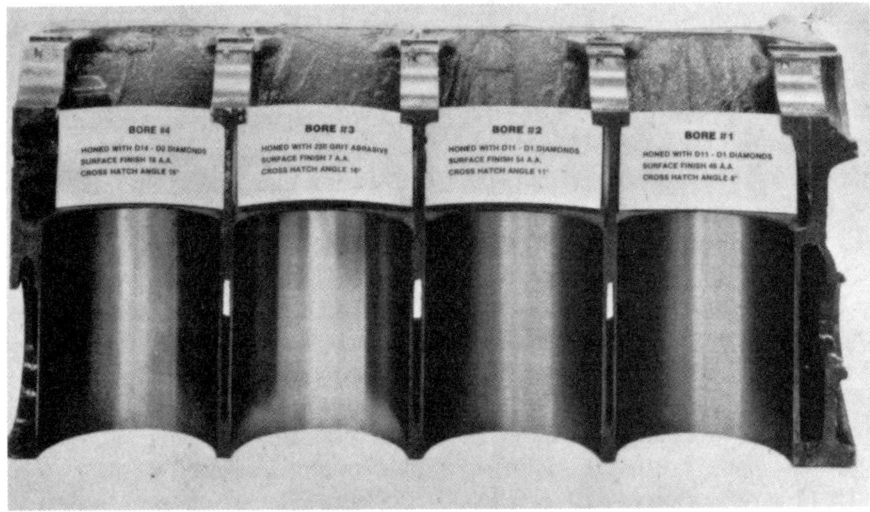

larger engines out of Detroit in the 60's and early 70's weighed upwards of 700 pounds without exhaust or intake, "plumbing" or engine-driven accessories. This only represented 12% of the weight of a 6,000 pound car, but would represent 25% of the weight of a modern, down-sized car. Little wonder that Detroit has phased out V-8 production, and big engines in general. All future powerplants will be (they say) in-line 6s, V-6s and possibly V-4's, in-line 4s and a scheduled 3-cylinder engine from GM.

And, while aluminum and other "exotic" alloys are gaining attention, thin-wall cast iron is still the favorite basic structural material, and the complex, strong, lightweight and inexpensive engine and head castings of today are a tribute to the advanced art of the modern casting plant. However, the thin sections which mean so much to the low weight of the modern powerplant have removed some of the strength of the cylinder walls. This thin section "problem" is *not* talked about much, but is easily seen in some reconditioning operations where older blocks, but thin-wall design, can cause problems in overboring for cylinder reconditioning. The top deck and bottom end of the cylinder jacket is rigidly fixtured, but the middle portion is supported only by itself. Thus, depending upon the depth of cut, rate of stock removal, etc., the finished bore can be beyond the finishing capability of the average honing operation. The problem is "taper," or rather, "barreling," with the top and bottom of the bore being well within tolerance, but the middle section (which is moving away from the boring tool,) of varying dimensions, depending upon varying strengths of differing wall thicknes-

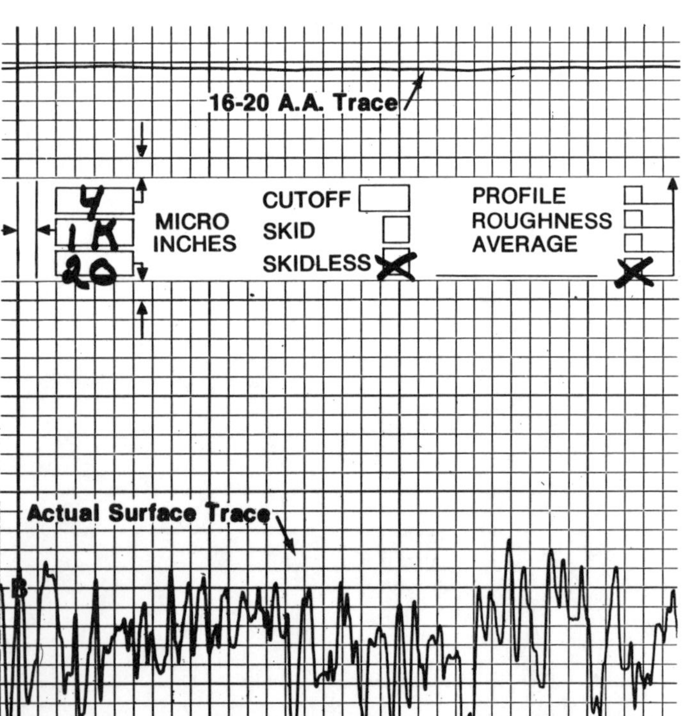

Figure 6. Laboratory trace of the 19 A.A. cylinder as described in Figure 6.

Figure 7. Very fine surface finishes are possible with diamond stones . . . at least fine from the automotive engine makers' point of view. This is a microphotograph (100X) of a bore of the same 6-cylinder block as in Figure 4, but after a change to the finish honing step. A.A. is 19.

ses. (This problem can be alleviated by reducing the cutting rate and other techniques, but the rate of production is slowed.)

According to Mark Estabrook, Chief Engineer, Barnes Drill Co. began its investigation into water soluble honing using vitrified stones. Use of water soluble coolants with common honing pressures can result in rapid glazing of the stones. (The use of kerosene as a cutting fluid prevented glazing . . . but did not meet the investigation criteria, of course.) Honing pressures were increased to prevent glazing, which was, indeed, accomplished even with the water soluble fluids. However, stone life was rapidly diminished, and the geometric tolerances were not controllable at the increased pressures, either. With vitrified stones proving unworkable with water soluble fluids, work proceeded with diamond abrasives.

A one-for-one replacement of diamond for vitrified without other changes was not possible. In fact, a look at the variables in cutting proves the degree of difficulty. The following were tested:

(1) speed (RPM)
(2) reciprocation rate (CPM)
(3) expansion pressure (PSI)
(4) coolant types and brands
(5) coolant concentration
(6) diamond stone configuration
(7) stone size.

In fact, each of these variables had a significant effect on the finished product, and indeed, experiments continue to find production optimums for some of them.

The overall problem, in machining terms, is to take a "rough" bored cylinder block and provide size, taper and out-of-round to less than .001" in any case, which amounts to an effective ±.0005" . . . especially critical on the bore size and out-of-round since we are dealing with a "class of fit," piston-to-bore. Also, the finished surface needs to be 20-50 AA.

The problem of "taper," or variations in size along the axis of the cylinder, is obviated by the nature of the honing process which spreads the load over a wide base . . . provided that the stroke of the honing machine is properly set.

The most difficult problem was out-of-round. Actual stock removal requirements are for .005" per minute minimum, and at this rate it was initially not possible to meet the out-of-round tolerance. Conversely, by lowering the pressures to meet the out-of-round tolerance (which *is* possible at the lowered pressures) the stock removal target could not be met.

Computer modeling and graphic techniques provided some clues to solving the stock removal rate/distortion problem. In general, it was found that greatly increasing the RPM of the machine was necessary. This increased speed, with greatly lowered pressures, helps to keep stock removal rates up while minimizimg bore distortion. This alone, however, is not enough to accomplish the total task.

Original assumptions on the proper R/C value (speed/stroke) were proven incorrect when checked by graph analysis of tests run. An analysis of the relationship of R/C (RPM/CPM) confirms earlier published data that R/C — N.417, and it is possible to construct a table of cross-hatch angles to conform to an R/C ratio of N.417 when the part dimensions are known, including of course the stroke length required for the hone. Working with this ratio, it was possible to improve the tolerances greatly, and this should hold true for all honing, not just diamond abrasive honing.

It was found that the cutting fluid used for a particular application is critical to performance. Initial tests, as said, showed that either stock removal rate *or* geometry would have to suffer. Revising procedures as outlined above did bring tolerance and stock removal requirements into acceptable areas. And, with the proper fluid, it was discovered that increased stock removal rates of are possible on a high production basis. In fact, the actual rate of removal was limited by the horsepower of the honing machine used to perform the tests. Not all coolants work equally well. Test programs for each application will probably be required.

There are two limiting factors yet to be addressed. First, the high speeds required for diamond honing tend to limit the cross-hatch angle . . . to somewhere around 10° or less. If a higher angle is desired by the customer, motor speed controls will be needed to lower the speed during the re-expansion cycle. Also, at this point in testing, diamond stones are not capable of achieving an ultimate surface finish below 20 AA, which means the need to use a vitrified stone to accomplish this final clean-up. The life of a vitrified stone in this point of the process would normally be high, anyway, which makes it economically acceptable within the framework of the process. Barnes Drill Co. is, however, looking to achieve 20-40 AA with diamond in the near future. (Incidentally, water is sufficient lubricant for this touch with the vitrified stone. A change to oil is not needed.)

European applications of this type have been successful, primarily due to their precision boring methods which result in far less stock removal requirements for the diamond. Thus, this application of diamond abrasive honing is a relatively significant achievement. It remains to be seen, of course, if the above application will actually see the production line, and whether or not the above conclusions will work equally well on other jobs, with different dimensions and cast iron compositions. However, if the higher speeds work with other metals (such as steels) and with other superabrasives (CBN) there is reason to believe that diamond abrasive honing has immediate benefits for the cylinder *liner* side of the automotive and truck engine business, and other areas as well, at a level of economic savings and production predictability that is not possible with the vitrified honing process.

• • •

The above article was prepared with the assistance of Mark Estabrook, Chief Engineer for Barnes Drill Co., Rockford, Illinois.

Reprinted from Cutting Tool Engineering, May/June 1981

A New Look at Water Soluble Cutting Fluids for Honing Operations

GENERALLY SPEAKING, for quite a period of time, the metal working industry has surely been in need of a good water honing and/or high quality micro finish grinding concentrate. Many companies have marketed some sort of a product along this line with nothing better than "reasonably acceptable" performance. This accounts for the continued use of kerosene in most all high demand honing/grinding operations. But kerosene is just not acceptable any more. OSHA standards, EPA standards, and normal health and safety standards just won't tolerate kerosene as honing/grinding fluid — not to mention availability and price.

Concepts and Requirements

For the purpose of this article, we will not take the workpiece material into consideration. Rather we will limit ourselves to the operation itself, and its relation to conversion from oil/solvent fluids to water/chemical fluids.

The oil/solvent-kerosene/fluids all offer a slight cushion between the cutting surface and the workpiece. It must be clearly understood that cushions will be considerably less to nothing when and depending on what kind of water fluid is in use. Emulsion type products can offer *some* cushion between the cutting surface and the workpiece while solution type products offer less to none.

This does not say there will not be a film between the interface. It does say, however, depending on the type of water fluid being used, that it will be less, and that the truer-to-true face of the cutting media *will* be traced into the workpiece.

Micro finishes are largely affected by this cushion, hence a change in cutting surface (grit size) may be required to achieve the desired micro finish. Cutting media bonding material can also become important, sometimes adversely affected by chemical components of water honing fluids. This, of course, can also cause changes in surface finish as well as show marked changes in cutting media service life. So to maintain a desired micro finish, it may be necessary to change cutting media grit size when changing to water fluids. It may also be necessary to select the wheel or stone bonding material based on the chemical components of the water honing material. Softened bonds mean premature wheel or stone breakdown which usually produce rougher micro finishes.

As we begin to review some of these operational concepts and requirements, it will become clear there have been as many, if not more, "people failures," as product failures in honing with water as the cutting fluid. As a good example illustrating this statement, consider Electromotive Force (EMF) and its ability to cause problems in water honing operations.

Corrosion is a problem normally associated with water honing/grinding fluids. It is not uncommon to find galvanic activity, meaning that EMF is causing the problem. This same force can and often does cause monumental problems in honing operations. Even though the honing stone cutting surfaces may stay clean and free cutting, the hone heads may not be operating properly, necessitating frequent cleaning of the heads.

If you are unaware of this factor, or choose to ignore it, all sorts of problems can be the result, including equipment failure due to stones not sliding in and out properly during operation cycles. It is very important that hone head metal surfaces, especially the slide areas, stay clean so the stone holders can expand and contract with use as they are engineered to do. But, EMF can cause accelerated attachment of metallic fines in this highly important area, thus restricting the required movement of the stone holders. All metals possess a measurable electric potential which varies with size and material. If the water honing fluid is a reasonably good electrolyte, there will be an electrical current flow and attraction between metal or metallic particles.

The rule that would apply is, flow is from least noble and anodic (this would be the scrap product of honing, metallic fines) to more noble-cathodic (hone equipment holders, heads, etc.), hence rapid fine build-up and attachment in the first area of opportunity . . . the slides and holders . . . thus restricting normal movement.

Oils/solvents being mostly *non*-conductors prevent both the flow of current and the attachment of fines to equipment surfaces in the immediate work area. Therefore, this problem does not exist with oils and solvent type materials. However, this is a major cause of product failure in the past use of water honing fluids, and is little discussed or appreciated as a major problem, varying largely with the type of operation even though it is present to some degree with all water honing fluids.

For continued good performance and long service life, bactericides, germicides and fungicides must also be incorporated as formula components to prevent product deterioration before planned.

Potential Solutions

Experience has taught that this EMF problem, when treated chemically, is best handled by an additive to an operating system on a daily or bi-daily basis. The reasoning behind this recommended frequency is quite simple: to prevent attachment of metallic fines in critical areas, such as hone heads and stone holders, a chemical coating or "plating action" is required. This coating on metal surfaces, or "plating action," is used up on a regular basis. Every *part* that is processed also is plated or coated by this material, hence, the plating material is used at a greater ratio than other formula components. To keep the supply of this coating component at the proper level, daily/bi-daily additions are required to maintain system balance.

Pillsbury Chemical & Oil, Inc. has developed a non-flammable water soluble solution for honing and for super-finishing which handles the EMF problem as above and provides fluid properties needed for cooling and lubricating many production applications. It is recommended for use at 10% by volume with water, however, the concentration can vary either way depending upon the pressure, type of stone and the stock that is honed or super-finished. The product is named Hone Rite. It offers the following features:

1. It is water soluble and can be used at varied concentrations.
2. It is non-flammable.
3. You can take deeper cuts, faster, and still obtain the desired micro.
4. It is a cleaner operation than when using the conventional mineral based products for honing or super-finishing.
5. It will repel all oils and greases and float them to the top. No smoking occurs during operation.
6. It contains a fluorescent dye to show honing and super-finishing pattern under the black light.
7. It will not foam.
8. The flushing action of the product keeps the stones clean and free from glazing.
9. It forms a clear green solution in water and, thus, is identifiable.
10. The product contains germicides to prevent rancidity and for the control of bacteria in self-contained units.
11. It holds the magnetic fines in suspension, giving greater efficiency for the removal of metallic particles with the magnetic separators.
12. It contains no phenol, no sulfur, no chlorine.
13. Contact should be made with local authorities on pollution regulations, but easy and economical disposal is common.

It is possible that when going to the water soluble Hone Rite from the basic mineral type solutions, that a stone change may become necessary. An innovation of the fluid is that on subsequent cleaning operations you will not contaminate your washers (such as you will with mineral base fluids), therefore allowing you to run washing equipment at lower temperatures, affording additional economies.

• • •

This article was edited from material supplied by H. R. Vahle of Pillsbury Chemical & Oil, Inc. For more investigation into the future of water soluble fluids for honing, read the article on diamond honing of automotive cylinder bores in the Diamond Directory Section of this issue of CTE.

Evaluation of cutting fluids for use with AMBER* BORON NITRIDE abrasives

BY **P D OATES** AND **H J BEZER** CASTROL RESEARCH & DEVELOPMENT LABORATORIES BURMAH-CASTROL HYDE CHESHIRE AND **A M BALFOUR** DE BEERS INDUSTRIAL DIAMOND DIVISION TECHNICAL SERVICE CENTRE CHARTERS ASCOT UK

Reprinted from Industrial Diamond Review, September 1977

This article presents the results of investigations into the effect on wheel performance of changes in wheel speed, coolant dilution, metal removal rate and the degree of coolant filtration, when grinding hardened T15, M2 and EN 31 steels with resin bond wheels containing AMBER BORON NITRIDE abrasive, using grinding fluids of different types. From the results of these studies, it is now possible to give clear recommendations on the best coolant formulae to use with these three representative workpiece materials, in order to obtain maximum efficiency in wet grinding with ABN wheels

Fig 1 *Jones & Shipman 540 surface grinder on which the tests were carried out*

Fig 2 *Standard Philips Philtrator unit*

Introduction

As part of De Beers continuing programme of applications development, an investigation into the effects of grinding with different cutting fluids was initiated as a joint venture with the Burmah-Castrol Company.

The purpose of this investigation was to determine the best water-based cutting fluid for use with grinding wheels containing AMBER* BORON NITRIDE abrasive. It is well known that neat cutting oils are the most effective coolants to use with ABN grinding wheels but because of the possible fire hazard and the environmental problems which neat oils create, their use is reducing and may well be prevented by legislation in both Europe and the USA. Thus, the investigation into means of obtaining a comparable performance with socially acceptable cutting fluids was important.

An evaluation was therefore made of the performance of a range of different fluids in a standard surface grinding test. The effects of wheel speed, coolant dilution, metal removal rate and the degree of filtration were studied using coolants of four basic types:

a) Neat oils.
b) Conventional soluble oils.
c) Semi-synthetic soluble oils.
d) Synthetic cutting fluids.

Test procedure

Tests were carried out using 1A1 resin bond wheels 152 mm diameter by 6·4 mm wide (6 in. $\times \frac{1}{4}$ in.) containing 80/100 US mesh ABN* 360. The machine was a Jones & Shipman 540 surface grinder and the coolant was filtered by a Philips Philtrator 7603/03 using a 25 micron nominal retention filter. The workpiece materials used were high speed steels (T15 and M2) and hardened ball bearing steel (EN 31).

*Registered Trade Mark

The T15 high speed steel was chosen as it is one of the most difficult steels to grind with Al_2O_3 abrasive, and from the coolant point of view it is interesting as it is particularly unreactive to the normal high pressure additive systems. The M2 was chosen as it is the most common type of high speed steel used in tool production, typically drills and milling cutters. Ball bearing steel (EN 31) was chosen as over 70% of production grinding is carried out on steel of this approximate composition, either in its through hardened form or as the case of case hardened parts which is of approximately the same composition. Typical components are roller bearing elements, car transmission parts and hydraulic pump parts.

The ideas that have been accepted up to now were that it is important to include a high pressure additive in the coolant when grinding with cubic boron nitride[1,2], and that due to the reactivity of cubic boron nitride with steam, it was important to have as large a percentage of oil as possible in the soluble coolant.

The investigations reported here were designed to discover how far these ideas could be exploited within the limit of using coolant formulations that were already available and already proven to have no harmful side effects, so that any results could quickly be applied. The results below show that the theories initially held have not proven correct, but that the initial objective of being able to recommend water based coolants and conditions of use that are advantageous, has been realised.

The results have been arranged to show the effect of a change in coolant specification on each parameter. As the absolute values of power and of grinding ratio particularly vary very widely with grinding conditions and steel specification, it is felt that the percentage format is the most explicit way of presenting the test results.

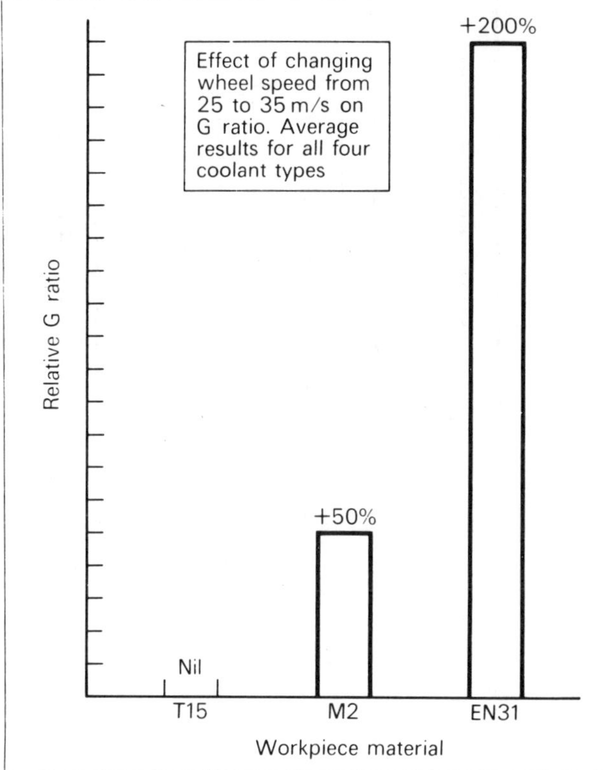

Fig 3 *The effect on G-ratio of changing wheel peripheral speed from 25 to 35 m/s (5000 to 7000 s.f.p.m.), when grinding T15, M2 and EN 31 steels with ABN 360. Results are the average for all four coolant types*

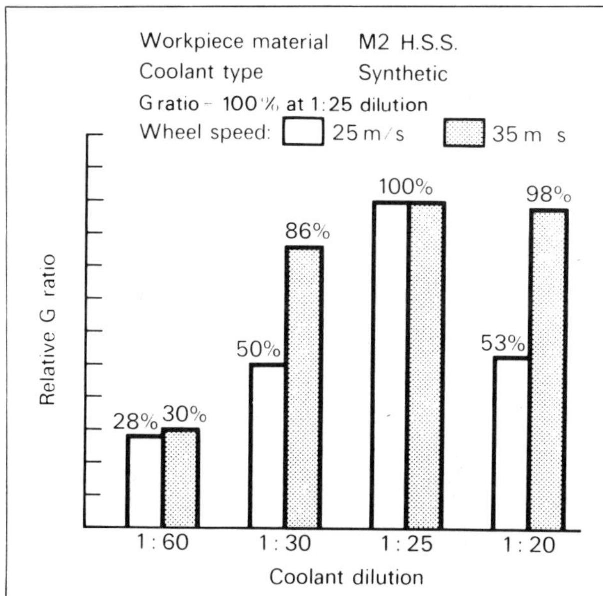

Fig 4 *The effect of coolant dilution (synthetic coolant) on G-ratio when grinding M2 high speed steel with ABN 360 at two different wheel speeds*

The effect of wheel speed

It is well known[1] that the life of a CBN grinding wheel increases with increasing wheel peripheral speed up to at least 80 m/s (16,000 s.f.p.m.). In the present investigation the effect of changing the wheel speed between 25 and 35 m/s (5000–7000 s.f.p.m.) was investigated to determine the effect of speed on performance when using different coolant additive systems.

Fig 3 shows the increase in grinding ratio with this 40% increase in wheel speed. These results are the average for all four coolant types as there was very little difference between them, and the results confirm that wheel speed is a significant variable in determining wheel life when grinding M2 and EN 31 steels but not when grinding T15 steel, although the result on T15 should be viewed with caution as it is possible that experimental error has masked a more limited benefit with higher speed. The other conclusion that can be drawn from these results is that a change in wheel speed will not affect the choice of coolant, as all four coolant types have approximately the same response to the wheel speed change.

The effect of coolant dilution

A detailed examination of the effect of dilution of a synthetic cutting fluid was carried out to examine the effect on power drawn and surface finish as well as grinding ratio.

Fig 4 shows that the effect of dilution of this type of coolant is very significant, and extremely critical if optimum results are to be achieved at 25 m/s (5000 s.f.p.m.) wheel speed. A reduction of sensitivity to dilution occurs at 35 m/s (7000 s.f.p.m.).

A comparison of Fig 4 and Fig 5 shows that as the grinding ratio increases the power requirement falls, which is against the normal trend. The indication therefore is that the lubrication properties of the coolant improve as the dilution is changed from 1 : 60 to 1 : 25.

There is, however, a reversal of this trend in line with the reduction in grinding ratio at 1 : 20 dilution. The relationship for both wheel speeds is substantially the same.

The relationship between coolant dilution and surface finish (Fig 6)—Ra value— is not as clear as for the other two parameters. At 25 m/s (5000 s.f.p.m.) wheel speed the surface finish does not show a minimum value but is lowest at 1 : 20 dilution, while at 35 m/s (7000 s.f.p.m.) a minimum value is shown at 1 : 30 dilution. However, this does not have a great significance as variations of up to 150% were recorded in the individual test results.

Fig 7 shows the dependence of G-ratio on the dilution

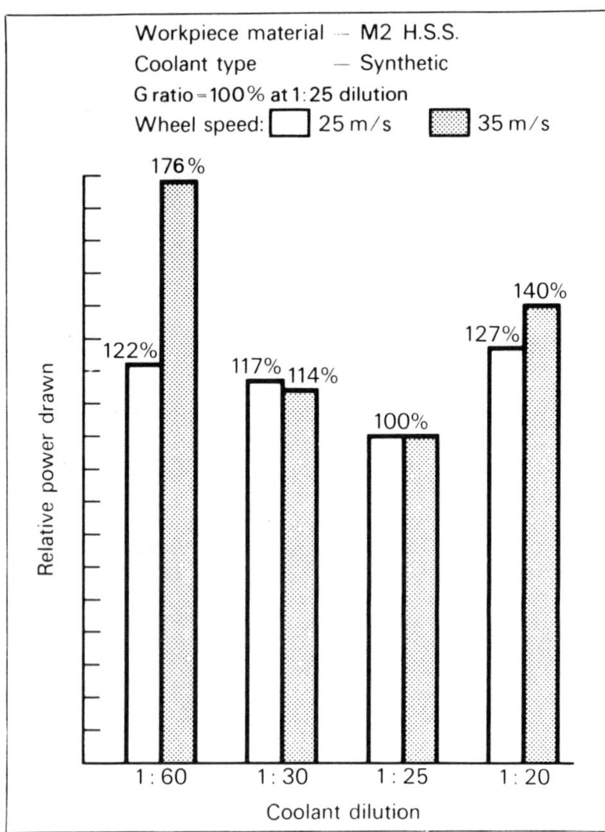

Fig 5 *The effect of coolant dilution (synthetic coolant) on power drawn by ABN 360 wheel when grinding M2 high speed steel at two different wheel speeds*

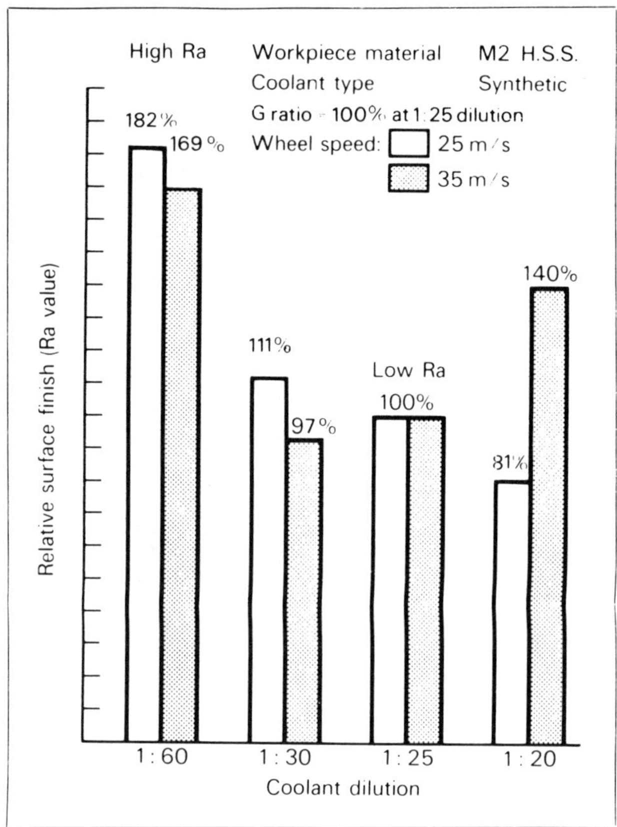

Fig 6 *The effect of coolant dilution (synthetic coolant) on workpiece surface finish when grinding M2 high speed steel with ABN 360 at two different wheel speeds*

of synthetic and semi-synthetic cutting fluids, on the three workpiece materials investigated. As can be seen, the dilution is not important on T15 H.S.S. (in the range 1 :20 to 1 :60) ; it is more significant on M2 steel and an important variable on EN 31 steel.

The other point of note is that the synthetic fluid is much more sensitive to dilution than the semi-synthetic type. This is of note as it has been found that it is also more difficult to control the concentration of the synthetic type, so that quite wide variations in wheel performance could be expected in practice. Tests were also carried out with dilutions of up to 1 :10 but no advantage was found in the use of solutions stronger than 1 : 20. This was contrary to our expectations, but a number of additional tests gave the same result. It was beyond the scope of this work to find out why this is so, but it is a point worth further investigation.

The effect of metal removal rate on 'G' ratio

Fig 8 shows the relative performance achieved at low (0·08 in.^{3}/min/per inch of wheel width) and high (0·4 in.^{3}/min/inch of wheel width) metal removal rates on M2 high speed steel and shows that at the low removal rate, the best soluble oil achieves only half the performance of the best neat cutting oil but that this difference disappears at the high removal rate.

Other tests showed that this effect is less on T15 and about the same on EN 31. The other feature of these results is that the cutting fluids giving a low performance at the low

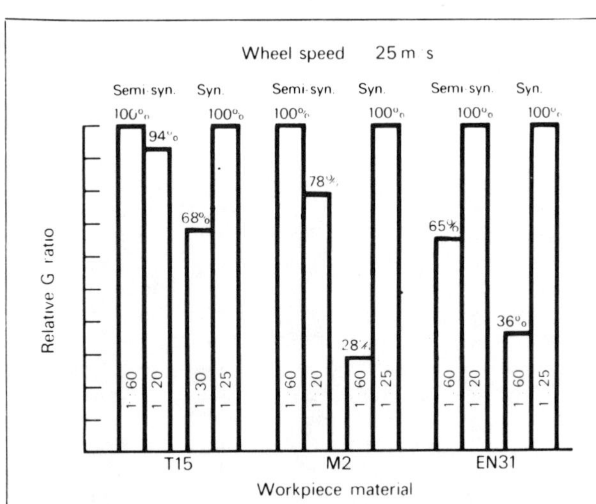

Fig 7 *The effect of coolant dilution (semi-synthetic and synthetic coolants) on G-ratio when grinding T15, M2 and EN 31 steels with ABN 360 at 25 m/s (5000 s.f.p.m.) wheel speed*

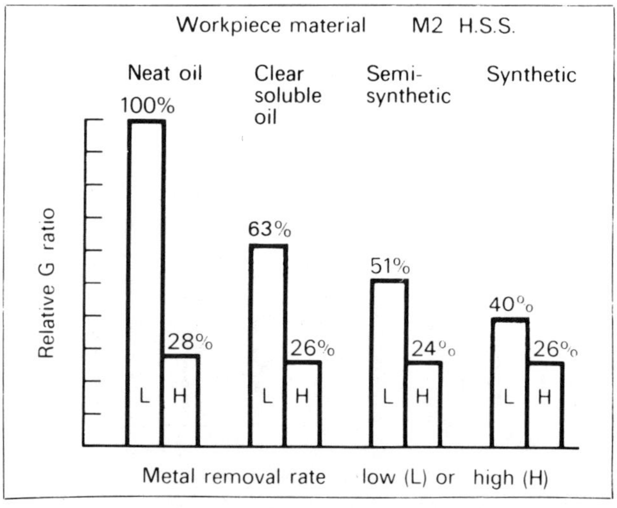

Fig 8 *The effect of metal removal rate on G-ratio when grinding M2 high speed steel with ABN 360 using different coolant types*

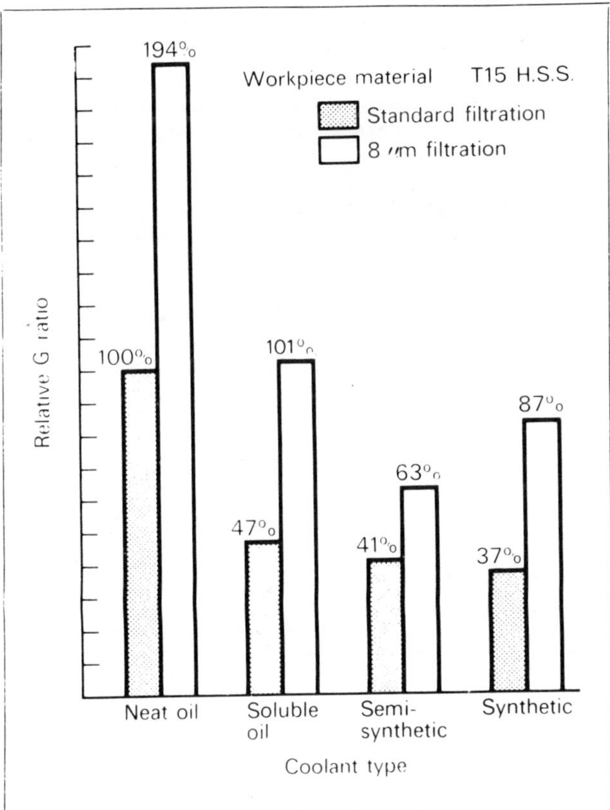

Fig 9 *The effect of coolant filtration on G-ratio when grinding T15 high speed steel with ABN 360 using different coolant types*

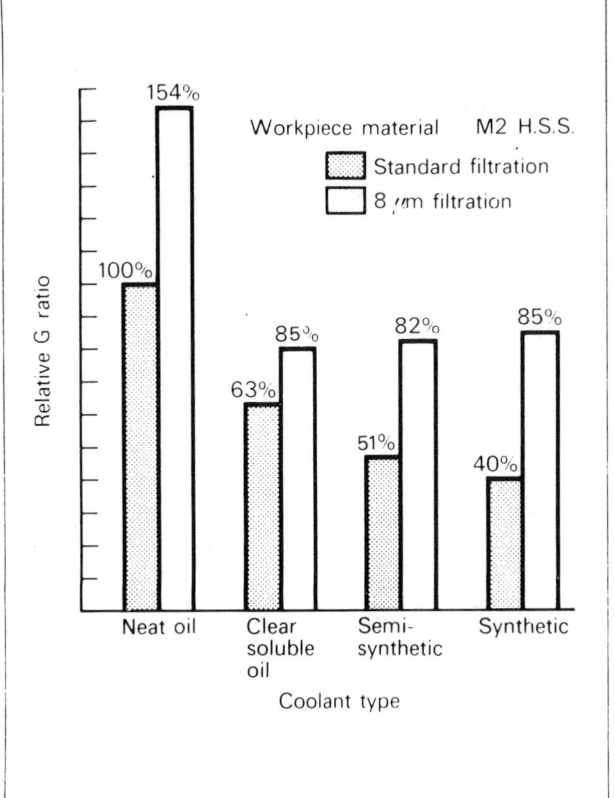

Fig 10 *The effect of coolant filtration on G-ratio when grinding M2 high speed steel with ABN 360 using different coolant types*

removal rate do not show the same loss of performance at the high removal rate (20% against 50%). This suggests that at relatively low metal removal rates the cooling and lubricating effects of the cutting fluid are important, but that at a higher removal rate the mode of wheel breakdown changes to one less dependent on these effects.

It is also worth noting that the method of increasing the removal rate is important. The tests reported here show that an increase in removal rate of 5 times reduced the grinding ratio by about 50% when the cross feed was increased. In a similar series of tests in which the removal rate was doubled (from 0·1 in.^{3}/min/inch wheel width to 0·2 in.^{3}/min/inch wheel width) by increasing the downfeed, the 'G' ratio was also reduced by about 50%.

The conclusion that can be drawn is that the metal removal rate does not affect the ranking of the coolants, but it does affect the percentage difference in performance.

The effect of coolant filtration on G-ratio

Figs 9–11 show the effect of introducing an improved filtration system for the coolant when grinding the three test materials. It is clear that the result on all three materials is a very useful increase in wheel life.

The initial objective of the test series is in part achieved by the step of adding extra filtration, in that the best water-based cutting fluids when used with fine filtration, perform as well as or better than the best neat oil used with standard filtration. However, the neat oil performance is also improved by the use of fine filtration by at least 54% (in terms of G-ratio).

A further interesting point about the change of filtration system is that the scatter of repeat results was reduced from ±20% to ±3% on T15, and from ±13% to ±1% on EN 31. The improved statistical significance of the results was achieved by reducing the contamination level of the fluid at the feed pipe from 40 mg/l to below 0·1 mg/l on an 8 μm filter.

Research summary

This investigation has shown that there are very wide variations in effectiveness between different types of soluble oil, and that the main influence which affects the choice of coolant type when grinding with ABN wheels, is the workpiece material. Other parameters such as wheel speed (between 25–35 m/s) and metal removal rate did not significantly affect the ranking of the coolants.

It will be noted that the presence of extreme pressure additive has not been found to have any beneficial effect. This point is contrary to our initial belief, but the results show that the physical characteristics of the oil had a measurable effect while the chemical additives did not. The detail of the coolants found to be most effective are given below.

(a) A thorough investigation of the performance of different specifications of neat oil revealed that neither sulphur or chlorine-based extreme pressure additives had any effect on wheel performance. The most important influence was viscosity. An oil with 110 centistokes viscosity at 20 °C gave the best performance under all conditions although it should be noted that thicker oils were not tested.

(b) For T15 H.S.S. the best soluble oil was a semi-synthetic type containing some mineral oil together with E.P. additives, surface active agents and inhibitors giving a clear emulsion with a small droplet size (Castrol Clearedge* EP 284) at a dilution of between 1 : 40 and 1 : 50.

A conventional soluble oil containing chlorine (Castrol Superedge* 4) at a dilution of between 1 : 20 and 1 : 30 also gave good results, but the semi-synthetic type of coolant is preferable as it tends to keep the wheel surface cleaner.

The synthetic coolant (Castrol Syntilo* 11) gave a

*Registered Trade Mark

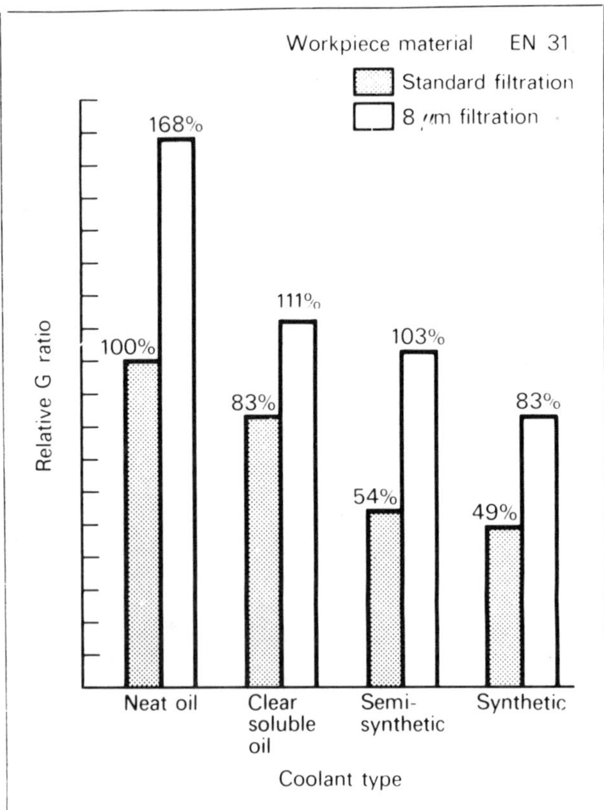

Fig 11 *The effect of coolant filtration on G-ratio when grinding EN 31 ball bearing steel with ABN 360 using different coolant types*

very good performance but as indicated it was more sensitive to dilution and suffered from high water evaporation rates, so that careful control of dilution is required if the best results are to be obtained.

(c) For M2 H.S.S. equally good results were obtained with the synthetic and semi-synthetic types described and a clear soluble oil containing equal parts of emulsifier and oil giving a droplet size of 0·5 to 1·5 microns (Castrol Clearedge E). The recommended coolants are therefore semi-synthetic oil (Castrol Clearedge EP 284), dilution between 1 : 40 and 1 : 30, or the clear soluble oil, dilution between 1 : 30 and 1 : 20.

(d) For EN 31 bearing steel the best results were again obtained with the clear soluble oil (Castrol Clearedge E), dilution between 1 : 25 and 1 : 15, and the semi-synthetic coolant (Castrol Clearedge EP 284), dilution between 1 : 25 and 1 :15.

(f) Finally it has been shown that the level of filtration is very important, and it is recommended that a filtration system that will at least remove particles larger than 8 microns be installed. Systems that remove particles larger than 5 microns are available for use with neat oil, but trouble can be experienced with them with soluble oil as the oil droplets tend to be filtered out with the grinding debris.

If the above recommendations are followed, results can be obtained with filtered soluble oils that equal or exceed those achieved with the best neat oil, conventionally filtered (particles larger than 25 microns). It remains true that a finely filtered neat oil performs better than the best soluble oil, and remains the best coolant to use when grinding with AMBER BORON NITRIDE when environmental considerations will allow.

Practical considerations of coolant use

It has been shown above that the correct choice of a grinding fluid can result in a substantial improvement in wheel life under a controlled laboratory environment. In practice whilst machining conditions can always be controlled, other factors integral with the operation cannot be neglected and may occasionally dominate.

Straight mineral oils are only environmentally acceptable on machines which are enclosed and are supplemented by good extraction facilities. The risk of fire from mist can be minimised by using higher viscosity products but since a secondary function of a grinding fluid is to transport debris, this feature will suffer if coolant flow is sluggish. Oils with 110 centistokes viscosity at 20 C have performed well but there is obviously an upper limit to the usable viscosity.

Synthetic soluble oils did not perform particularly well in these tests on metals which are difficult to grind. In a sense this is fortunate since there are some practical disadvantages associated with this range which are not apparent under laboratory conditions. When synthetic soluble oils (i.e. those containing no mineral oil) are in constant use, machine corrosion protection is provided by chemical additions but evaporation of the pure water during a hot day, followed by condensation onto dry unprotected machine parts at night, may result in rusting of finished parts, or of the machine itself, with results that can be very costly.

Semi-synthetic soluble oils contain some mineral oil and under all conditions there is protection for the machine. The products are tailored to give good detergency, low foaming, protection from bacterial attack, excellent extreme pressure quality and stable emulsions[3]. For AMBER BORON NITRIDE on the metals tested it was shown that extreme pressure qualities are not essential but these other features, not measured in the tests, are important for general workshop situations.

Conventional soluble oils offer a very wide range of properties and costs. The cheaper products contain no E.P. additives, poor microbial resistance and smaller amounts of emulsifier. They are completely satisfactory when the metal being worked has low tensile strength, the system is kept relatively clear of bacterial infection, there is no hydraulic oil leakage into the coolant and effluent treatment processes are adequate. If any or all of these conditions are abnormal, more sophisticated conventional soluble oils must be used.

Conclusions

As a result of the work reported in this article it has been possible to give clear recommendations on the best coolant formulae to use when grinding three representative workpiece materials with AMBER BORON NITRIDE. The results have thrown up three points that are contrary to previous ideas: it has been shown that E.P. additives are not effective, that high concentration soluble oils are not beneficial, and that the difference in performance between neat and soluble oils reduces at high metal removal rates. No analytical study of the reasons for the observed phenomena was possible, but such an investigation could well throw valuable light on the grinding process.

Finally the beneficial effect of fine filtration does not need explanation, but the very large effect recorded is again surprising.

It is hoped that these results will prove, as they were intended, to be a useful practical guide to coolant selection when grinding with AMBER BORON NITRIDE abrasives, and that the present lack of a clear understanding as to why the effects occur, will not be seen as detracting from the validity of this necessarily empirical investigation.

References

1. DE JAGER, P. W. G., BUSCH, D. M., BALFOUR, A. M. Latest developments in grinding steels. *IDR* (1976), **36**, pp 6–14 (January).
2. General Electric Company (USA) brochure No. SMD3–152.
3. OATES, P. D. Advances in grinding fluids. *IDR* (1972), **32**, pp 12–14 (January).

Solid Lubricants Can Reduce Wear and Increase Cutting Tool Life

L. Stallings and Maj-Britt K. Gabel

Reprinted from Cutting Tool Engineering, May/June 1980

WORK done at the Naval Air Development Center shows that coatings of solid lubricants can reduce wear and extend life of tools made from high speed steel. In addition, reduced machining and tool dressing times can be realized.

Besides the obvious and more immediate reasons for such tests, there is the less obvious one of this country's heavy dependence on imported tungsten. It has been estimated that 75% of the tungsten used in this country is imported and if imports were cut off, it would pose serious problems. Therefore, anything that can be done to extend the life of cutting tools can be to the metalworking industry's advantage.

The increase in the number of drilling operations using molybdenum disulfide coated drills is dramatic, as shown in Table 1. Failure criterion was the appearance of burrs around the hole. The coolant used with the uncoated drills was sulfurized oil. The coated drills showed a very significant increase in the number of holes drilled without the use of a coolant. In these tests, the drilling was done on a piece of 3.2 mm (0.125 in.) thick titanium alloy (Ti-6Al-6V-2Sn) and the cutting speed was 300 rpm.

The results of tapping tests performed at the Naval Air Engineering Center are shown in Table 2. In this test a MIL-L-8937 solid film lubricant containing MoS_2 applied by a special process was used. A radial drill was used with water soluble oil on forged AISI 4340 steel. Workpiece hardness was 34-38 HRC. Normally, this process required the use of two or more taps to prevent tearing of the threads. The criterion of tool failure was either an undersized hole or a rough finish. In two runs, the coated tap required only one pass with two to four times as many holes tapped as the uncoated tool would accomplish.

The Naval Air Engineering Center evaluated indexable inserts and a twist drill coated with MIL-L-8937 solid film lubricant. The inserts were used on 17-4PH (AMS 5643) stainless steel heat treated to a tensile strength of 1070 MPa (155,000 psi). The coated inserts were able to cut 50 deeper, resulting in a considerable saving of machining time.

A 36.5 mm (1.44 in.) twist drill was used to drill forged AISI 1020 steel. The coated drill only increased the number of machined pieces by one; however, its wear area was one-third the size of that of the uncoated drill, and it had no metal cold-welded to the tip. This would reduce the amount of time necessary to resharpen the drill.

The Naval Air Rework Facility at North Island, California, evaluated taper locks, drills and reamers coated with the MoS_2-containing MIL-L-8937 lubricant applied by a proprietary process. A four-fluted 10-32NF tap was used to tap holes in the exhaust frame of a T-64 engine. The frame is made from Inconel 722 (AMS 5541). Here, the savings were in time formerly taken to remove broken taps. The coated taps completed seven times as many holes as the uncoated ones.

This same coating was also used on ball-shaped milling cutters for manufacturing the F-4 wing fold lugs made from forged AISI 4340 steel. The results are shown in Table 3. The machine was a Cincinnati Hydro-Tel tracer mill using both closed angle and open angle ball-shaped cutters. The cutting fluid was water soluble oil. The machine operated at 145 rpm with a feed speed of 50 mm/min (2 in./min) to produce a cut measuring 2.5 x 50 mm (0.10 x 2 in.). • • •

L. Stallings is the chief of the Lubricants Section at the Naval Air Development Center, Warminster, Pa.

Maj-Britt K. Gabel was a lubricant chemist at the Naval Air Development Center and now is retired.

This is an edited version of the original paper which was presented in the AMAX Molysulfide Newsletter, August 1979.

Table 1. Drilling Tests on Ti-6Al-6V-2Sn Alloy Plate

Tool Coating Designation	Number of Drilling Operations
None	2
MIL-L-23398 (MoS_2, air dried, organic)	70
NADC-23A (MoS_2, heat-cured, inorganic)	100

Table 2. Tapping Tests on AISI 4340 Stock

Run	Tool Life	
	Coated	Uncoated
1	1 pass 12 holes	2 passes 3 holes
2	1 pass 24 holes	3 passes 12 holes

Table 3. Milling Tests on AISI 4340 Stock

Tool	Number of Parts Machined	
	Coated	Uncoated
Closed ball-shaped cutters	280	100
Open ball-shaped cutters	280	100

CHAPTER 4

EXTENDING USEFUL FLUID LIFE

Reprinted from Modern Machine Shop, March 1979

That clear, cool, water that comes from the tap is anything but ideal for water-soluble cutting and grinding fluids.

Water Is Water... Or Is It?

By WILLIAM A. SLUHAN
General Manager
Master Chemical Corporation
Perrysburg, Ohio

With the ever-tightening environmental regulations facing today's metalworking industry, managers of American metalworking plants are recognizing what those in the aerospace industry, the plating industry and the electronics industry have known for a relatively long time—*water is not just water.* Rather, the quality of the water used in a manufacturing process can dramatically affect the quality of the product, the efficiency of the process, the maintenance costs of the process machinery and the useful life of any process chemical bath.

It is primarily in relation to the useful life of chemical baths that managers are beginning to recognize the effect of water quality; generally, the purer the water the greater the useful life of the chemical bath. And the greater the useful life, the less material which must ultimately be processed for disposal.

Cutting Fluids

Water soluble cutting and grinding fluids are chemical baths; that is, they are composed of fluid concentrate diluted normally from one to ten percent with water. Water, then, amounts to 90 to 99 percent of the cutting or grinding fluid in the machine coolant sump. Poor quality (high mineral content) water dramatically affects the performance of the fluid, as follows:

- Hardness minerals (primarily calcium and magnesium chlorides, carbonates and sulfates) affect chemical, semichemical and emulsion type fluids, and result in gummy, sticky, residues, separated emulsions and tend to foster microbial growth.
- Nonhardness minerals (primarily sodium and potassium chlorides and sulfates) affect all types of water-miscible fluids, and are frequently the cause of corrosion problems associated with these fluids.
- Sulfates (sodium, potassium, calcium and magnesium) all act as oxygen sources for the sulfate-reducing bacteria, which are responsible for liberating hydrogen sulfide gas—commonly referred to as "Monday-morning stink" in metalworking shops.

Virtually all water-miscible fluids are formulated to overcome some of the detrimental effects of minerals dissolved in the water with which the concentrate is mixed. Generally, cutting fluid emulsifier systems contain both anionic and nonionic wetting agents; the latter are unaffected by hardness salts present in the water. But nonionic wetting agents, while able to overcome some effects of hardness salts, make poor cutting fluid lubricants and tend to form stable foams, which can cause operational problems with some coolant systems.

Anti-corrosion systems in the fluid will suppress the corrosive effects of the acid radical (Cl^- or $SO_4^=$) of the dissolved minerals in the water supply. However, the corrosion inhibitors will tolerate only certain concentrations of these salts before their effectiveness is overwhelmed. The effect of these chloride and sulfate ions in the coolant is perhaps most noticeable when machined parts are stacked wet and allowed to dry in tote boxes. Staining or corrosion is frequently found on mating surfaces of these parts after a relatively short time and this staining is usually directly attributable to the presence of these ions in the coolant solution trapped between parts.

This latter point is important because a machine tool coolant sump acts like a still at room temperature and much of the daily coolant usage is actually replacement of water lost by evaporation. Whatever salts were present in the evaporated water are left behind in the coolant solution in the machine sump. Consequently, although a coolant solution starts off with a relatively good water the accumulation effect rapidly converts the water in the sump to a poor quality water.

Figure 1 shows the increase in mineral content of a coolant sump for a 90,000-gallon central system run three shifts per day, six days per week. Coolant makeup required was 10,000 gallons every twenty-four hours—8,800 gallons of water lost by evaporation and 1200 gallons that adhered to removed chips and parts. In 40 working days the mineral content of the coolant solution increased from 250 ppm (14 grains) to 1000 ppm (56 grains). This increase is typical of most coolant systems.

Water Improvement

Since water quality affects the performance of water-miscible cutting and grinding fluids and the performance of the fluid affects the efficiency of the manufacturing operation, what options does the manager have in regard to the quality of water?

To date, the majority of metalworking plants have done nothing to improve water quality. Typically, they continue to use raw or untreated water supplies, either private wells or public water systems. The quality of such sources ranges from very good to very bad. One metalworking plant uses a private artesian well, which produces water containing about 9 ppm total dissolved solids, about a half grain, and is almost as pure as distilled water while another plant's water system contains over 110 grains per gallon total dissolved solids. The majority of metalworking plants mix coolant concentrates with waters varying from four to five grains (considered to be moderately hard) to 15 to 20 grains (considered to be very hard).

In some metalworking plants with extremely poor quality water, managers have chosen to improve water quality by installing water softening equipment. Water softening is a process in which hardness mineral ions (calcium and magnesium ions primarily) are exchanged for nonhardness mineral ion (sodium) by passing the water through an ion exchange resin bed. The ion exchange bed is a pressure-tight tank filled with ion exchange resin (tiny, porous, plastic beads which carry a negative electric charge), and the necessary plumbing and controls to affect water flow through the bed as well as periodic regeneration of the bed. As water flows through the bed, calcium and magnesium ions adsorb onto the resin particles and in so doing, replace sodium ions present on the resin particles. Thus, calcium and magnesium ions are exchanged for sodium ions.

Periodically, the ion exchange bed is regenerated with a saturated salt (sodium chloride) solution. The highly concentrated sodium ions replace the calcium and magnesium ions previously removed from the water, and the bed is rinsed to remove excess salt. The resin bed is now recharged or regenerated with sodium ions and ready to soften more water.

The total amount of dissolved solids present in softened water is not appreciably different from the hard water. But the nature of the water is different in that the calcium and magnesium salts have been exchanged for sodium salts and the

Fig. 1—Mineral accumulation versus time in a water-miscible metalworking fluid.

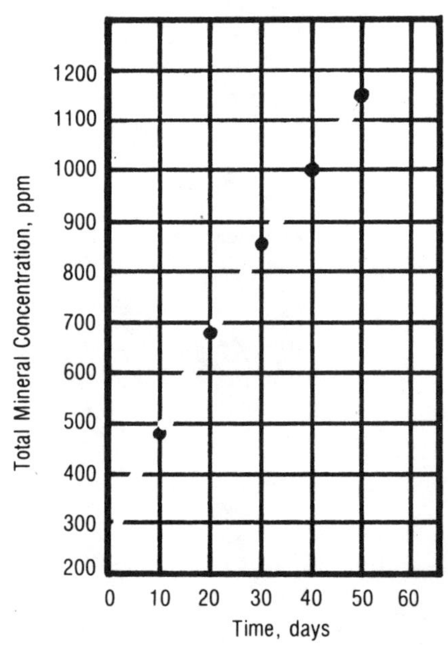

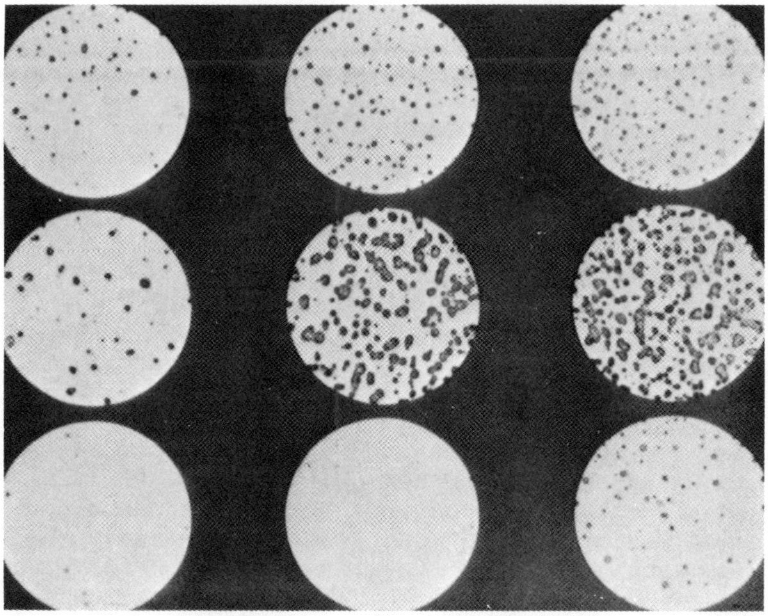

Fig. 2—Results of cast iron corrosion tests at various concentrations of the same metalworking fluid in waters with different mineral contents.

sticky, gummy residues which result when mixing coolant concentrates with hard water are no longer a problem. However, coolants mixed with softened water have a greater tendency to cause corrosion than coolants mixed with either hard water or demineralized water, as shown in Figure 2.

Three processes are available to remove dissolved minerals from water: (1) distillation, (2) reverse osmosis, and (3) deionization.

Distillation is the process by which dissolved minerals are removed by first evaporating the water (thereby leaving the dissolved solids behind) and condensing the water vapor. Distillation is extremely effective in producing high quality water but has the drawbacks of requiring high initial investment and being both energy- and maintenance-intensive. The cost of water produced by distillation is relatively high.

Reverse osmosis is a technique whereby relatively pure water is produced by forcing water through a semipermeable membrane under high pressure. Water molecules pass through the membrane while the majority of dissolved ions are filtered out by the membrane. While the process does improve water quality, it does not produce water of sufficiently high quality for use with water-miscible fluids; typically, only 90 percent of the minerals are removed from the water supply. Further, the membranes have relatively unpredictable lives and relatively high replacement costs; and approximately half of the water fed to the system goes down the drain as waste.

Deionization is the process by which dissolved minerals are removed from water by passing the water through ion exchange beds. Both negatively and positively charged ions are removed to produce the equivalent of distilled water with much lower installation, operating and maintenance costs than distillation equipment.

The Deionization Process

Deionizers, shown in Figure 3, are similar in operation to the water softeners described previously. The major difference is that softeners consist of a single ion exchange bed wherein sodium ions are exchanged for calcium and magnesium ions, whereas deionizers are composed of two ion exchange beds:

1. A cation exchanger wherein hydrogen ions are exchanged for all cations present in the water supply, normally sodium, potassium, calcium, magnesium, iron and aluminum.
2. An anion exchanger wherein hydroxyl ions are exchanged for all anions present in the water, normally sulfates, chlorides and carbonates.

Whereas softeners are regenerated with sodium chloride, deionizers normally utilize hydrochloric acid to regenerate the cation exchanger and sodium hydroxide to regenerate the anion exchanger, although other regenerating chemicals can be used in certain instances.

Generally, deionizers can produce water equivalent in quality to distilled water for about one-fourth to two cents per gallon (½ cent per gallon is about average). This cost includes the equipment purchase price amortized over five years and the cost for regenerant chemicals used over that same period. The cost per gallon varies with the relative quality of the water being deionized (the better the quality, the lower the cost) and with the amount of water required. Since a deionizer has a fixed minimum cost, the greater the amount of water that is deionized the less the equipment cost per gallon.

The average turret lathe requires approximately 500 gallons of water per year and, assuming an unusually high cost of one cent per gallon, deionized water for that turret lathe would cost only five dollars per year. Deionized water can reduce coolant consumption as much as 80 percent and normally 30 to 40 percent compared to mixing coolant concentrate with raw water. If that average turret lathe used $40.00 worth of cutting fluid concentrate per year, the use of deionized water would save three to six times the cost of the deionized water yearly.

Deionized water can also greatly extend the sump life of water-miscible fluids. Theoretically, if the coolant sump can be kept relatively free of tramp lubricating and hydraulic oils and other contaminants, the fluid could have virtually unlimited sump life. However, let us assume deionized water would only extend fluid sump life from three months to four months. Such an extension would save one machine pump-out and cleaning per year. If machine cleaning is done during production time and requires three hours for each cleaning, it would cost $36.00 to clean a manual turret lathe, assuming the machine carries a burden rate of $12.00 per hour. Again, assuming deionized water for this average lathe costs $5.00 per year but saves one cleaning and recharging per year, the net savings would be $31.00 per year.

In addition to lowering coolant

Fig. 3—Deionizers can produce water equivalent in quality to distilled water for an average of one-half cent per gallon. Of course, the cost varies with the initial quality of the water being deionized and with the amount of water required for plant usage.

consumption and reducing machine cleaning and pump-out costs, deionized water can reduce machine corrosion problems. Simutaneously, it will reduce workpiece corrosion problems and reduce the tendency for bacterial growth in coolants. In summary, a small expenditure for improved quality of the water used to dilute water-miscible coolants can produce major savings in overall plant maintenance costs as well as major increases in overall plant efficiency. **MMS**

ABOUT THE AUTHOR

As General Manager of Master Chemical Corporation's Systems Equipment Division, William A. Sluhan directs the design, manufacture and marketing of Master Chemical's closed loop coolant systems. Upon graduation from Ohio Wesleyan University in 1964 with majors in chemistry and business administration, Mr. Sluhan joined the Master Chemical Corporation as a salesman. After serving two years in the U.S. Army, he returned to Master Chemical Corporation as Assistant to the Sales Manager and Director of Product Service and Evaluation, responsible for the field testing and evaluation of research and development products. In 1973, he was promoted to Vice President - Operations, and assumed his present responsibilities in 1976.

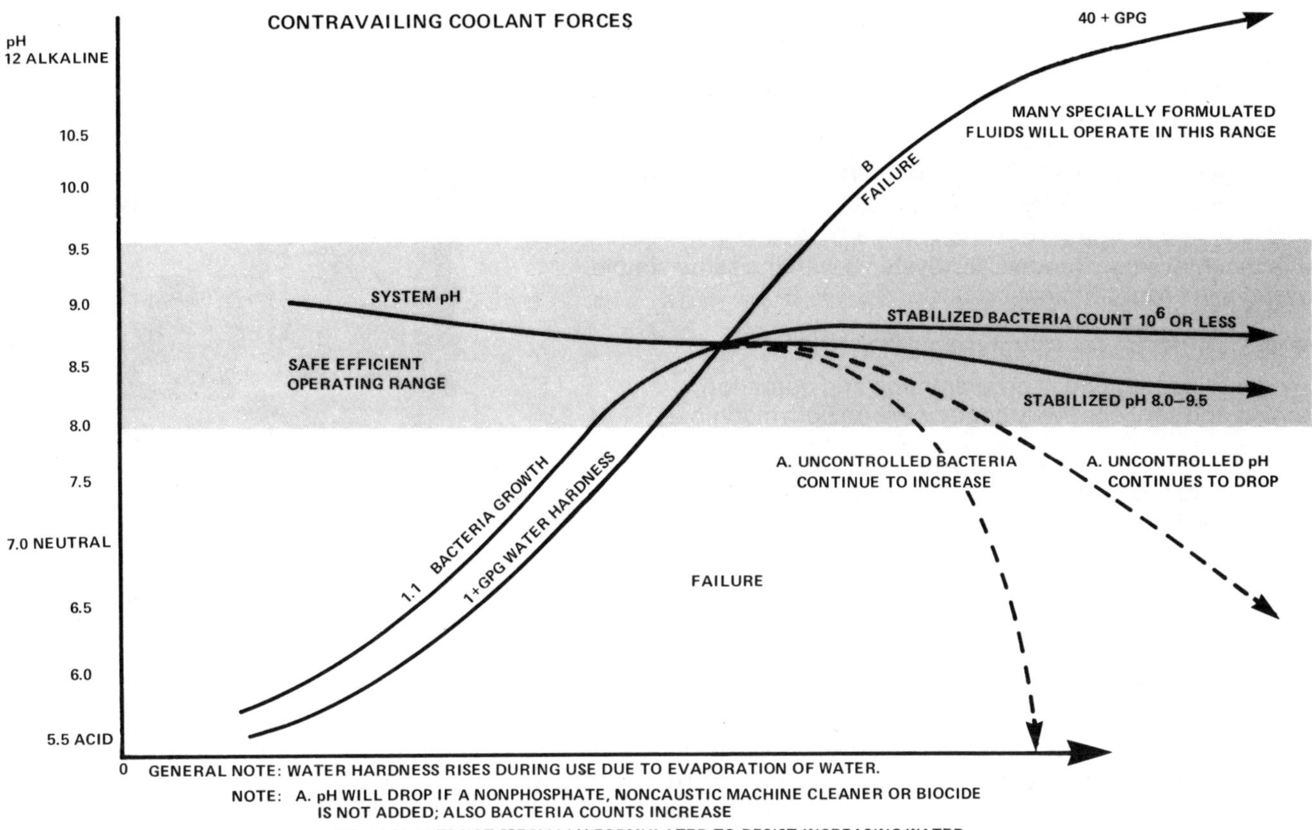

GENERAL NOTE: WATER HARDNESS RISES DURING USE DUE TO EVAPORATION OF WATER.

NOTE: A. pH WILL DROP IF A NONPHOSPHATE, NONCAUSTIC MACHINE CLEANER OR BIOCIDE IS NOT ADDED; ALSO BACTERIA COUNTS INCREASE

NOTE: B. MOST COOLANTS NOT SPECIALLY FORMULATED TO RESIST INCREASING WATER HARDNESS FAIL AT THIS POINT BY SEPARATING INTO TRAMP OIL

NOTE: C. AS THE pH GOES BELOW 8.0 AND/OR THE BACTERIA COUNT EXCEEDS 10^6 AND IS INCREASING, RUST APPEARS AND RANCIDITY (SMELL) OCCURS. DERMATITIS CAN OCCUR. FLUIDS SEPARATE. TOOL, WHEEL AND BATCH LIFE DECREASE.

Reprinted from Manufacturing Engineering, October 1980

Contravailing Coolant Forces

DONALD R. HIXSON
Product Manager
International Refining and Manufacturing Company

MANY FORCES OPERATE in your plant to promote water diluted coolant failure. The accompanying chart is used by our company in an attempt to visualize these effects. Through years of research and experience the petroleum industry has established some fundamental facts which have become apparent:

1. As water hardness increases, measured in gpg (grains per gallon), solution and emulsion stability drop
2. As gpg rises, more tramp oil, rancidity, rust and coolant failures occur
3. While not directly related to rising gpg, pH tends to drop as additives are removed by the increasing hardness in the form of insoluble hardwater soaps. The rate of pH and gpg change are affected by initial gpg, coolant ratio, coolant formulation, type of operation and outside contamination
4. In addition and also influenced by gpg rise and pH fall, are the occurrence and rate of growth of bacteria.

These facts are usually treated in isolation. The interaction of these forces is shown in the chart. By controlling these forces, productivity can increase by improved machine availability and reduced fluid consumption due to spoilage.

The shaded portion of the graph is looked on by our industry as the safe or efficient operating range. The areas above or below this range usually exhibit accelerated objectional conditions.

Water hardness can be controlled. Various methods of softening, deionizing and distilling it can be used, but they add to cost and maintenance effort. Some hardness control is especially beneficial in long running central systems. In many cases, however, a properly formulated product operates very nicely on available water.

The control of pH pays large dividends in extending fluid life and performance. A simple method is to control the addition of concentrate. Constant small additions of concentrate and water are better than irregular massive additions of water and/or concentrate. Regular additions also help stabilize particle distribution more uniformly and control bacteria growth. The lower the number and kind of tank side additive additions made, the more consistent the finish results obtained.

Each ecological system establishes a different need for make-up, pH control, bacteria growth, percent of tramp oil, and increase in water hardness. This is effected by system turbulence, lube hydraulic oil leakage rate, make-up water hardness, slushing oil or rust preventive contamination from the raw material, air-borne contaminants, temperature, effectiveness of the filtration and, most of all, operator and plant hygiene practices.

Once the individual sump or system stabilizes, the actions your supplier recommends can be taken to keep your fluids in the operating ranges for the longest time, consistent with your particular ambient environment. ■

Presented at SME's Westec '79 Conference, March 1979

Modular Coolant Approach And Diagnosis For Better Efficiency And Economy

By S. Bob Vora
Johnson Wax & Son, Inc.

With the introduction of new machining techniques, various metals and manufacturing processes, it is becoming increasingly difficult to select a proper fluid for grinding or machining that will provide maximum efficiency and economy.

There is no "universal coolant," and yet many of you tend to rely on one coolant to perform all jobs - reduce tool wear, provide proper lubricating and cooling properties, rust protection and long tank life.

In order to achieve your goals, and obtain the maximum mileage, a modular coolant approach and analytical method for diagnosing and correcting your problems is of great value.

In today's production-oriented world, high-performance and speed are key factors. Your cutting fluid is an integral part of the production formula in your shop. The average user tends to ignore the importance of the coolant while spending a great deal of time and money in selecting tools. He does not realize that the proper selection and maintenance of a coolant will extend tool life, improve finishes on his parts and provide a better working environment for his employees.

With these facts in mind, let's take a look at some key factors in your coolant's character!

FUNCTIONAL OBJECTIVES OF A COOLANT

A great deal of R & D effort is spent in synthesizing coolant formulations to keep pace with the demands of modern machining needs. Considerations are given to:

(1) Safety

Operator safety is of great concern. Use of phenols or nitrites in the coolants, due to their alleged carcinogenic properties, has become a major issue. Other factors which are taken into consideration are properties which will reduce causes of dermatitis, foul odors and provide easier coolant maintenance.

(2) Cooling

Effective cooling properties are necessary to cool the workpiece and tool. Efficiency in heat dissipation is often the primary measuring stick in coolant evaluation.

(3) Lubricity

To reduce friction between the tool and chip interface; and provide for adequate machine lubrication.

(4) Corrosion

To give good protection to the workpiece, tool and machine.

(5) Removal of Debris

Provide for larger particles to sink or settle, and smaller particles to wet partially and float for easy filtration.

(6) Microbial Contamination

Ability to resist rancidity, mold and fungus.

(7) Disposal

Ability to prolong coolant life so that frequent disposal is not necessary. Also, ease of separating oil and contaminated portion for acceptable disposal.

I am part of an industry which is extremely cognizant of its customers' needs and problems. Our customer contact doesn't end with the sale. Rather, it starts! It is this type of supplier-customer relationship that provides coolant manufacturers the information necessary to constantly improve their products and services to you. One of these services is help in selecting the proper coolant for your particular need.

COOLANT SELECTION

Generally speaking, there are two broad categories:

(1) Straight oil; (2) Water soluble.

As a rule of thumb, straight oils impart higher lubricity and, thus, are used for slow-speed operations.

In the water-soluble category, there are three basic types:

(1) Soluble oils;

(2) Chemical emulsions or semi-synthetics; and

(3) Chemical solutions or synthetics.

The basic difference is that soluble oils contain approximately

80% oil; chemical emulsions or semi-synthetics contain 20-25% maximum oil; and chemical solutions or synthetics do not contain any oil.

Unlike the oil type, water-based coolants impart rapid cooling and thus are used frequently in high-speed operations.

The trend of using water-based coolants is on the rise due to machining operations using tools at higher speeds and feeds. Water, being an excellent conductor of heat, (cooling effectiveness of water is approximately 4-5 times greater than oil), it makes sense that water-based coolants, with proper wetting agents, lubricity agents, corrosion inhibitors, etc., function exceptionally well in many of the high-speed applications in today's machine shops.

Once your basic needs are determined and a specific coolant selected, a second important service can be supplied to you by our industry. That is, monitoring and controling the cutting fluid in your equipment. A sou d control program will eliminate many of the problems so frequently associated with coolants; such as, corrosion, rancidity, poor tool life, etc.

Many large manufacturing plants have their own chemical laboratories, and conduct daily tests on their coolants. They try to detect problems before they occur and implement remedial action promptly. For those who do not have this "in-house" facility, a portable test kit can be used almost as effectively. Most coolant manufacturers will help you set up a diagnostic/analysis program to fit your individual needs. To illustrate, I would like to discuss some of the more commen test procedures with you today.

DIAGNOSTIC METHODS

Test #1 - Refractometer

This is a simple test of determining the concentration ratio of coolant to water. In this test, one or two drops of coolant from the tank are placed on a prism. The light transmitted through the prism will give a reading of light refraction on a scale of 0-30. These readings can be compared to a set of calibrated readings supplied by the manufacturer of the subject coolant.

Heavily contaminated coolant with emulsified tramp oils, or dissolved salts will result in false readings. When this occurs, a titration test will be needed to determine coolant concentration.

Test #2 - pH

pH is a measure of an acidity or alkalinity of a coolant. Readings below 7.0 indicate acidity, while readings above 7.0 indicate alkalinity. The pH of the coolant will give you an indication of possible contamination by detergents, bacteria, acid carryovers, etc. (i.e. anerobic bacteria are acid producing and will cause low pH readings.)

A piece of litmus pH paper is dipped in the coolant and the color developed on the wetted paper is matched with the accompanying color chart to give a direct reading. If a pH meter is available, the readings can be made much more accurately.

Test #3 - Concentration by Cationic Titration

This method is used when the refractometer reading is inconclusive. In this type of test, cationic titrating solution is used to titrate anionic emulsifiers. These emulsifiers such as soaps, fats and oil-based materials, are used to impart boundary lubrication. By this method, you can find out if your coolant is giving you the necessary lubrication. Boundary lubricant depletion can cause poor tool life and finish, particularly in the more difficult operations.

When using the cationic solution, the color change or end point is reached which makes it a rather simple test to run in your shop. You simply count the number of drops to cause the color change, and then read the concentration ratio from an accompanying chart.

Test #4 - Corrosion Inhibitor Level

Corrosion is a common and expensive problem in our industry. Its presence is due to a number of reasons. To name a few - improper mixing of coolant/water ratio. acid carryovers, tramp oils, bacteria, etc.

In all of the above examples, the chemical ingredients will deplete, thus causing poor corrosion protection.

To run this test, highly corrosive cast iron chips are used as standard material. Chips of equal quantity are soaked with the subject coolant on the filter paper using a petri-dish. Three mixes are used:

(1) A fresh standard of the known ratio for comparison;
(2) The coolant from the tank;
(3) The coolant from the tank plus rust inhibitor.

After soaking the chips in these samples for 20 minutes, the excess is drained off and the chips left to stand overnight. The next day, these filter papers are checked for corrosion.

Test #5 - Total Alkalinity

In this test, all alkaline constituents are neutralized to determine the coolant concentration. This also gives you some indication of corrosion level and bacterial activity. The concentration is determined by titrating the coolant and reaching the end point by color change. Low alkalinity can indicate bacterial contamination, possible corrosion and unstable emulsion.

Test #6 - Tramp Oil Content

Excess tramp oils can cause severe problems such as rancidity, corrosion control, poor wetting, machine buildup, poor filterability, etc.

This test will determine what percent of tramp oil is present in the coolant. In this method, used coolant is split by excess acid and the percent of oil is calculated.

Test #7 - Bacteria Contamination

This test will indicate whether the coolant is contaminated with bacteria and what action must be taken to handle a specific problem. In this method, a bio-stik is dipped in the coolant and left to incubate for 48 hours. The bacteria counts are read from a chart prepared for this purpose.

Test #8 - Yeast/Fungus Contamination

This test will determine fungus and mold contamination. A bio-stik is dipped in the coolant and left to incubate for 48 hours. Colonies will develop on the stik if it is contaminated. A chart will read the severity of the contamination.

The above tests should be conducted on a regular basis. A monitoring chart should be established and all test results logged for future reference.

Diagnostic analysis is only part of the story. The other part is equally as important and, in fact, would negate the benefits of the former if left out.

What I am referring to, of course, are the actions to be taken as a result of a systematic analysis. Where the tests prescribe normalcy, we, of course, will do nothing. However, where deficiencies are evident, we will want to take positive, corrective action.

This is where a modular-coolant concept becomes a reality. Based on the results of a diagnostic analysis, we will precribe treatment of your coolant utilizing a number of key additives.

To illustrate this, let me walk you through a few common examples:

PROBLEM	POSSIBLE CAUSES	STEPS TO FOLLOW	CORRECTIVE MEASURES
Corrosion	A. Improper dilution	A(1) Use Test #1 (Refractometer) or (2) Use Test #2 (Acidity/Alkalinity) or (3) Use Test #3 (Lubricity Titration) or (4) Use Test #4 (Corrosion)	A. Add concentrate to correct to proper dilution range.
	B. Acid contamination	B(1) Use Test #2	B(1) Eliminate source of contamination. (2) Recharge sump, or (3) Neutralize with suitable alkali.
	C. Oil contamination	C(1) Check wetting, then (2) Use Test #6 (Tramp oil)	C(1) Eliminate source of oil if possible (2) Recommend equipment for oil removal.
	D. Bacterial contamination	D(1) Check for odors, if present (2) Use Test #7 (Bacterial)	D(1) Drop product if possible, (2) Clean machine. (3) Treat with biocide (4) Recharge with fresh Product - IF NOT - (2) Add suitable bactericide. (3) Monitor routinely.

PROBLEM	POSSIBLE CAUSES	STEPS TO FOLLOW	CORRECTIVE MEASURES
Tool Life	A. Coolant application	A. Consult product application.	A. If current application is in error, suggest the correct.
	B. Speeds & feeds	B. Brochures & machinist handbook.	B. Product & working conditions such as speeds & feed, and also tool geometry.
	C. Delivery of coolant	C. Check for adequate delivery to workpiece.	C. If coolant flow is inadequate, make suggestions to correct deficiencies.
	D. Product concentration	D(1) Use Test #1 (2) Use Test #3 (3) Use Test #5	D. If coolant dilution is too lean or too rich, suggest the proper working range.
Buildup on Machines	A. Water hardness	A. Use test sticks.	A(1) Suggest the use of deionizer or similar equipment. (2) Suggest the use of chelating chemicals.
	B. Tramp oil	B(1) Visual inspection should indicate if problem exists. (2) Run Test #3 (3) Run Test #6	B(1) Eliminate source if possible. (2) Recommend oil removal equipment. (3) Use product with higher emulsifying properties (i.e. Trampol-X)
	C. Mold	C(1) Check for musty odors. (2) Use Test #8	C(1) Calculate the amt. of moldicide necessary. (2) Suggest proper cleaning procedures.

PROBLEM	POSSIBLE CAUSES	STEPS TO FOLLOW	CORRECTIVE MEASURES
Buildup on Machines - Cont'd.			(3) Recharge with fresh product if possible.
	D. Bacteria	D(1) Check for "sewerlike" odors. (2) Use Test #7	D. Follow procedures outlined in "Odors" (a).
	E. Improper dilution (too rich)	E(1) Use Test #1 (2) Use Test #3	E. Correct the dilution with water.

Coolants are truly chemical cutting tools subject to wear and abuse which, if left unchecked, cost your company money.

My mission today was to give you an insight to the analysis of these costly coolant problems, and recommend corrective measures therefor, through the perception of a chemical company supplying your industry.

Presented at SME's Fabricability Conference, December 1975

The Effect Of Water Soluble Cutting Fluids On Operating Conditions In Machining

By Robert Ciesko
Cincinnati Milacron

Cutting Fluids affect productivity and, therefore, the total operating costs of machining systems. Part quality, machine tool performance, and general operating efficiency are affected by cutting fluids. Maintaining cutting fluids through proper control and maintenance pays back handsome dividends. The effects of improper control (dirt and oil) are discussed. An in-depth control program is recommended for achieving maximum cutting fluid life.

INTRODUCTION

I think we can safely say that the majority of the people in the metalworking industry do not consider the cutting fluid a significant factor in the metal removal process.

An enormous amount of engineering time is spent in developing a new production line. The machine tool must meet rigid specifications in terms of dimensional accuracy and productivity, and tooling must be correct for the operation. Even though a cutting fluid will be used when the production line becomes operational, little attention goes to its selection, use, and control.

This is a sad fact because the cutting fluid is a vital part of the total metal removal system. It can mean the difference between profit and loss. It has a direct bearing on the quality of the parts produced, the amount of wear on the moving parts of the machine tool, and operational costs. Even when some thought is given to selecting an appropriate cutting fluid, it is then often promptly forgotten.

In order to function effectively, a cutting fluid requires maintenance just as the machine and the tool require maintenance. The purpose of this talk is to discuss the effects of water-soluble cutting fluids on the machining operation. Inevitably, this depends on cutting fluid maintenance, which affects productivity and, therefore, the total operating cost of the production system. The effect of a cutting fluid can be illustrated in three production areas: part quality, machine tool performance, and operational efficiency.

Part Quality

Cutting fluids, when properly controlled, protect the parts and machine tools from rust and help to achieve a satisfactory finish. The key, of course, is proper control.

If a cutting fluid mix becomes too lean, rust forms on the parts. When a part must be scrapped because of rust, this is a loss of material and the machining cost. Since so much is invested already, parts are generally reworked which incurs additional expense. Obviously, this has an adverse effect on production costs.

The same is true for scrap or rework because of poor finish. Cutting fluid that is too lean cannot provide adequate lubrication, if lubrication is a requirement, to obtain good surface finish. Tramp oil contamination of the cutting fluid reduces cooling ability and sometimes causes premature tool failure. Finally, fines and grit that are allowed to recirculate mar finishes and cause other problems as well.

Scrap and rework are production losses that can be minimized by using the proper cutting fluid and good control practices.

Machine Tool Performance

Cutting fluids flush chips from the cut zone, keep machines clean, and minimize wear of the moving parts. Poor maintenance of a fluid, however, seriously limits its ability to do these things. Dirty fluids are abrasive and accelerate wear of moving parts. They deposit residues on machines that are not only unsightly and unpleasant to operators but also cause wear and even sticking of gages, slides, etc. Excessive dirt can clog cutting fluid return lines, necessitating machine shutdown for cleaning to prevent poor tool life because of inadequate fluid application.

Keeping a cutting fluid clean is well worth the effort. Replacing slides, ways, pumps, and tools, and contending with operator dissatisfaction can skyrocket production costs.

Operational Efficiency

Cutting fluid characteristics are determining factors in the efficiency of an operation and, therefore, its cost. Such things as rust, rancidity, and foam control, mildness, useful life, and the kind of treatment required for disposal are properties often ignored in the total cost of machining. Desirable properties are taken for granted and it is only when problems interfere with production and raise costs that these properties are appreciated. Few realize that control is necessary to maintain the cutting fluid.

Rancidity not only causes objectionable odors but eventually can lead to poor tool life as the bacteria deteriorate the lubricating value of the fluid. Also, machines are cleaned out more often leaving less available time for production.

Excessive foaming leads to fluid losses.

Skin irritation will not only cause operator complaints but will lower their morale and decrease efficiency.

Any of these problems can lead to short cutting fluid life and this is where hidden costs add up rapidly. The following example shows what kind of hidden costs are involved.

EFFECTS OF PROPER MAINTENANCE
ON CUTTING FLUID COSTS

Machine: Turret Lathe
Sump Capacity: 25 gallons

	Cutting Fluid A	Cutting Fluid B
Purchase Price/gallon	$3.00	$3.00
Concentration	1:25	1:25
Cost to Fill Sump	$3.00	$3.00
Cutting Fluid Life	4 weeks	12 weeks
Number of Cleanouts per year (50 weeks)	12.5	4.16
Labor Costs (1 man, 2 hours @ $4/hr)	$ 8.00 per cleanout	
Machine Burden Rate (1 man, 2 hours @ $14/hr)	$28.00 per cleanout	
Disposal Cost ($0.10 per gallon)	$ 2.50 per cleanout	
Additional Maintenance/week (weekly concentration checks)	----	$1.50
TOTAL FLUID COSTS/YEAR		
Cutting Fluid Cost	$ 37.50	$ 12.48
Labor Costs	100.00	33.28
Cleanout Downtime	350.00	116.48
Disposal Costs	31.25	10.40
Maintenance Costs	---	75.00
TOTALS	$518.75	$247.64

Figure 1

In figure 1, we have ignored cutting fluid daily makeup cost because this is the same fluid in both cases and carryoff and evaporation are identical. As you can see, by increasing the life of the same cutting fluid three times, through proper maintenance, a $271.11 yearly savings is achieved. This is a 1/2 reduction in costs; or another way of saying, this is a three-fold return on the invested additional maintenance cost.

Since the cutting fluid and the operating conditions under which the cutting fluid is maintained can affect the part quality and the performance of the machine tool, as well as the operational costs of manufacturing a unit, how can we best control operating conditions in machining? Before we can discuss controls, we need to identify what factors affect performance over an extended period of time. There are many factors which can affect the performance and, therefore, the operating conditions of the cutting fluid in machining, but the most important are:

1) keeping the cutting fluid clean (free of dirt and oil) and,
2) maintaining the recommended concentration.

Dirt

Dirt or metal fines act as a breeding ground for microbes which will destroy the lubricating and corrosion inhibiting properties of the cutting fluid. Bacteria and fungi are everywhere--in the water, the air, on the operators hands, the parts, and the machines. It is only natural that they get into cutting fluid mixes. As soon as we set out to remove metal, bacteria are present. These metal fines, if they are allowed to accumulate in the system, act as a breeding ground for the bacteria. Recirculating dirt in the cutting fluid leads to poor part finish and eventual wear out of tools which results in high regrind costs and low production per tool. Recirculating fines build up and plug coolant lines. Machine shutdown is the final result. As the dirt content in the fluid builds up, the flow of fluid to the tool and work-piece is being reduced, resulting in less lubrication and cooling. Recirculating fines wear out the vanes and the bearings in the fluid sump pumps. The ways and slides of the machine tool can wear, causing inaccuracy in position, thereby affecting the quality of the part. Also, the metal fines act as a sponge, depleting the important performance characteristics of the cutting fluid; in reality, wearing it out and leading to short fluid life.

Oil

Hydraulic oils, way oils, spindle oils, etc., are all used on machine tools. Many of these oils contain extreme pressure agents. These extreme pressure agents are usually sulfur- or phosphorus-based. These additives are a food source for bacteria. Oil leakage continuously feeds the bacteria in water-soluble cutting fluids. Rancidity and short life result. Oil makes machines dirty. The tramp oils end up coating the machine surfaces and attracting fines and shop dust. Not only is this unsightly and unpleasant to the operator and lowers his morale, but in most cases, management wants a good, clean working environment and maintenance costs in keeping the machines clean are increased.

As oil builds up in the fluid, the parts and floor get slippery -- safety hazards. The machining operation may start smoking, especially if it is a high-speed, high-feed operation thereby decreasing the quality of the air in the shop. Cooling, which is an important characteristic of a cutting fluid, is lost as the oil content increases. We know that water is the best cooling media and possesses a specific heat of 1; oil has a specific heat of 0.5 and is not as effective as water in cooling the part. In very critical operations, loss of the cooling ability of the fluid creates size control problems which lead to rework or scrap.

Some very interesting studies conducted at Cincinnati Milacron, Inc. recently showed that if oil levels are not controlled in water-soluble cutting fluids, it is very possible that the skin irritation potential of the cutting fluid may increase. These tests were conducted in the laboratory on animals. These animals were immersed in fresh mixes of cutting fluids with and without typical oils found in the shop. By measuring the irritability of these mixes, Cincinnati Milacron found that some cutting fluids and certain oils can make a normally mild cutting fluid irritating. It is important to add that much more work needs to be done in this area because not all cutting fluids are similarly affected. In fact, this is a function of the compatibility of the oil and the cutting fluid. However, I point this out because it is another reason to control the amount of oil that leaks into the cutting fluid system.

One of the most timely concerns related to oil is the recent Occupational Health and Safety legislation. Oil mist levels in shops are directly influenced by oil contamination of the cutting fluid. Tests conducted at Cincinnati Milacron, Inc. showed that excessive oil levels from machine hydraulic and lubricating systems played a very significant role in the misting problem. In these tests, the researchers at Cincinnati Milacron selected a centerless bar grinder because this represents the most severe misting condition found in the metalworking shop. The graph in figure 2 shows the results of these determinations. The grinding was done in a closed room with low turbulence conditions. Sampling devices were positioned at operator nose level, on the sides and to the rear of the grinder.

The test conditions were:

Mist Measuring Time	3 hours
Actual Grinding Time	1-1/2 hours
Room Temperature	80-85°F
Relative Humidity	72-84%

Measurements were conducted on four different types of water-soluble cutting fluids. A conventional soluble oil containing a high amount of oil, a premium soluble oil containing a similar amount of oil, a chemical-emulsion product which contains a small amount of oil, and a chemical-solution product which contains no oil.

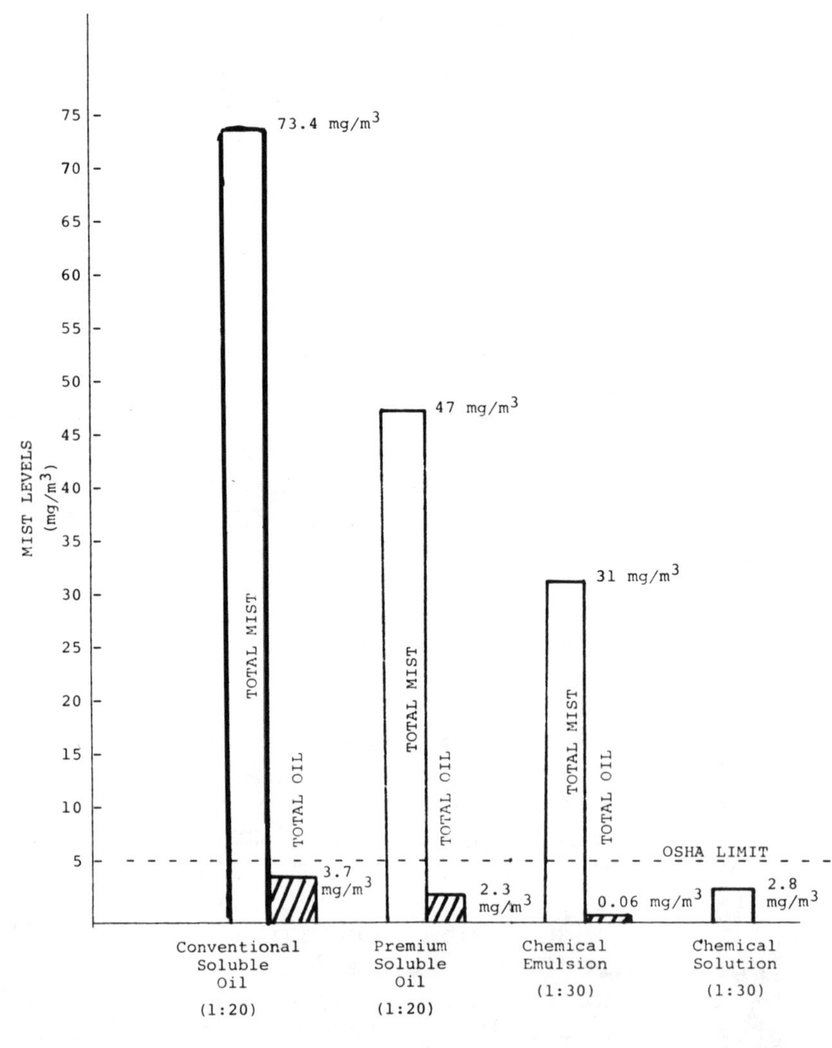

Figure 2

Plotted on the graph is total mist, which includes water plus product plus oil, and oil mist levels. As you can see, the graph shows that the oil content of the products affects mist levels. As the amount of oil in the fluid decreases, so does total mist. Also, the oil mist levels decrease. Since the solution product contains no oil, there can be no oil mist. But the important conclusion from this test is that if oil levels are controlled, misting will also be controlled.

Concentration Control

Maintaining the recommended concentration has a significant effect on the operating condition in machining. Proper concentration control prevents rust which means no rework or scrap for this reason.

Rancidity is often a result of a lean concentration. Maintaining the recommended concentration prevents odor and the resultant loss of tool life caused by deteriorating cutting fluid.

Cutting fluids are formulated to provide settling and filtration properties. But if the concentration is too lean, dirt recirculation occurs resulting in loss of surface finish, poor tool life, machine wear, and short fluid life.

On the other hand, if the concentration is maintained too rich or too strong, heavy residues become a problem. This causes difficulty in operating machine chucks, slows down machine slides, and attracts dirt and dust, leading to machine tool wear.

Skin irritation may also occur with rich concentrations, leading to operator dissatisfaction, loss of efficiency, and the associated cost of treating this medical condition.

Foam can result because wetting and cleaning properties of the fluid are too strong, creating housekeeping and safety problems. Foam suspends the metal fines and leads to recirculation with its attendant problems. Suspended dirt, in turn, depletes vital cutting fluid ingredients, which can contribute to the generation of ammonia in some types of cutting fluids. Foam interferes with the proper functioning of positive control filtration systems having cloth or paper media. It can cause rapid indexing with high media costs and inefficient dirt removal. And, perhaps most important, concentrations that are too strong or too rich increase cutting fluid costs.

All these problems can be controlled by a good cutting fluid maintenance program. One of the most important factors in setting up an overall cutting fluid program is to assign one person in the plant the responsibility for cutting fluid maintenance and control. The overall control program consists, basically, of daily and weekly checks of the cutting fluid system.

Daily concentration checks are recommended for central filtration system applications. When one considers that a central filtration system may be connected to 50 machines, the importance of controlling concentration is dramatically brought into focus. If we do not control concentration, we will have the rust, rancidity, loss of tool life and settling properties, residue, skin irritation, or foam problems that we talked about. If the production unit must be shut down to change the fluid because of these problems, we generate a tremendous nonproductive cost. With individual machines, it is impractical to check the concentrations daily. Concentration checks should only be made as problems are reported at the individual machine. However, one very good way with both central systems and individual machines to avoid concentration control problems is through the use of a premix makeup fluid. Every time the volume of the fluid must be made up, it should be made up with water plus concentrate at the proper ratio. This will maintain a steady-state concentration condition. For if water alone is added to make up the volume, a lean mix will occur, even if only temporarily, and this is when many, many rust, rancidity, tool life, and settling problems occur.

pH is another important factor to check. pH is one measurement of the effectiveness of the rust control, cleaning ability and bacterial control. Cutting fluids are designed so that bacteria are most dormant at a pH range above 8.8. (See figure 3.) If one maintains the pH of the cutting fluid within an 8.8 to 9.2 range, minimum problems can be expected. pHs higher than 9.2 result in low bacteria activity, but above 9.5 skin irritation becomes a concern.

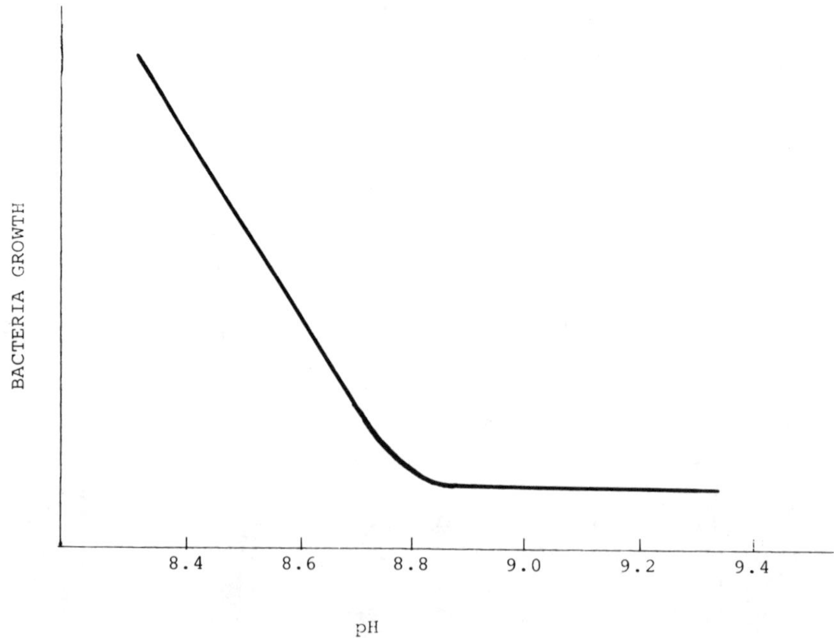

Figure 3

Filter operations should be checked daily. With central systems (whether they be pressure or vacuum), check the filter setting. Also, insure that adequate media is available when using positive filtration. Some individual machines may be attached to magnetic separators or cyclonic filtration units. Their operation should be checked daily to insure that these units are functioning properly.

Oil removing devices such as skimmers or centrifuges should be checked daily to insure that they are functioning properly and removing the tramp oils.

Weekly determinations consist of an analysis either made in your own facilities or by the cutting fluid supplier. Oil determinations which measure the amount of extraneous oil and free oil should be checked weekly. Extraneous oil levels should be kept below 3% through the use of oil removing devices such as skimmers and centrifuges. If this is not done, you will have the associated problems of rancidity control, high misting, dirty machines, slippery floors and parts, smoking, loss of cooling ability, and, possibly, increased skin irritation.

The amount of dirt and metal fines should be checked weekly. This is an indication of the operating efficiency of the filter or the settling characteristics of the cutting fluid. Dirt levels should be kept below 0.5% on volumetric basis or less than 50 ppm on a gravimetric basis. Otherwise, one can expect odor problems, poor finish, coolant line plugging, coolant pump wear, machine tool, slide and way wear, as well as the short useful life of the cutting fluid.

Perhaps the most important weekly determination is the micro-organism count of the cutting fluids. It is strongly recommended that total plate counts and mold counts be made for central system cutting fluids. This is the only way of determining the true biological activity of the cutting fluid. This test indicates the need for corrective action based upon sound scientific fact rather than guess.

It is a practice in some plants to add bactericide on a weekly or periodic basis to prevent high bacterial and mold counts. However, this is not a very wise approach because excessive use of bactericides can lead to skin irritation. This method of overkill can cause many problems. The only way to determine if bactericides are really needed is through bacterial counts. Many suppliers now provide this service to their customers. Recently, test kits have been introduced for determining bacterial and mold counts in the plant. These are little plastic plates or paper strips which have a bacterial or mold growth media attached. The plate or strip is dipped into the cutting fluid sample and then incubated for 24-48 hours. Comparison with photographs to determine the approximate count along with other considerations determine if there is a need to add a bactericide.

Shown in figure 4 is a weekly monitoring sheet for determining all the necessary chemical and physical characteristics of a cutting fluid. As you can see, these sheets show concentration, pH, the % dirt, the amount of free oil, extraneous oil and total oil, as well as the total plate counts of bacteria and mold. By keeping records of this sort, we are in essence developing a preventive maintenance program. We can see changes in any one of these particular parameters from week to week, and as these parameters start heading toward danger areas, corrective action can be taken before a problem presents itself on the production floor. Obviously, this type of control is ideally suited to central filtration systems but is impractical in individual machine applications. In individual machines, perhaps the most important control parameter is concentration control, which can be easily achieved through the use of premix.

In conclusion, I want to emphasize that cutting fluids are an important part of the metal removal system. They must not only be properly selected, but they must also be properly maintained. Through proper selection and maintenance, we can avoid many of the production losses that are involved in the machining operation. Parts scrapped, machine tool-related problems, and the total operational considerations can be controlled by a program of effective maintenance of the cutting fluids. If these operating conditions for cutting fluids are improved, profit will be improved through increased productivity per unit cost. And, this is no small consideration in today's economic climate.

DISTRIBUTION

ANALYSIS OF **FLUID MIX SAMPLE**

COMPANY ____________ CITY ____________ SYSTEM NO. ____________

PRODUCT ____________ REC'D CONC. 1: ____________ CAPACITY ____________

PROGRAM ____________ OPERATION ____________

MATERIAL ____________

SYSTEM TYPE ____________

SYSTEM PARAMETERS													
DATE		SAMPLE NO.	CONC.	pH	% DIRT VOLM.	% - OIL			PLATE COUNT (1ML.)	MOLD COUNT (1ML.)		DATE CHARGED - - ACTION REQUIRED	TECH SPEC.
TAKEN	REC'D					FREE	EXTRANEOUS	TOTAL					

> = GREATER THAN < = LESS THAN M = MILLION

Figure 4

Presented at SME's Industrial Fluids Clinic, November 1980

Operating Procedures For The Conservation Of Hydraulic Fluids

By Joseph Ivaska, Jr.
Tower Oil & Technology Company

The conservation of hydraulic fluids has received greater attention and emphasis due to several factors. The marked increase in the cost of petroleum based fluids and FR fluids, along with disposal costs and a greater emphasis on health and safety restrictions.

In this technical paper we'll show how good fluid conservation practices can result in added productive gains, such as improved system performance, greater hydraulic component life, and a marked reduction in overall hydraulic system maintenance costs.

To achieve this added dimension of "productivity" via fluid conservation we must first address ourselves to a systems approach in upgrading and improving the present techniques in dealing with conservation.

A complete program must include in-plant training of shop personnel, elimination of sloppy and undesirable shop practices, greater emphasis on maintenance by management, and a thorough understanding of the different type of fluid conditioning equipment available.

Let's break down the conservation program into a nine point outline in order to fully realize that an in-depth analysis of fluid conservation techniques must first be made in order to achieve some degree of success. The points to consider are

1. Value Analysis
2. Proper Fluid Selection
3. Storage and Handling of Fluid
4. Contamination Control
5. Learning from your Breakdowns
6. System Design Consideration
7. Fluid Reclamation Procedures
8. Leakage Control
9. Keeping Fluid Conservation Records

The most important portion of any fluid conservation program is communication between fluid supplier, management, shop personnel, and any regulating agencies that are involved in pollution control, health and safety regulation.

From looking at the above outline one also realizes that the greatest underlying factor in fluid conservation is contamination control. To be successful in conservation one must understand how contamination of hydraulic fluids occurs, and what procedures are necessary to extend or maximize fluid life.

VALUE ANALYSIS

Before choosing a hydraulic fluid from an energy conservation standpoint, there are several important considerations that must be reviewed and analyzed before the proper fluid choice can be made. Proper evaluation is actually mandatory in order to provide your facility or process with trouble-free fluid conservation procedures necessary for successful operation during the 1980's and beyond.

1. Determine what fluids are available for your hydraulic components that meet the hydraulic fluid specification requirements.
2. An economic impact study should be made for each fluid as follows:
   a) Initial cost
   b) Leakage rate
   c) Reclamation alternatives
   d) Disposal cost
   e) Health & Safety requirements
   f) Maintenance Procedures
   g) Availability of Fluid

It now becomes quite apparent that it's no easy task to select a fluid that fulfills all of the above requirements and still conserve energy along with meeting the new health and safety requirements.

There are several options open in determining what is the best procedure to follow in fluid conservation.

The first and most obvious is to select a premium grade hydraulic fluid, minimize leakage, use conditioning equipment to keep the fluid clean (using filters, magnets, centrifuges and even reclamation equipment). This procedure can be quite feasible if you have the proper maintenance procedures and the system design enables you to operate for optimum fluid conservation.

However you may find in your analysis that you have hydraulic systems with excessive leakage, inferior maintenance, or poor system design that allows high levels of contamination, etc. Here again there are several choices. For the long term it certainly could be economically feasible to clean up the act, rework the system and bring it up to operating standards that provide for optimum fluid conservation.

The other alternative could be to operate with lowest possible cost fluid, and then try to condition or reclaim the fluid by using conditioning equipment such as auxiliary filters or centrifuges. When the volume of hydraulic fluid used becomes substantial, a re-refining process by a reputable refiner might be in order.

In the final analysis the overall economic realities of system operation (including fluid cost), disposal cost, and health and safety limitations may be of such magnitude that a high water based hydraulic fluid should be considered. These new fluids are certainly economical, safe, and are quite easy to dispose of.

Last but not least, high water based fluids certainly fill the bill on energy conservation. These "HWB" fluids contain only about 5% volume of petroleum based ingredients; the remainder is water.

Another possible fluid choice might be an invert emulsion which again conserves energy. Inverts are usually 60% petroleum by volume and the remainder is water.

Actually some of the water based fluids usually designated as "FR" fluids should actually be identified as "EC" fluids (energy conservation fluids). Both HWBF fluids and invert emulsions indeed conserve our valuable resources. They are also safe and economical to use.

PROPER FLUID SELECTION

If you've done your homework on value analysis for fluid conservation, you'll find that the choice of fluid has already been made for you. The economic facts of your particular situation have determined what family of hydraulic fluids are best suited for you, namely petroleum based, HWB fluids, invert emulsions, etc.

The important factor to consider here is to understand the maintenance procedures for each family of fluid. Your fluid supplier can be of invaluable assistance in determining if any component changes should be made, such as better inlet conditions, pressurized inlets, larger heat exchangers, improved filtration, different types of seals, higher quality fittings and hoses, and capital investment in reclamation equipment.

STORAGE AND HANDLING OF FLUIDS

Hydraulic fluid can be contaminated before it is even added to a hydraulic system, causing problems from the start. Improper storage of fresh hydraulic fluid either outside or inside the plant, where the contaminanats can collect on the exterior of drums, can result in harmful dirt and other contaminants being introduced.

Drums stored outside should be stored on their sides instead of on their bases. If this is impossible, shims can be used to tilt the drums with the bungs in a horizontal line so water will drain away from them, rather than puddle around the bungs.

Storing fluid drums on their sides, with the bungs horizontal, will also keep moisture and water from being sucked into the drums. (This can be brought about by breathing because the contents of the drum expand and contract with changes of temperature.)

Moisture in a drum not only contaminates the fluid, but also forms rust on the drum interior in the air space above the fluid level. The resulting rust can flake off into the fluid. For this and other reasons it is advisable to use an auxiliary filtering unit to pump all fresh hydraulic fluid into your hydraulic equipment. Let's look at the overall subject of contamination control of hydraulic fluids.

CONTAMINATION CONTROL

It was stated earlier in this paper that contamination of the hydraulic fluid and system components is the chief source of costly downtime and repairs on most hydraulic equipment. For this important reason a positive system of contamination control for hydraulic fluids and system components is mandatory. Actually controlling contamination is the heart of any fluid conservation program for hydraulic fluids.

Up to this point in the program the proper hydraulic fluid was selected and stored safely to make sure that only clean, fresh fluid is readily available.

Let's look at the different sources of fluid contamination and see what procedures should be followed to either eliminate or sharply reduce the chances of fluid and system contamination.

There are three major sources of hydraulic fluid contamination. They can be broken down into (1) external contaminants which are related directly to the plant environmnet, (2) internal contaminants which are generated by the hydraulic system itself, and (3) contaminants introduced during system repair.

EXTERNAL SOURCES OF CONTAMINATION

A good place to start is the initial filtering or adding of fresh hydraulic oil. It is a wise investment to use an auxiliary filtering and pumping unit to add all fresh oil to the hydraulic system as indicated earlier. If a system is using electrohydraulic servo valves, then the "extra clean" fluid requirements are positively a necessity. For these hydraulic components the filtering and pumping unit should have a filtration capability of 10 microns or less in order to insure the cleanliness that is required for servo systems.

Even when adding small amounts of fluid, using some type of secondary containers can be dangerous. Dirty containers that may have housed such items as oil absorbent, lubricating oils, or solvents will also contaminate fresh hydraulic fluid. Other containers such as fill cans, collecting tanks or lubrication carts which may hold hydraulic fluid are not sufficiently clean to permit adding of those fluids directly to a hydraulic reservoir. All fluids should be filtered first from any kind of container before being added to a system, usually by means of an auxiliary filter and pumping unit.

Another common and costly source of contamination is the unintentional use of drained oil and spent fluids. Never store used fluids of any type in the same colored barrels as fresh hydraulic fluid. Improper identification of drums in storage can also result in the use of an incorrect fluid. It is quite easy and very costly to contaminate an entire system through the addition of either drain oil or an incorrect fluid. This occurence can be avoided by marking drums clearly and keeping other oils and liquids away from fresh hydraulic fluids.

Other external sources of contamination can be the result of improper equipment design or poor planning. If your equipment is operating under unusual environmental conditions, make sure the hydraulic system is properly protected from these sources of contamination. A good example is a plastic regrind machine which is generating large amounts of plastic fines and dust. This type of equipment should be installed away from the hydraulic components of injection molding equipment. Too often rapid fluid contamination is discovered only by rapid hydraulic component wear, and then it is too late.

Other external sources of contamination are foreign particles that are allowed to enter hydraulic systems through loose reservoir covers, through an open fill hole where the cover is left off, or when protective screens or inlets are either broken or missing. Dirty air breathers can also be a source of external contaminants.

Now let's look at the possible sources of internal contamination.

INTERNAL SOURCES OF CONTAMINANTS

When a new hydraulic system is started up, there can be a diverse array of foreign particles present in the hydraulic circuit. Examples are: core sand dislodged from castings, metal fines and particles from pumps, valves and fittings, particles from packings and seals, lint, moisture, rust, pipe joint compound, paint, etc. During the initial run-in period all intake filters, strainers, and full flow in-line filters should be kept clean and checked periodically to make sure they are not plugged or loaded, and therefore by-passing the contaminants through the new hydraulic circuit components, causing sticking of pressure and flow control valves. Spools don't shift properly and pump wear is also accelerated.

Even extremely fine particles that are actually smaller than the clearance of any hydraulic components can cause problems. These fine particles can accumulate and packup (sometimes called silting) to cause erratic movement of controls which can be a source of considerable downtime. This hydraulic control problem due to sticking of valves can occur especially on machine tools furnished with servo systems. On this type of equipment it is important to keep all of the fluid conditioning equipment (in-line filters, return line filters and auxiliary filters) in the Class IV range of fluid cleanliness.

Another harmful side affect of contamination is poor performance of even the best of hydraulic fluids. Contamination by foreign and metal particles increases the heat, friction, and even the oxidation rate of hydraulic fluids. Contaminants don't allow the fluid to reach the critical spots in your hydraulic system, such as valves and servo systems. Contamination can also raise the system operating temperatures, especially if accumulated particulate form sludge, or varnish deposits from oxidized oil are present in the hydraulic system reservoir. Last but not least, contaminants bring about accelerated pump wear.

For the above reasons auxiliary filtration with an extra clean capability of 10 microns or less is advisable. The filter elements ahead of the servo circuit should be kept clean. During the initial run-in period these filters may have to be checked and changed quite frequently. Under normal operating conditions servo filters should be cleaned every 500 hours or every 60 days.

On conventional high pressure hydraulic systems filters should be cleaned or replaced every 1,000 hours or when the by-passing indicator shows that the filter needs cleaning.

Another source of internal contamination is sludge and oxidation deposits resulting from the oxidation of hydraulic oil. The resulting gummy residue on valves and spools is one of the leading contributors to poor operations of hydraulic equipment. Let's discuss in some detail the causes of this eventual breakdown of hydraulic oil.

These deposits usually occur because systems have been run too hot. Some of the causes for this are dirty or plugged heat exchangers, inadequate cooling capacity, or letting reservoir fluid levels run too low. Heat exchanger water turned off completely can also result in accelerating overheating and oxidation of the hydraulic fluid.

An important fact to remember is that the oxidation rate of hydraulic oil doubles for every 18 degree rise in temperature. There is four times more likelihood of sludge and varnish developing when oil is operated at 140°F. than when the system is operating at 104°F.

Foam is another internal "contaminant" in hydraulic fluid. It signifies that air is entrained in the oil, and compressible air in oil causes erratic response during cycling. Further, the lubricating properties of the oil are reduced by the amount of air present, and accelerated pump wear can be brought on by foaming.

Foaming or aeration of hydraulic oil generally occurs for one of the following reasons:

1. Oil is too thick
2. Operating temperature too low
3. Oil level is too low
4. Return line is above the oil in the pump
5. Contamination with another fluid such as solvent
6. Filters are clogged or not sealed properly
7. There are bad seals on the suction or pressure side of the pump
8. Restriction in the inlet piping.

Besides visually noticing the foaming taking place in the hydraulic fluid, aeration can be detected by its characteristic sound, as if the pump were pumping marbles, or a stick was being moved along a picket fence.

CONTAMINANTS INTRODUCED DURING HYDRAULIC SYSTEM REPAIR

Here are some helpful hints that can keep you from introducing needless contaminants into your hydraulic system during repair. (1)

1. If you have just had a pump breakdown, make sure that the system is cleaned and flushed properly before a new pump or other component is installed, and before fresh, clean hydraulic fluid is added to the system.
2. All openings in the reservoir, including filter housings and filter caps, should be kept sealed or covered.
3. If any welding or grinding has to be done on any system component, make sure this work is performed away from hydraulic system components.
4. If you must use shop air for cleaning fittings and other system components, make sure the air is clean and dry.

5. Check all cylinders, sub-assemblies, pipe and tube ends, to make sure that burrs, dirt or scale are not present. Check all pipe and tube ends for any restrictions that can cause problems later.
6. When using teflon tape or compound on pipe threads, always leave the first inside thread bare.
7. Last but not least, if you have introduced water or some other fluid contaminants into your hydraulic system, make sure that all of the improper fluid has been completely flushed and drained.

It is a good policy to call your fluid supplier for the proper techniques and required system cleaners to use in order to properly purge and clean your hydraulic system.

LEARNING FROM YOUR REPAIRS AND BREAKDOWNS

In life we usually learn from our mistakes. This analogy somehow doesn't seem to apply in hydraulic component repairs and maintenance. Why do hydraulic equipment users continue to repair the same type of breakdown over and over without determining how the costly problem can be eliminated, and then taking the proper action to prevent it?

When a failure occurs in a hydraulic system, such as pump failures that show up as accelerated ring wear, shaft failure, bearing failure, cocked vanes, etc., there is usually a very logical and reasonable cause for the failure. You will note that many of these breakdowns are directly related to improper handling or treatment of the hydraulic fluid, poor contamination control and improper shop repair practices.(1)

One of the most common causes of pump failurs is cavitation. Cavitation occurs when the components involved in the hydraulid circuit don't completely fill with fluid. This can be caused by poor inlet conditions due to a restricted pump inlet, sharp bends in the pump intake piping, clogged intake strainers. The pump inlet can be too high above the fluid level making it difficult to draw enough fluid to fully prime the pump. The viscosity of the hydraulic fluid may be too heavy, or the operating temperature too low, causing the fluid to be too heavy to pump.

In brief, the main causes of cavitation are (1) poor inlet conditions, which may or may not be designed into the sytem, or (2) the fluid viscosity is not right -- either too heavy or too cold.

What is actually happening in this type of failure, is that the system pressure has been reduced to the vapor pressure of the fluid. In essence you are pumping a vacuum. The resulting explosions of the voids present in the fluid brings on a violent reaction which literally tears metal particules out of the pump components.(2)

SYSTEM DESIGN CONSIDERATIONS

There are many ways to conserve or extend the life of hydraulic fluid through good system design.

Let's start out with the system reservoir. Improper design can cause the fluid to overheat. First, the fluid capacity of the reservoir in gallons should be three times the pump capacity, or rating in gallons per minute. If you're using "HWB" hydraulic fluids the reservoir capacity need only be twice the pump capacity in gallons.

Reservoirs should be designed so that the drain plug is at the lowest point, with the drain plug fitting flush with the inside so that all the oil can be properly drained. Good reservoir design should provide baffles between the pump inlet and return lines. Make sure the return line is not above the fluid level, since this will bring about excessive foaming. Most reservoirs are vented in order to maintain atmospheric pressure; this provides the pressure necessary to force the fluid into the pump. Air breathers should be cleaned regularly. In very contaminated air environments it may become necessary to elevate the breathers sufficiently to prevent the drawing in of dirty shop air.

On machine tools, where coolants are being used, special care should be taken to separate the hydraulic fluid reservoir from the coolant tank. Sealing out cutting fluids and coolant from the reservoir area is imperative.

Another design consideration that helps control contamination and leakage are exclusion seals and devices such as cylinder rod wipers, scrapers, and boots that keep out both solid and liquid contaminants. These devices can prevent contaminant particles in the fluid from being carried into a system by a retracting cylinder rod and then becoming trapped between the seal and reciprocating member. To make matters worse these particles, when abrasive, can actually gouge leakage paths in the rod or cylinder wall.

The design and shape of the seal is also a factor in controlling the exclusion or flushing of contaminants from dynamic contact areas. Square section seals have the ability to wipe off particles before they arrive at the sealing area. Thinner lip seal design provides an opportunity for the fluid to flush out any contaminants during the low pressure stroke.

Another factor is the actual finish of the wearing surface on the rod and cylinder wiping areas. Recent studies have shown that dynamic surface finishes that interface with the seals can become too smooth. Dynamic surfaces below 10 RMS can actually increase friction and wear and shorten seal life.

During the design stages of your hydraulic system, it's very helpful to determine what requirements you may need in reservoir design, seal design, the type of sealing compounds, exclusion devices, and the proper finish of the dynamic metal surfaces. Your seal supplier along with your fluid supplier can provide guidance and the necessary technical data before the system is fully engineered and built. (4)

Last but not least, make sure you have sufficient cooling or heat exchange capacity to maintain an operating temperature of 120°F. or less. Proper operating temperature is imperative in reducing the rate of oxidation and the build up of damaging varnishes, acids, and residue related to it. In all applications during winter months you may require heating coils to maintain proper start-up temperature, especially when fire-resistant fluids are being used.

RECLAMATION PROCEDURES

In any conservation program for hydraulic fluids there are certain conditioning fluid methods that can be employed to minimize fluid waste. For the sake of simplicity let's separate the methods used to purify the fluid into conditioning equipment that can be used (1) to reclaim fluid outside the system, and (2) equipment that operates while the system runs. In the latter methods the fluid is cleaned while the system is in operation, using micronic filtration, magnetic separation and auxiliary filtration. These techniques can remove metal and foreign particles only and generally cannot remove water, oxidation derivatives, solvents, etc.,

The conditioning methods for purifying or reclaiming hydraulic fluids outside of the hydraulic system are settling tanks, activated earth filtration, centrifuging, and reclaiming or re-refining.

Let's first discuss the merits of "in system" filtration. Micronic filtration is now a way of life. Hydraulic equipment manufacturers will generally provide your components with sufficient micronic filtering capacity, providing that you give them the necessary information pertaining to environmental conditions, type of filters you desire and what grade of cleanliness is required. The use of auxiliary filters or pumping units was discussed earlier in this paper. A typical auxiliary filtration unit is shown in Figure 1.

Figure 1.
Typical Auxiliary Micronic Filtration Unit
(Courtesy Marvel Engineering Co.)

These units are extremely helpful in providing micronically clean fluid when adding fresh fluid to a system. They also can be used to aid or accelerate the filtration of fluid while the system is in operation. This type of unit can be extremely helpful in cleaning up a system after a major breakdown. These auxiliary filters can provide accelerated filtration on the start-up of new systems by removing the unique type of foreign particles that are generally present in new hydraulic systems.

Now let's discuss the fluid conditioning techniques that are performed outside of the hydraulic system, namely settling, activated earth filtration, centrifuging, and vacuum distillation.

Settling tanks are an economical and easy way to drop out large foreign particles. The bottom portion of the settling is drawn off and disposed of. The remainder of the tank can be filtered by means of an auxiliary filter or centrifuge to further purify the fluid. It must be stressed that settling tanks won't settle small particles, and they definitely do not remove water effectively from hydraulic fluid. A centrifuge is one of the best ways to remove water from hydraulic fluid.

The centrifuge can remove both solid and liquid contaminants, including water, from hydraulic fluids. Centrifuges are accelerated means of gravity settling. They create forces much greater than gravity, thereby separating contaminants from hydraulic fluid, quickly, continuously, and completely.

Centrifuges operate automatically with very little maintenance required, and no filter elements are used in this process, hence no changing of filters. Centrifuges normally remove all solid particles above one micrometre in size providing they have a specific gravity greater than the fluid being centrifuged.

A typical medium sized automatic centrifuge can be seen below in Figure 2. (Centrifuges are available in capacities of 5 gallons to 1,000 gallons per hour).

Figure 2.
Medium Sized Automatic Centrifuge Separator
Courtesy of Alfa-Laval, Inc.

There is another more sophisticated method of filtration that can be performed on fluid outside of the hydraulic system, and this involves the use of activated clay or fullers earth filters. In this method the filtering media absorbs the residual contaminants derived from oxidation, and also screens out fine foreign particles. The fullers earth filter may be of the cartridge type, or in some instances the clay is mixed with oil and then removed from suspension by micronic filtration. A word of warning in using this technique: make sure you contact your fluid supplier to find out what oil additives can and should be replaced.

Another method of reclaiming hydraulic fluid is vacuum distillation. The advantages of vacuum distilling are settling, filtration, plus the ability of this process to eliminate both water and solvents from the hydraulic fluid. Large volume users of hydraulic oil have turned to this process in order to sharply reduce their oil consumption.

There are two types of reclamation units available, either a batch unit or one that is designed for continuous operation.

In a batch unit, the dirty oil is mixed with an activated clay. This mixture is then drawn by vacuum into an electrically heated chamber. Under these conditions of vacuum and high temperature, all moisture and solvents are removed. The high temperatures also serve as a catalyst to aid the reaction between the clay and the oil. After a sufficient digestion period, the warm oil is then filtered through a filter press containing both cloth and paper filters. The vapors which were drawn off in the vacuum chamber are condensed and collected in an auxiliary tank. The filtered oil is collected and appropriate additives are mixed in to bring the oil back to the required additive level.

While many plants have installed their own reclamation facilities, there are others who send their oil to local facilities who specialize in reclamation work. The cost per gallon may be higher than doing it yourself, but it is still much cheaper than buying new oil. (3)

LEAKAGE CONTROL

The leakage of hydraulic fluid is somewhat like death and taxes. It will never go away or be completely eliminated. The only exception to this statement on leakage is the aircraft and defense industry, where their ultimate goal is <u>zero leakage</u> to provide for reliability and repeatability. One fluid power executive has a unique way of introducing the importance of controlling leakage in his talks at seminars and workshops: "If you believe in leakage don't fly home from your next business trip or vacation."

Why isn't leakage control higher on the priority list of both management and operating personnel involved with fluid power equipment? The answer to this question could very well be that someone is not doing his homework.

The conservation of hydraulic fluid is tied directly to the control of leakage. There are two sobering statistics that illustrate how important leakage really is. First, it's a known fact that approximately 300,000,000 million gallons of hydraulic oil are used each year in the U.S. The disturbing point is that this figure is over four times the capacity of all the hydraulic systems in this country. This simply means that we consume over four times the capacity of each hydraulic system by either poor contamination control or leakage.

The next point is even more disturbing; where does all of the drain oil go? You will note from the figure (3) that it has been estimated that over 50% of all the waste hydraulic oil is simply dumped; most of the rest is burned.

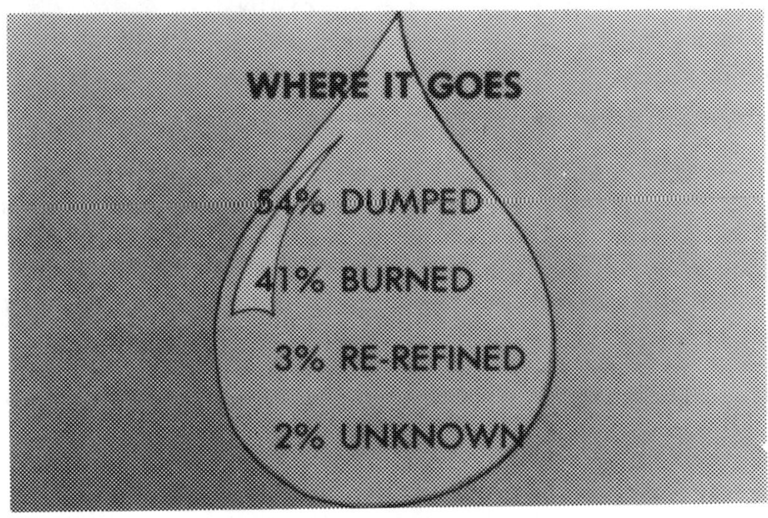

Figure 3.
Breakdown of Hydraulic Fluid Consumption
Courtesy Rexnord, Inc.

The dumping problem has now reached a new degree of importance. Government, state and local agencies will simply not permit this indiscriminate dumping of hydraulic fluid to continue. We have already polluted both surface and water areas with chemical wastes and toxic substances, so why add hydraulic fluids to the list?

Remember earlier we discussed value analysis as a method of choosing your hydraulic fluid. The cost of disposal plus the labor involved becomes an operating cost that is directly attributed to leakage.

The next cost is the actual cost of the lost hydraulic fluid. One drop of fluid lost every five seconds equals a consumption of approximately eighty gallons per year. A drop every second amounts to about 400 gallons per year. This may be only at one fitting. What if you had three or four leaky joints on the same system, which is not uncommon? This could translate to about $180.00 a year for a leak of one drop every 5 seconds, or $900.00 per year for a leak of one drop every second. (The basis for this cost is premium anti-wear hydraulic oil @ $2.25 a gallon - a very realistic figure for the early 1980's, and this is bound to go even higher.) (3)

The cost of leakage is still higher when using fire-resistant fluids such as water glycols and phosphate esters. Acute leakage of fire resistant fluids is a good reason to consider the feasibility of using invert emulsions, and high water based fluids.

Let's investigate the cause for leaks. There are actually three main areas: seals, elastomers, and poor system maintenance procedures.

In selecting fluid system seals and elastomers the first consideration should be fluid compatibility. Some fluids will cause seals to swell excessively, causing excessive friction and wear. Other fluids may cause physical breakdown of the compound used to manufacture the seals, and others may harden the material to a degree where sufficient resilience is no longer present to provide sufficient sealing.

When ordering seals make sure your supplier is aware of the precise application and environment that the seals will undergo; namely, temperature, stroke, frequency, and pressure. All these points have to be considered in choosing the right sealing compound for a particular fluid.

Enough is not said about how high temperature can and does affect seal life, and therefore brings about leakage. In choosing seals the heat compatibility of the sealing compound may be the overriding factor.

Heat resistance is a time-temperature function. The higher the temperature seen by the seal, the shorter life the seal will have. With a given temperature, most sealing compounds will expand or contract about 10 times as much as the metal materials they seal. But the more serious heat problems are related to resilience and hardness. High heat tends to diminish the resilience and increase the hardness. At the lowest temperatures expected seal materials must remain resilient.

Most seal compound compatibility charts also give maximum and minimum temperature ranges within which the seal material can function. It should be noted that this temperature range is not always applicable for all fluids. Fluids containing water are generally very harsh on most seal polymers, with the exception of ethylene propylene compounds. Generally, a significantly reduced temperature is recommended for water base fluids with many compounds.(4)

Remember we discussed temperature control; poor temperature control contributes to leaks and contamination. Seals and gaskets deteriorate rapidly at excessively high temperatures. They become embrittled and portions of the sealing material are eroded into the fluid stream. Ultimately oil or fluid leaks past the damaged seals, resulting in reduced system performance, and bringing about even higher operating temperature due to the increased friction, causing even more leaks, etc.

Hose failures with catastrophic leaks can also be attributed to high temperature.

Poor maintenance practices also contribute to leakage. Careless or improper installation procedures can promote leaks on dynamic joints when rods and shafts have been abused by contamination, or the finishes roughed up or scarred by sloppy handling.

On static joints the same type of problems exists: poor quality of threads, the wrong type of pipe sealant, too much tightening of fittings, loose fittings due to improper installation, abrasion failure on hoses when a rotating member rubs against a hose, twisted hoses, or fittings and hoses that are not properly clamped to resist system vibration. All of these poor maintenance procedures contribute greatly to leakage.

Most hydraulic component suppliers conduct training sessions on proper use of fittings, hoses, seals and sealants. Sending maintenance people for this type of "Hands on" training is a wise and profitable investment. (Many maintenance people haven't been given the opportunity to learn the proper techniques for maintaining fluid power components.)

FLUID MAINTENANCE PROCEDURES

(1) Select the viscosity and type of hydraulic fluid recommended by the component and hydraulic equipment manufacturer. (Remember this may include fluids which aren't necessarily petroleum based.)

(2) Be sure that fluid is clean to the degree required by the component or equipment manufacturer. On certain machine tools and other types of production equipment with servo systems, the filtration requirements are generally 10 micrometres (Microns) absolute or less. This means that when adding fresh hydraulic fluid in any quantity, the fluid must be filtered by some auxiliary means to the degree recommended for the equipment. This same procedure of extra clean or ultra-fine filtration follows for systems with electrically modulated hydraulic valves.

The same filtration applies for fluid being transferred from holding tanks, lubrication carts, and partially opened barrels of fluids.

(3) Temperature control can be a very effective way of increasing fluid life. The operating temperature should be held to from 100°F - 120°F. for best results if at all possible. Heat exchangers should be periodically cleaned to make sure they are functioning properly.

(4) Check fluid condition for both foreign particle contamination and chemical condition every 90 to 120 days. On systems requiring ultra-clean filtration a contamination check may be required every 60 days or less, depending on the hours of operation.

If your hydraulic system has to be located in an environment where dust or fine foreign particles are encountered, make sure all hydraulic components are fully protected against this kind of contamination. This also means that any special covers that may be installed should always be kept in place and not removed. Fluids can be easily contaminated by loose covers or dirty breathers.

(5) Change or filter oil when inspection and/or analysis shows it is necessary. Good maintenance procedures can be rewarding; changing oil on a chronological basis is wasteful and unnecessary.

Many systems have run and can run with the same hydraulic fluid for extended periods of time, as long as good maintenance procedures are being followed to help extend the fluid life. Actually avoiding fluid changeovers saves money, energy, and it also reduces the possibility of introducing contaminants into the machine's hydraulic system. It is not uncommon to obtain fluid life of from 10,000 - 20,000 operating hours or more through the aid of fluid conservation measures.

Periodic inspection and tests of hydraulic fluid on a regular schedule are vital parts of any effective fluid conservation program. It is extremely important to take samples for inspection. Laboratory analysis is often necessary to determine whether hydraulic fluid is safe for continued use. Sampling should be done every three to six months, depending upon the operation, unless laboratory inspections indicate otherwise.

Sight glasses should be observed daily, and special checks of the fluid should be made after any of the following occurrences:

(1) After starting up a new piece of equipment, it's wise to sample hydraulic fluid for the various types of contaminants which could have been introduced during the run-in period. Usually these contaminants can be cleaned up through the use of auxiliary filtration and through the use of magnets to pick up ferrous metal particles. Remember, if you are operating a servo system, or electrically modulated valves, then extra clean filtration is necessary, especially during the start-up period. Make sure you follow the equipment and component manufacturers' filtering recommendations.

(2) Whenever any appreciable amount of water contaminates the oil, the sight glass will show a light brown or cloudy yellow color. The oil should then be checked immediately when these signs are observed, and the source of moisture investigated. (Remember a centrifuge separator can help you reclaim oil contaminated with water.)

(3) If a particular system has been operated at an excessively high temperature for an extended period of time, an oil sample should be taken and tested in the laboratory to discover whether excessive oxidation has depleted the inhibitors, making the oil unsafe for continued use.

(4) Fluid should be checked for metal particles or other related contamination following a major breakdown of hydraulic component. This is especially true if pump failure occurs. All this investigation should take place <u>before</u> the system is started up again, to make sure the new or rebuilt pump is not affected by contaminants from the previous breakdown.

All of the above procedures concerned the type of contaminants that are harmful to the properties of hydraulic fluid. For this reason, it is important to recognize the entrance of contaminants that can appreciably shorten fluid life, and remove them before any permanent damage occurs. Remember, it isn't necessary to change fluid simply because it is six months or a year old, as explained earlier.

The last important step in any hydraulic protective maintenance program is to keep maintenance records. If at all possible, each piece of hydraulic equipment should have its own record, showing the following data:

1. Complete equipment description, including location of machine.
2. Type and grade of hydraulic fluid required, including reservoir capacity.
3. Records and dates of all protective maintenance done on the hydraulic fluid, including such things as dates of auxiliary filtration, cleaning of filter elements and magnets, results of laboratory analysis, and actions taken.
4. Materials used to clean and flush system. Include dates of fluid change.
5. If any reclamation work has been performed on the fluid, such as centrifuging, filtering thru fuller's earth, or the fluid was subjected to re-refining, make sure all of the pertinent data is noted, such as viscosity, fluid condition, acid number, and particle count. If the fluid has been filtered with fuller's earth or a similar process - the fluid condition should show if the anti-wear agents and other additive properties are still present in the filtered fluid.
6. Full information on replacement or repair of hydraulic components.

This paper has been written to bring into focus the crying need for education, training and orientation of shop personnel to current maintenance requirements of molding machines. On the other hand management should understand the productivity and profit that will be gained by the use of the conservation procedures and expertise outlined in this paper.

BIBLIOGRAPHY

1. "Identifying Causes of Hydraulic Component Failures" Rexnord Hydraulic Components Division Form S-121
2. Sperry Vickers - "Hydraulic Hints - Diagnosis and Correction of Aeration and Cavitation" Bulletin MH-15 4-1-69
3. "The Case of the Costly Drip" Paul Schact Rexnord, Inc. 4-76
4. "Leakage and Seals" Ralph E. Peterson Parker Packing Co.

Lubrication System Aids Equipment Uptime

"I would estimate that 80 percent of machine downtime is caused by a lack of lubrication . . . "

The above quote by Arthur Lorenzen, Ford facilities maintenance engineer, shows the importance of a lubrication system. Consequently, when the Ford Motor Co. transaxle plant was being built in Batavia, OH, every piece of equipment having more than six lubrication points was specified with an automatic lubrication system.

The plant uses Trabon lubrication systems and monitor/control panels by the Lubriquip Div., Houdaille Industries, Cleveland.

Machine tool lubrication systems are of the series-progressive type, based on divider valve manifolds that positively displace metered amounts of lubricant. Most of the systems are controlled by Trabon maxi-monitors, discrete programmable units that provide the status of lubrication cycling at a glance. Machines with lubrication cycle intervals of a week or longer are controlled by electromechanical panels. The lube cycles are based on elapsed time of machine operation.

"We specified the lubrication control panels be installed to shut machines down after completion of the full cycle, when a lubrication fault occurs," Lorenzen explains.

Each maxi-monitor panel contains LEDs that indicate one of the following conditions: 1) the last lubrication cycle was completed successfully and things are working properly; 2) a lubrication cycle is taking place; and 3) the last lube cycle was not properly completed.

In addition to these standard indicators, Ford specifications called for high-pressure switches and low reservoir level-sensing switches.

The master divider valve manifolds are fitted with performance indicators. When pressure builds above a preset point in a lubrication line, a disc ruptures causing a pin to protrude from that performance indicator.

Both terminating and circulating types of lubrication systems supply metered amounts of lubricant to slides, ways, pillow blocks and spindles. Circulating oil systems lubricate such things as drive gearboxes.

Lubrication of free-standing equipment is achieved basically by a control panel, reservoir, pump, master divider valve and secondary divider valves that supply lubricant via steel lines.

Transfer Line System. Automatic lubrication of large transfer lines requires a more sophisticated approach, especially when more than 1000 lubrication points must be monitored on one machine. On these applications, Ford specified a posiflex system consisting of a conventional series-progressive, master-secondary divider valve system at each workstation fed by a single line reversing loop system. This system utilizes the lube lines themselves to convey information about each lube point back to the central panel.

The series-progressive nature of both types of systems facilitates troubleshooting. Divider valve manifolds in the conventional and loop systems are ported so that each piston must complete its full cycle before lubricant is directed to the next piston for further divisioning.

If a line blockage or tight bearing raises line pressure to the point that the piston cannot cycle, this is reflected by a stoppage of cycling at the master divider valve manifold in the conventional system or at the reverser valve of the loop system. This condition is sensed by a microswitch which causes the monitoring panel to shut the equipment down or sound an alarm.

The closed-loop system feeding transfer line workstations has positive displacement-type divider valves with a single piston feeding each workstation's series-progressive system. As oil is pumped in one direction through the loop, each piston meters some lubricant to the workstation and then permits the lubricant to flow to the next divider valve. When all of the valves on the loop have received their lube, a reversing valve directs the lube through the loop in the opposite direction. A single valve delivers both shots of lubricant to the same workstation and a twin valve dispenses lubricant to two workstations.

With this design, if a single lube point becomes blocked, its secondary feeder will build pressure and stop cycling. This in turn produces a signal which can be used to shut the line down.

Lorenzen explained, "I like the maxi-monitor type of lube system because it tells you when you have low lube or a blocked line, and you know where to look for the problem. We can go down the line, master block by master block, tracking the fault, hopefully correcting lube faults before they result in equipment downtime." □

CHAPTER 5

DISPOSAL OF SPENT FLUID SYSTEMS

Presented at SME's 1974 International Engineering Conference, April 1974

The Influence Of Safety, Environmental Controls And The Availability Of Raw Materials On New Cutting And Grinding Fluids

By George E. Barker
VanStraaten Chemical Company

The growing demand for cleaner air, water, working conditions and the emergence of OSHA has created a new environment for the development, in-plant use, and disposal of cutting and grinding fluids. The demands require more careful selection of materials for new products as well as the reformulation of many existing products.

More importantly these demands have necessitated compromises for the manufacturer as well as the user. Unfortunately, some are not always well received, i.e., elimination of phenolic germicides and the resulting slime problems.

There are many factors, which influence the characteristics and composition of cutting and grinding fluids in today's production facilities. Some of these factors are even overriding the importance of the influence of the cutting and grinding fluids on tool life and wheel wear. They are of vital importance to the user and present problems to the new product development chemist and to the manufacturer of metalworking fluids. Although safety was always a prime consideration among the manufacturers of quality cutting and grinding fluids, it has taken on a new significance and in some instances, perhaps even a more prominent place. The effect of all manufacturing operations on the environment is becoming very apparent in the plans for the development of products as well as in the designs for production.

The scarcity of many raw materials is limiting the breadth of formulation in the hands of the research chemist in developing all types of new chemical products. The development of new cutting and grinding fluids is no exception. It is necessary to be certain that ingredients are selected,which are available widely in the quantities needed, and of a uniform quality. Lack of raw material has also manifested itself in the withdrawal of basic suppliers from some markets.

Today, I am going to take you behind the scenes in order to show some of the restrictions and desireable features which are being incorporated into new cutting and grinding fluids to meet the modern-day requirements in our high productivity society. I shall dwell mostly on those qualities which are required in addition to the very important characteristics of the cutting and grinding fluid in increasing the rate of production and reducing the consumption of tools and wheels.

SAFETY:

Safety is a concern of all responsible manufacturers. It has been a major concern of all quality producers of cutting and grinding fluids. It is rapidly becoming mandatory for all producers of these fluids.

Listed below are the chief properties which affect the safety of cutting and grinding fluids in use:

1. Combustibility:

   a. Flash Point.
   b. Fire hazard.
   c. Explosion hazard.

2. Misting:

   a. Misting of straight oil fluids.
   b. Misting of water-miscible fluids.

3. Smoking.

4. Evolving fumes due to decomposition.

5. Irritating or toxic ingredients.

6. Allergenic ingredients.

7. Contact dermatitis.

1. Combustibility:

A flash point of a fluid is the temperature at which a spark will cause a flash. A continuation of the same test carried out through raising the temperature of the fluid will reach a point where the flash will become a sustained flame and that is called the fire point. It is desireable to have a minimum flash point of 150°F for any cutting and grinding fluid, and in the interest of real safety, the flash point should be above 200°F. In order to be safe, any mist or vapors should not form an explosive mixture with air.

The hazards from flash, fire and explosion are completely avoided by the use of water emulsions or solutions. They are also pretty much avoided by the use of high flash ingredients and certain ingredients which tend to make the fluid non-combustible.

2. Misting:

Misting is caused by either the fluid or the operation or a combination of the two. A fluid can be formulated so as to reduce the misting of a straight oil by as much as 90 to 95%. Developments in this area have been slow and erratic, largely because of the absence of a suitable laboratory test. Grinding operations, and particularly high speed grinding operations, are prone to develop mists, because of the tendency to develop very small drops of the fluid and disperse them in air. It is undesireable since it is particularly irritating to the nose and throat. Even plain water will irritate the nose and throat, and sometimes the eyes, if it is in a very finely divided mist. If the mist is from straight oils, it increases the fire hazard because finely divided organic particles, even with a very high flash point, may become combustible or even explosive if mixed with the proper proportion of air. An excessive inhalation of petroleum oils, for example, may cause lung irritation, consequently, there are definite limits placed on the concentration of petroleum mist which may be contained in air. The most recent figures for the limits on this are 5 milligrams per cubic meter of air.

3. Smoking:

Although in many people's minds, smoking and misting are closely related, there is a distinct difference. Smoking is generally caused by a partial decomposition or oxidation of organic matter at elevated temperatures. The smoke may contain carbon particles or other particles of organic matter. Smoke is annoying, disagreeable, and can be detrimental to health if it is inhaled in sufficient quantities. Smoking of an oil composition used for cutting or grinding can be reduced by cooling the oil so that the temperature, which it reaches on the cutting and grinding operation, is not sufficient to produce a smoke. The smoking is also reduced by the proper formulation of the oil system for the particular type of operations and conditions of temperature which are reached during the cutting or grinding operation.

4. Evolving Fumes Due to Decomposition:

It is possible that fumes or vapors may be evolved during a cutting or grinding operation due to decomposition of some of the ingredients in the fluids. These fumes may or may not be toxic, but it is necessary to be certain the ingredients used in the fluid are stable under the conditions of use so that no toxic gases or vapors are produced. None of the fumes from present fluids are believed to be toxic.

5. Irritating or Toxic Ingredients:

It is essential to avoid the use of irritating or toxic ingredients in the fluid. The developer of fluids must be very cautious in the selection of ingredients used in the formulation of his new fluids. These are not easy decisions. Toxicity is relative. A chemical which is toxic when inhaled may not be toxic when ingested. Likewise, a chemical which is toxic when ingested may not be toxic when applied to the skin.

No attempt will be made to list all of the chemicals which are toxic, and which might be used in cutting and grinding fluids, but some of them are listed. First, there should be none of the toxic heavy metal ingredients such as, lead compounds, chromates, mercury compounds, beryllium compounds, etc.. Recently a list of carcinogenic compounds has been released by the Department of Labor under the Occupational Safety and Health Standards Act as amended May 3, 1973. This list is given below:

2-Acetylaminofluorene
4-Aminodiphenyl
Benzidine (and its salts)
3,3'-Dichlorobenzidine (and its salts)
4-Dimethylaminoazobenzene
alpha-Naphthylamine
beta-Naphthylamine
4-Nitrobiphenyl
N-Nitrosodimethylamine
beta-Propiolactone
Bis-Chloromethyl ether
Methyl Chloromethyl ether
4,4'-Methylene (bis)-2-chloraniline
Ethyleneimine

None of the chemicals in this list should be used in cutting and grinding fluids.

6. Allergenic Materials:

It is possible to find individuals who are allergic to all sorts of common materials, even to toilet soap. There are some types of materials to which a relatively high proportion of the population is allergic. These materials should be avoided. Examples of these materials are Mercaptobenzothiazoles, Dithiocarbamates, some amines, and sometimes formaldehyde.

7. Contact Dermatitis:

Approximately 800,000 cases of skin disease are reported to Workmens Compensation Insurance Companies annually in the United States. Most of these cases result from contact with a chemical substance. A small number of the chemical substances with which workers come into contact are potentially skin sensitizers. Approximately 80% of all cases of occupational contact dermatitis result from contact with primary irritants. It should be said at this point that no responsible manufacturer includes ingredients in his cutting and grinding fluids which are primary skin irritants. The principal dermatitis caused by straight oils is a condition called folliculitis. This is caused by the plugging of the hair follicles in the skin with oil or grease. Oil-soaked clothing will also cause this condition. Occasionally bacteria invade the irritated area, and may result in severe skin infections. Chloracne is another condition which had developed from straight oils in the past. One of the principal causes of chloracne is the use of certain chlorinated naphthalenes and diphenyls. The chlorinated naphthalenes are never used now, and the chlorinated diphenyls have in recent years been disqualified for use in cutting fluids.

The water soluble cutting fluids produce an eczematous dermatitis from their defatting action on the skin. This is quite common, particularly with those people who have a dry skin. This type of dermatitis is especially noticeable in winter, when the skin is dryer and when the worker is exposed to conditions outside of employment, which are conducive to an irritation of the skin.

In order to avoid dermatitis caused by the defatting of the skin, the fluid developer is careful to reach a balance between wetting, detergency and the cleanliness required in the fluid for application in the cutting and grinding process. The very clean fluids, and particularly some of the newer synthetic fluids, are prone to cause this defatting of the skin. Some people are so effected that they cannot work with these very clean products or they must use protective creams or other protection for the skin which will prevent the defatting action.

Dermatitis is reduced drastically when the operators are convinced to practice good personal hygiene. The hands and forearms should be washed at least twice a shift, before eating and at the end of the shift. A mild hand soap should be provided and the operators instructed to rinse well and dry the skin thoroughly. In the cases of those individuals who appear to have sensitive skins, the use of protective creams is sometimes helpful.

In severe cases the dermatologist should determine diagnostically whether the dermatitis is of occupational origin. There are many conditions in the home, in conjunction with sports and the pursuit of hobbies, which cause dermatitis.

ENVIRONMENTAL CONTROLS:

1. Air Pollution:

a. Smoke from straight oils:

As indicated under the discussion of safety, smoke sometimes is generated in the use of straight oils for cutting and grinding. This occurs generally when the stock removal is made at a very high rate so that high temperatures are generated. Even though the ventilation system is such that the concentration of smoke particles in the working environment is not excessive or does not cause any discomfort, the exhaust of this smoke with the air to the atmosphere is not permitted. This is usually prevented from entering the atmosphere by electrostatic precipitaters or a scrubber or some other means which is adequate for removing it. Good practice demands the formulation of oils of a type, which do not smoke excessively under the conditions of use. Some reduction is gained through using better quality products, which offer better friction reductions, less heat, and therefore produce less smoke.

One automotive manufacturer reports they are spending $30.00 per car to ventilate working areas. A reduction in the causes and cost would provide significant savings.

b. Mist from either straight oils or water solubles:

The generation of a mist and its exhaust to the atmosphere presents problems similar to those occurring with the exhausting of smoke to the atmosphere. In the case of straight oils, the mist may consist of small particles of oil as well as the other ingredients in the formulation. As with the smoke, the mist can be removed by electrostatic precipitaters, sometimes by scrubbing, which is particularly effective in removing the mist from water solubles, and by burning with an auxiliary fuel, such as natural gas. Through experimentation and proper formulation, it is possible to reduce mist to a very low concentration so that generally it is acceptable and presents no major problems with air pollution. Naturally, there are some instances where operations generate such turbulence in the air and fluid that the fluid becomes dispersed in the atmosphere to a degree that cannot be prevented without the auxiliary equipment and processes to remove the mist from the air before it enters the atmosphere.

c. Gases, volatile ingredients:

There are some gases and many vapors or volatile ingredients which could be generated during the use of some cutting and grinding fluids, whether water solubles or straight oils. The famous rule 66 which has been established by the City of Los Angeles for the control of atmospheric contaminants is one of the most severe and is being adopted by more and more cities as well as some states. These regulations prohibit the entrance into the atmosphere of most aromatic hydrocarbons, of many chlorinated solvents and many chemicals, which could be produced during the use of cutting and grinding fluids, especially if they become overheated. It is best to avoid the use of these ingredients in cutting and grinding fluids, but if they must be used or if the conditions of employment of the fluid are such as to generate unacceptable byproducts which might enter the atmosphere, it is generally necessary to burn these exhaust vapors, usually with the employment of an auxiliary fuel, such as gas or oil. If oil is used, it must be of low sulfur content and be used under conditions which will lead to complete combustion.

2. Water Pollution:

The principal source of water pollution from cutting and grinding fluids occurs during their disposal after they have served their purpose and are no longer useful. Fluids become inoperable due to the excessive accumulation of dirt, contaminants, and breakdown of some of the ingredients caused by biological or chemical processes. Generally, the straight oils present no particular problem for disposal since they are either burned or are sold to an oil reclaimer. The water soluble types present problems because they may contain ingredients which are not acceptable for entrance into natural bodies of water or in most instances, even into sanitary sewers.

a. Chemical Pollutants which can be derived from metalworking fluids:

(1) Metallic Ions:

There are two sources of metallic ions in used cutting and grinding fluids. One source is the composition of the fluid concentrate itself and the other is the accumulation of metal ions from continued but very slight corrosion of the metals being worked in the particular operation. Most water soluble cutting and grinding fluids contain only alkali metal ions and in particular, only sodium or potassium ions. No quality fluid should contain the heavy metal ions such as, lead, mercury, chromium, nickel, etc.. Two other ions which are not generally recommended for use in metalworking fluids are cadmium and zinc, the cadmium being the most undesireable.

(2) Inorganic anions:

Perhaps the most commonly known inorganic anion to which there is an objection is phosphate. This is not too commonly employed in cutting and grinding fluids but there are some which contain high percentages of phosphate. The nitrate anion is another one which is generally regarded as undesireable in natural bodies of water due to its tendency to promote the growth of algae as does the phosphate. Nitrates are not too commonly used in cutting and grinding fluids. Nitrites are present in many cutting and grinding fluids, but they do not contribute to undesireable properties to the water, since they are readily biodegradable.

(3) Hydrocarbons:

Hydrocarbons are present in most cutting and grinding fluids, because even in those fluids containing no petroleum based ingredient in the fluid itself, they are almost always contaminated with way oils, hydraulic fluids, and other hydrocarbon based lubricants used in lubricating the machine tools.

Not only does tramp oil contribute to problems of disposal of used fluids, it necessitates much better control of concentration, and leads to residue problems, rust problems, and sometimes to the development of disagreeable odors. It usually requires the installation of special equipment such as a centrifuge, to remove the excessive tramp oil in order to avoid a much shortened fluid life.

(4) Phenols and other Biocides:

Another source of pollutants from cutting and grinding fluids is the class of chemicals usually termed biocides. This includes bacteriacides, fungicides, slimicides, etc.. Biocide is a general term which describes all chemical agents, which kill microorganisms, i.e., bacteriacides kill bacteria, fungicides kill fungi and slimicides kill slime forming microorganisms regardless of their type. It should be interjected here that metalworking fluids are often controlled by the use of bacteriostats, which keep the bacteria from reproducing, but do not kill them.
The only class of biocides which are generally banned from natural bodies of water are the phenolic and mercurial biocides. It is the personal opinion of the author that the only reason the other biocides are not banned is because their characteristics and their effects on the receiving bodies of water are not well known. In many instances they may be more detrimental to the natural waters than are the phenolics. Some phenolic derivatives do have the very unappreciated ability to contribute undesireable taste to water, which is used for potable purposes. This, and the fact that phenols are easily detected probably are the main reasons for isolating this class for particular exclusion from water entering the natural environment.

Many times, there are additives used, especially with central systems, to suppress microorganisms growth, and maintain other properties of the system, and these can be sources of pollution. Among the most common additives are biocides, alkalies to adjust the pH and chemicals to maintain rust protection. These additives must be carefully chosen and used in judicious quantities to avoid water pollution, when the used fluids are discarded.

(5) Organic chemicals from the classes of amines, alcohols, acids, esters, and surfactants:

All water miscible cutting and grinding fluids, whether they are the emulsion type, or the synthetic type, contain some or all of the above listed organic chemical ingredients. Some of these are biodegradable, and so can be disposed of in the normal sanitary sewer, while others are not, and therefore, need to be removed before the used fluid is discharged to a natural body of water, or a sanitary sewer. The particular compounds of the above classes must be carefully selected, and the combinations must be chosen so as to present a minimum or problems in the disposal systems which will be discussed.

c. Characteristics influencing the disposal of metal-working fluids:

(1) Biodegradability:

Although every manufacturer produces some biodegradable fluids, most of the fluids on the market are not totally biodegradable. There are both advantages and disadvantages to biodegradability. As pointed out earlier, all used fluids contain some nonbiodegradable contaminants and consequently, these nonbiodegradable contaminants must be removed. It is just as easy to remove the nonbiodegradable ingredients of the fluid as it is the nonbiodegradable contaminants present in it. There is another disadvantage to highly biodegradable products and that is the excessive oxygen demand usually expressed as biological oxygen demand required to digest these ingredients. Some of the sanitary sewers are demanding high fees for the acceptance of biodegradable products which have a high B.O.D. The alternative to this is for the user of the fluids to put in a biological digestive system to treat his fluids before discharging them into a sanitary sewer. This is expensive and requires a considerable amount of space.

(2) Toxicity to microorganisms used in sewage disposal plants:

It is necessary to avoid the incorporation of any materials which tend to interfere with the sludge digestion process in sewage disposal plants, if the fluid is discharged to a sanitary sewer. If it is not discharged to a sanitary sewer, it will probably be necessary for the user to treat it in some way to remove the biodegradable and nonbiodegradable ingredients. The supplier must be careful to avoid the incorporation of biocides which either are not biodegradable at low concentrations or kill the microorganisms used in the sewage digestion process at the concentration which they are received. It is particularly necessary to avoid the introduction of toxic metallic ions, since these can accumulate in the sewage digestion plant and present problems over a period of time. Such metallic ions as indicated earlier are chromates, copper, nickel, lead and zinc salts. Besides interfering with the sewage digestive process, the accumulation of these metals in the sludge from the digesters makes them unacceptable for use as land fill or as fertilizer to be placed on agricultural land.

(3) Ease of Removal of Prohibited Materials:

(a) Insoluble oils (hexane extractables):

Insoluble oils present a special hazard to the disposability of used cutting and grinding fluids. The oils generally are not biodegradable at a reasonable rate so they must be removed by a physical or chemical procedure before disposal of the used fluid. As low as 35 ppm. of the oil can be seen upon the surface of the lake or river as a colored film. As small a quantity as 1 pint of oil produces a slick of about 1 acre in size when floating on water. One difficulty with the hexane extractable test is that in addition to removing or determining the insoluble nonbiodegradable petroleum oils, it also removes certain fatty acids and greases which are biodegradable and therefore, acceptable to sanitary sewage treatment plants.

(b) Phenols:

Most phenolic derivatives present no difficulty if they are disposed of in sufficiently dilute form through sewage treatment systems. The phenols, fortunately, at low concentrations are easily biodegradable in a sewage treatment plant.

If the emulsions are split with acid, generally the phenols go into the oil phase and are removed with it. If a few ppm. of phenol remains in the water phase, it can be removed by filtering through activated carbon.

RAW MATERIALS:

Everyone involved in the manufacturing industry today is confronted with raw material problems. These problems are of various types and sources, but they are very definite restrictive forces on the types of products being manufactured. Some of these restrictions are government originated, such as the restrictions on the use of certain materials which may be derived from so-called endangered species. Other restrictions, which are government derived, depend on the prohibition of the use of certain materials because of their alleged toxicity or environmental incompatibility.

1. Government Restrictions:

A good example of the effect of government restrictions is the ban on the importation of sperm oil, which is derived from an endangered species, the sperm whale. Over the past 50 years, probably every producer of cutting and grinding fluids has used sperm oil or some derivative of it in at least one of its products. Sperm oil has certain characteristics which are different from other fatty oils derived from either animal, vegetable or marine sources. It is composed principally of the esters of a long carbon chain fatty acid and a long chain alcohol. There are ways of devising products which have adequate properties and perhaps even equal properties without using sperm oil and its derivatives, but the changeover has required an enormous amount of research and development work on the part of the suppliers of cutting and grinding fluids.

Another area in which government restrictions limit or define the ingredients which may be used in products coming into contact with people, is the group of materials alleged to be toxic. A list of forbidden materials, because they are alleged to be carcinogenic, was presented in a previous section of this paper.

Fortunately, to the best of the author's knowledge, no cutting and grinding fluid manufacturer uses any of these materials at the present time. Some of these restrictions are not too well founded. Take an example from another area, that of a food additive, the cyclamates, which is an artificial sweetener and was banned here some years ago because it was alleged to be carcinogenic. There have been very extensive investigations to show whether the cyclamates do cause cancer in man or animals. Some of these investigations, particularly in Germany, have been very exhaustive and have not shown any indication, that when used in reasonable quantities, they cause any cancerous condition. In fact, Abbott Laboratories, which was the principal producer of this sweetener, has done considerable research on its own and has submitted a new application to the Food & Drug Administration requesting them to reverse their original decision.

2. Scarcities:

Manufacturers, generally, are aware of scarcities in the area of raw materials, intermediates, and subassemblies. The scarcity of raw materials is also hitting the manufacturers of cutting and grinding fluids. Some of the basic ingredients, including the fats and oils and derivatives of them, such as fatty acids, soaps, synthetic esters, etc., are scarce and very expensive. It is interesting to note that within five or six months, the price of tallow and other similar fats went from 7¢ to 30¢ per pound, and even then it was difficult to obtain. It has since gone down in price to a more reasonable level. The desireable lubricating oil fractions are becoming scarce and they are being refined from different crudes from month to month, which contribute some different properties to the oils, particularly to the characteristics involved in emulsification.

Another condition which contributes to the raw material problem is the shortage of transportation. This condition is particularly severe with the railroads, where boxcars, gondola cars and tank cars are in short supply. It necessitates manufacturers carrying much increased inventories to allow for the lengthened lead times on deliveries of raw materials.

Another group of ingredients becoming scarce and used very extensively, particularly in the water-miscible fluids are the surface active agents, including both the anionic and nonionic types. The nonionic ingredients are particularly difficult to obtain because of the scarcity of ethylene oxide, used as an intermediate in their manufacture. In turn, the ethylene oxide is scarce because there is a scarcity of ethylene at the present time, which is a raw material used in its manufacture. The whole scarcity situation goes back originally to the scarcity of petroleum oil and the changing in the product mix which the refineries are turning out at present. In addition to the surface active agents, there are a number of synthetic organic chemicals used in the manufacture of cutting and grinding fluids which are becoming scarce due to the inadequate supply of raw materials and intermediates for their manufacture. Every supplier or manufacturer of cutting and grinding fluids must have good sources of these materials and must avoid the scarcities through selection of his formulas and the constant surveillance of the raw material markets.

CONCLUSIONS:

In addition to using metalworking fluids which decrease the tool wear and wheel wear, as well as increase the rate of production, it is necessary to use fluids which are safe, and do not endanger the environment. These fluids are available from the major quality manufacturers and as users, you should make yourself aware of the advantages and limitations of the fluids which you are now using. If they do not meet modern standards, you should discuss your needs with your supplier to be certain that he recommends the best fluid for your purposes under conditions which you use.

There are some things which you can do as a user to help yourself, and among them are the following:

1. Select proper fluid for your conditions of operation.
2. Apply the fluid correctly.
3. In the case of water-miscible fluids, control the concentration, pH and microbiological attack.
4. Keep fluid clean by filtration.
5. Prepare used fluid properly for disposal.

Presented at SME's Economics of Environmental Pollution Abatement Seminar, December 1970

Control Of Oily Waste Pollution By Recycling And Re-refining

By John W. Swain, Jr.
Swain Corporation

The problem of pollution of water by oily wastes needs no emphasis at this meeting. Through personal experience and incessant bombardment by the news media, we are well aware of the effects, if not the details of the causes. The ecological effects, rightly or wrongly, have been widely discussed. The degree of rightness or wrongness is not really important at this point. Pollution of water by oily wastes is bad. To the degree that much of this is caused by individuals and households is not important. The fact that approximately seventy percent of water pollution is caused by individuals is interesting but is not germane to industry.

Industry is the bad guy--it wears the black hat. Industry is obvious. Industry produces large and obvious quantities of waste. Industry does not vote. Industry is taxable. Industry must consider its customers--the individual who votes, pay taxes, and buys the products of industry.

Therefore, despite any ameliorating factors, industry must eliminate its pollution of water, air and land...not reduce, but eliminate. The American people, and yes, the people of the world expect, and rightly so, that the scientists, the engineers and the industries of America can, and will, eliminate pollution by industry.

The fact is--industry can eliminate its own pollution of water, air and land.

That this will require dedication, wisdom, improved technology, managerial skills, a high degree of engineering and money is obvious. But the job can and will be done.

It will be done because industry, in some cases, will make the necessary effort as part of the society in which it exists. In other cases, industry will, of necessity, meet the current stringent regulations which are harbingers of more stringent regulations to come.

As an example of some of these regulations and the standards which may be imposed, we mention the following either enacted into law or in the process of being enacted at this time. Remember that pollution is a hot political issue and politicians are--politicians.

In 1969 a waste oil collector dumped seven thousand gallons of waste oil into the Clinton River, a small river near Detroit. This resulted in the enactment of "The Phantom Dumper's Bill", now Public Act 136 of the State of Michigan.

This bill requires that all haulers of liquid waste, and each of his vehicles, be licensed by the State of Michigan and that the hauler be bonded.

It further requires the maintenance of records which clearly show the origin, description and ultimate disposal of each load of waste. This is a good law. Larger and more governmentally aware industry now requires of its waste haulers a certification that the waste will be disposed of in a suitable manner.

This does not, however, relieve the waste producer of either direct or contingent liability for the proper disposal of the waste. Contingent liability can cover the "public relations responsibility"--as an example, a plant of a major manufacturer sold its oily waste, under contract, to a waste hauler. The hauler in turn contracted for the use of a farmer's field which had a natural dike. In due course, well over a million gallons was pumped into this natural resevoir. One day the oil started to burn. The black, rolling smoke and the possible danger was enough to earn very noticeable headlines.

Joe Waste Hauler was not mentioned. The headline was "XYZ Corporation Oil Burns." This is known as bad public relations.

A second Bill, Public Act No. 127, recently passed by the Michigan Legislature is what I call the "Little Old Lady With the Tennis Shoes Bill."

Under the terms of this bill, "The attorney general, any political subdivision of the state, any instrumentality or agency of the state or of a political subdivision thereof, any person, partnership, corporation, association, organization or other legal entity may maintain an action in the circuit court have jurisdiction where the alleged violation occurred or is likely to occur for declaratory and equitable relief against the state, any political subdivision thereof, any person, partnership, corporation, association, organization or other legal entity for the protection of the air, water and other natural resources and the public trust therein from pollution, impairment or destruction."

The implications of this Bill are enormous. Suit can be brought as a personal vendetta. Suit can be brought by members of the so-called "lunatic fringe." Suit can be brought by well meaning but uninformed individuals or groups, and on and on.

Since suit can be brought in Circuit Court, and with the great interest in pollution in mind, we may very well find the dockets of our courts overloaded with these suits. Since very few judges are qualified experts in pollution, or its control, the testimony of experts will be necessary. This will require, not only the experts hired by the defendant, who is deemed guilty, but also the testimony of the only unbiased experts, representatives of the governmental regulatory agencies. The interesting question arises, "When will the government people do the work assigned to them?"

The State of Michigan legislators have proposed two more interesting and indicative Bills.

One requires that every "Commercial Enterprise" submit a list of every material used in its processing and every material being discharged as waste. The enormity of this requirement is almost beyond comprehension, despite its necessity or apparent necessity. We say apparent because, for example, a laundromat operator, as an operator of a commercial enterprise, will be required to list the various materials used in his establishment and the specific materials contributing to his waste. Assuming that the housewives who use the laundromat were forced to use cleaners supplied by him (an unlikely possibility) we must assume that this entrepreneur is qualified to select cleaners which are non-pollutable. He must, if we carry this to its logical conclusion, be aware of the type of soil in the clothes which combine with the cleaner and the water to become his effluent.

Let us further assume that the husbands of most of the women, who bring clothes to this laundromat, work in a plant which makes products high in phenol. The standard on phenol in waste waters is 5 parts per billion...not million...billion. This is not much. The poor guy is in violation and doesn't or can't know it.

This is obviously ridiculous. However, the operator of a manufacturing plant is assumed to know what is going on in his plant.

Since industry is the "bad guy", and is vulnerable, the implication is obvious. Joe Laundromat probably has no great problems but the Joseph F. Manufacturing Company does. Since Joseph F. makes widgets using 1010 steel, we now arrive at pollution of water by industry with major emphasis on machining wastes.

We eliminate oily wastes from food processing, textile, transportation, agriculture and oil refining industries.

The magnitude of the problem is greater than many realize.

In 1962 the country produced over 150 billion of gallons of petroleum products. In the more familiar motor oils and industrial oil markets the United States used---

Motor Oils and Greases..............987 million gallons
Industrial Oils.....................847 million gallons

In 1969, it is estimated that in these categories the consumption was---

Motor Oils...............1.4 billion gallons
Industrial Oils..........1.7 billion gallons

The total consumption increased by 70%, but significantly the increase in industrial oil was much greater. 98% for industrial oils but 42% for motor oils.

We will not dwell on the significance of these figures, but only note the rather startling increase in the use of industrial oils.

The increase is due largely to the increase in machining of metals and other materials used in products used by our society. It is significant to note that the states using the largest amounts of industrial oil are those in what we call the "Fertile Crescent"---New England, New York, Pennsylvania, Ohio, Michigan, Indiana, Illinois, Wisconsin and parts of Tennessee and Kentucky, the heavy industrial states.

Therefore, oily wastes from machining are an important factor in water pollution. They are also, far more than is realized, an important factor in air pollution.

It is not part of this presentation to discuss air pollution but we will mention here, the effects of machining on air pollution.

In the machining of metals, in the forming of metals, in the casting of metals, in the manufacturing of metals and the cleaning of parts, various lubricants and machining fluids are used. Since most of these operations either require or generate heat, there is a consequent volitilization of the lubricants and fluids which then become air pollutants. The government air pollution people are aware of this problem and effective steps are either being taken or are planned to eliminate the problem.

However, let us consider now the problem of handling oily wastes generated by machining operations.

The problem is more complex than appears on the surface.

We will use, as an example, a plant that produces automatic transmissions. Not only are these produced in large quantity, but almost every machining method, a wide variety of materials and high standards are involved. Also, the process requires sophisticated waste treatment facilities and allows for control of wastes during the manufacturing process.

An operation such as this will use:

Type I	Hydraulic oils
	Machine Oils
	Compressor oils
	Quench oils
	Air Conditioner oils
	Transformer oils
Type II	Mineral Seal oil
	Automatic Transmission fluid
	Straight cutting oils
	Emulsifiable or soluble cutting oils
	Drawing compounds
	Honing oils
	Rustproofing oils
	Die Casting parting agents
	Greases
	Assembly lubricants

Type III Solvent emulsion cleaners
Alkaline base cleaners
Tumbling fluids
Acids and alkalis
Floor cleaning compounds

Type I materials are those which have petroleum oil bases with varying, but relatively minor amounts of additives. The range is 0 to 20% additive content. These oils, in service, are subjected to heat, air, moisture and the fine solids with the result that they become physically or chemically contaminated. The physical contamination derives from free or emulsified water and fine dust particles. In some areas of the plant, however, the solid particles can be relatively large, 25 microns or more. The chemical contamination is the result, to a major extent, of oxidation with some possible polymerization.

From a pollution control aspect the major problems with this contamination are:

(1) Loss in useful life and increase in waste oil.
(2) Difficulty in processing in a waste treatment plant.
(3) Increased difficulty in reprocessing or rerefining.

Type II materials, used in the production of the parts are composed of a wide variety of petroleum oils with many chemicals or chemical compounds present either as reaction products or as additives.

In Type II materials, in addition to petroleum oil, we find such substances as:

Sulfur
Chlorine
Phosphorus
Zinc
Lead
Copper
Iron
Tin
Fatty oils
Soaps and emulsifiers
Petroleum oxidates
Glycols
Wetting agents
Polymers
Pigments, such as graphite, talc, whiting, molybdenum Disulfide
and many others

These present many difficulties in waste treatment and subsequent reclaiming, reprocessing and rerefining.

Type III compounds are widely varied but will contain:

Soaps and emulsifiers
Wetting agents
A wide variety of alkaline compounds
such as-Sodium hydroxide
Potassium hydroxide
Sodium silicates

Phosphates
Carbonates
Acids
such as-Phosphoric
Sulfuric
Hydrochloric
Solvents
such as-Trichlorethylene
Perchlorethylene
Benzene
Mineral Spirits
Kerosene

All in all, there are quite a few problems in handling of oily wastes.

However, these wastes can be handled and are being handled with varying degrees of success. We will cover this subject in two areas--large plants and small plants. We will remain in the machining waste area with the automatic transmission plant and a small supplier of transmission parts.

Pollution control should, and must, start at the front door--not the back door. Let us start at the back door, however, and discover why this is so. The automatic transmission plant will have a waste treatment plant capable of handling an influent of 300 to 500 gpm. This influent will be largely water but will produce between 100,000 and 200,000 gallons per month of oil in its final oily waste. This waste will run from 5% to 50% water and solids. The wide variation is due to:

(1) Inadequate engineering of the plant.
(2) Inadequate processing methods.
(3) Indiscriminate dumping by the manufacturing area with a consequent overload.
(4) Inadequate waste treatment plant operators.
(5) Mechanical breakdown.

The waste water will contain more than oils and chemicals from the manufacturing process. It will be the vehicle for implementing the "Garbage Can Philosophy." This philosophy--or "Swain's Law" is:

"Anything the mind of man can contrive will end up in a waste treatment plant."

Needless to say, this philosophy does not enhance the efficiency of a waste treatment plant.

While there are several variations, the waste treatment process has the following steps:

(1) Gravity separation of oil, water and solids.
(2) Treatment of oil.
(3) Treatment of water.
(4) Disposal of water to sewer, stream or lake.
(5) Disposal of oil.

The gravity separation of oil, water and solids may be, and should be, assisted by acid, demulsifiers, wetting and dispersing agents and heat.

The treatment of the oil ranges from further settling with or without heat to such operations as centrifuging, filtration, and processing with heat and acid.

The treatment of water is more complicated. In general, the water, which contains 50 to 1000 PPM of oil and possible fine solids, is mixed with materials which will form a floc which traps the oil and solids. The floc which either floats, if it is formed from aluminum sulfate, or sinks, if formed from ferric chloride or ferric sulfate, is removed. The water is neutralized with caustic soda, soda ash or lime and after settling or filtration is disposed of to a sewer or stream.

Because of the increasing cost of water and its disposal, there is a growing recognition of the desirability, even necessity, of re-using rather than dumping the water.

The oil is disposed of to a waste oil dealer or rerefiner. We will return to this later.

We earlier mentioned the "Front Door Approach" to pollution control. In order to achieve this, management must give more than lip service, must give more than dollar service to Pollution Control.

The "Front Door Approach" comprises five steps:

(1) Management Organization
(2) Engineering and funding
(3) Designation of adequate authority and responsibility.
(4) Selection and use of processing materials for control of pollution.

    (a) Prolong the life of oils, coolants and cleaners.
    (b) Re-use the materials in other operations.
    (c) Segregate the materials at the source for easier processing.
    (d) Use of materials compatible with the waste treatment process.

(5) Re-use of waste materials recovered in the waste treatment operation.

    (a) Filter and use as cutting or slushing oil.
    (b) Process for plant fuel.
    (c) Contract for reprocessing or rerefining of waste oil to products usable in the manufacturing operation.

This same procedure can be followed in smaller plants. The equipment may be as simple as a settling tank. Here a gross separation of oil and water is achieved. The water quantity may be unacceptable by large company and government standards, but, at least, it is far better than waste now being dumped.

The next step is the use of heat and demulsifiers and/or acid.

The third step would be the use of polyelectrolytes with the water to further improve its quality.

The oil derived should be stored in tanks of 1000 gallon capacity or more for more economical disposal.

Again, the same management philosophy must prevail.

Now we have large quantities of waste oil to be handled. There are five methods of disposal:

(1) Incineration
(2) Use as fuel
(3) Road oiling
(4) Ground disposal
(5) Re-use through reprocessing and rerefining

Incineration, if properly done, is an effective method of disposal. However, there are a few problems.

(1) The cost - $0.03 to $0.08 per gallon
(2) High capital and operating costs
(3) Noise
(4) Possible air pollution
(5) Destruction of a natural resource and useful product.

Burning as fuel is satisfactory if the oil is processed to remove or reduce such materials as--sulfur, chlorine, phosphorus, fatty acids, organic acids, alkaline matter and solids and water. The effect on these materials on corrosion, coking and clinkering is obvious. In this regard, Consolidated Edison of New York, a large fuel user, has approved the use of crude fuels containing up to 25% of crankcase oil-used motor oil. No study has been made by them, or the American Petroleum Institute on the use of industrial waste oils. However, we processed 5000 gallons of industrial waste oil for use in a major automotive manufacturers boiler. The results were good. However, we have had a great amount of experience in use of similar oils in our own boilers and stills with the problems noted above.

One of the larger markets for waste oil in the past was sales to residual fuel oil dealers. As industry has converted more and more to natural gas, this market has steadily declined. With the lower air pollution effects of gas, we can expect no increase, and probably, a continuing decrease in this market.

Oiling of roads has used as much as 200 million gallons of waste oil in the past. But with the effects on ground and water pollution and the hard surfacing of suburban streets, this market is steadily declining.

Ground disposal has two aspects...deep well disposal and in-soil bacterial degradation. Earlier, it appeared that many of our wastes could be disposed of in deep wells, 2000 to 5000 feet. However, the governmental regulatory agencies are very chary of this. We don't know enough about the underground waterways. We also have noted that there are distinct possibilities that we may affect sub-surface structures with possible earthquake potential.

In-soil degradation is being tried in the South-West United States. It appears that it is reasonably successful if the climactic conditions are right--moisture and temperature. However, this method probably can't handle the total volume even if we ignore the cost of shipping to the proper areas.

The last method of disposal-recycling for re-use offers many advantages. These include immediate economic gains as well as the long-range gains of conservation of matter, or a conservation of a not inexhaustible natural resource.

We have used the terms reclaiming, reprocessing and rerefining.

We define these:

(1) Reclaiming is that process by which waste oil is settled, dehydrated and filtered.
(2) Reprocessing is that process by which the waste oil is chemically treated to remove various unwanted constituents and is then filtered.
(3) Rerefining can, and usually does, use both of the above process steps, but, also subjects the oil to high temperature. (550° - 700°F.) distillation to produce a high quality lubricating or hydraulic oil.

Reprocessing and rerefining can produce oils which cover 90% of the industrial usage. These oils have specifications equal to and, in many cases, superior to oils currently being used.

Ignoring the readily apparent conservation of a natural resource by recycling through rerefining, let us look at the immediate economic gains.

(1) The cost of disposing of waste oils, as well as other waste material, is increasing. Where waste oil dealers, not too long ago, paid for waste oil, there is now a charge from $0.01 to $0.10 per gallon. This also applies to other methods of disposal such as incineration.
(2) The increasing cost of crude oil, refining and additives will continue to advance the price of oil products.
(3) Rerefining with a more stable cost base can supply useful products with savings in the 20% plus range. These savings will more than amortize the capital and operating costs of waste treatment.

The challenge is here--the dollars are available--now we only need the will to do the job.

Presented IIT's Lubrication Challenges in Metalworking and Processing Conference, June 1978

ULTRAFILTRATION FOR DEWATERING OF WASTE EMULSIFIED OILS

Steven D. Pinto
ABCOR, INC.
Wilmington, Massachusetts

Abstract

Ultrafiltration of emulsified oil in water wastestreams from waste cutting and grinding operations, waste rolling oil solutions, and alkaline degreasing baths can produce an effluent water phase that is acceptable for discharge to sewer systems. Since ultrafiltration membranes act as a physical barrier between the wastestream and the clean water effluent, operator or flow caused upsets are prevented, ensuring no carryover of oil to the effluent. The effluent (permeate) from the ultrafiltration system is usually of sufficient quality for discharge to a sanitary sewer system. The permeate, as a rule, contains less than 100 mg/l of freon-extractable oil and grease and less than 10 mg/l suspended solids. The ultrafiltration process also produces a liquid oil concentrate containing up to 60% oil and solids, which can be disposed of by hauling or incineration. The concentrate will be only 3 5% of the original volume. No sludge containing large amounts of chemicals and heavy metals is produced in ultrafiltration because no such chemicals are required as in other methods of treating oily emulsions. Ultrafiltration has been found to be the most cost-effective treatment method for treatment of oil/water emulsions in many applications.

Introduction

Recent developments in ultrafiltration membrane technology have made an alternative treatment method available for wastewaters containing emulsified oil. Prior to the development of ultrafiltration, the standard method for treatment of emulsified oily wastes was chemical demulsification followed by secondary clarification of the water phase. The system design for chemical treatment requires the use of a variety of chemicals including sulfuric acid, waste pickle acid, alum, iron sulfate, and proprietary chemicals which usually include polymers.[1]

All chemical treatment methods produce a sludge in which the dirt, floc, and trapped water remain in the oil phase. The water phase from chemical treatment of emulsions may need additional treatment to meet the quality standards for discharge to a sewer system. Some final polishing of the water phase is generally required. The sludge phase almost always requires further stabilization by cracking or filtration before it can be disposed of.

Ultrafiltration, on the other hand, produces a water phase that can usually be discharged to a sewer with no post treatment, and an oil phase that can generally support combustion. Thus, hauling of oily wastes will not be necessary when using ultrafiltration and incineration. If the concentrate cannot be burned, then only 3-5% of the original waste volume need be hauled.

Principles of Ultrafiltration

In an ultrafiltration (UF) process, feed solution of wastewater containing emulsified oils is introduced into and pumped through a membrane unit (Figure 1).

Water and some dissolved low molecular weight materials pass through the membrane under an applied hydrostatic pressure. Emulsified oil droplets and suspended solids are retained by the membrane while water and dissolved salts pass through the membrane (the water phase is called permeate).[2] The emulsified oil is concentrated in this manner up to about 60% oil and solids content.

The pore structure of the ultrafiltration membrane acts as a filter, passing small solutes (i.e., salts) and water while retaining larger emulsified oil particles and suspended solids. The pores of the ultrafiltration membrane are much smaller than the particles rejected, and thus the particles cannot enter the membrane structure and plug the pores. Pore structure and size (less than 0.005 microns) of the membrane are quite different from those of ordinary filters, in which pore plugging results in drastically reduced filtration rates and requires frequent backflushing, etc., which may produce extra solid or liquid wastes.

In contrast to a conventional filter, an ultrafilter separates the water from the emulsified oil mixture rather than the oil from the oil-water mixture. The oil remains in solution and is retained by the ultrafiltration membrane. The high velocity of the solution over the membrane prevents buildup on the membrane surface itself. Thus, the rate and amount of oil collection on the membrane surface is very small compared to conventional filtration.

Design and Operation of Ultrafilters

The important design criterion for operation of an ultrafiltration system is the capacity of the membrane to pass water. This rate of filtration is termed flux, and is the volume of water permeated per unit membrane area per unit time (gallons per square foot per day, GFD). Since both membrane equipment capital costs and operating costs increase with the membrane area required, it is highly desirable to maximize membrane flux.

For a wide spectrum of ultrafiltration applications, membrane flux has been found to be described by the following relationship:

$$J = k \ln \left(\frac{C^*}{C} \right) \qquad (1)$$

where J is the flux, k is the aqueous phase mass transfer coefficient, C^* is a constant and C is the feed concentration. The mass transfer coefficient (k), and correspondingly flux, increases with increasing feed velocity through the membrane system. An optimum pumping rate is generally determined by a cost balance involving power and membrane area costs. Flux also increases with temperature. The normal operating temperature is the temperature at which the feed is discharged to the treatment system, which is usually between 30 and 40°C.

The model leading to Equation (1) assumes that the resistance to flux lies in a concentrated boundary layer at the membrane surface as is the case in the ultrafiltration of oil/water emulsions. The relationship in Equation (1) does not, however, appear to be the dominant reason for flux decline in ultrafiltration of oil/water emulsions until the oil and solids content is above 10-15%. The primary reason for flux decline in low concentration oil/water solutions is buildup of oil, calcium complexes, graphite, or ferric hydroxide on the membrane surface resulting in a time related fouling. The surest indication of this type of flux reduction is that the water flux of the membranes is very low, and dilution of the feed does not return the flux to its original level.

The reason that membrane systems foul in this manner is that as a dispersed phase in water, the oil or solids are inherently unstable with respect to agglomeration into a continuous second phase. In a quiescent state, this inherent instability leads to oil floating and most solid particles sinking. Under most ultrafiltration conditions, sufficient turbulence exists to prevent this type of gravity enhanced phase separation. Nevertheless, membrane surfaces do become covered with oil or solid particles due to the operation of the ultrafiltration process. The concentration of particles or droplets is higher near the membrane surface than in the bulk of the fluid because the solution that permeates through the membrane surface carries with it rejected particles to the membrane surface. These rejected particles are transported away from the membrane surface by various diffusion mechanisms; the net result is a higher concentration of the rejected species near the membrane surface than in the bulk of the fluid. The increased concentration of particles at the membrane surface will result in agglomeration of droplets at the membrane surface -- that is, the sticking together of particles into large entities; thus created, the large entities diffuse away from the membrane more slowly than their constituents. This leads to a still higher concentration of the dispersed phase at the surface of the membrane and thus, the solids will continue to build in size. Unless the surface energy relationships are favorable, these large, slowly diffusing entities will stick to the surface of the membrane and create an additional hydrodynamic resistance, thus lowering the flux.

In an ideal ultrafiltration operation there would be no build-up of oil on the membrane surface. An oil film, however, does eventually form on the membrane surface necessitating periodic cleaning of the membrane. Cleaning consists of washing the ultrafiltration system with a soap solution, which is displaced back to the feed tank after the cleaning is completed. This method results in no additional solid or liquid wastes that would need hauling, as the soap solution is ultrafiltered along with the feed solution.

In large oil/water systems, it is anticipated that cleaning of the membranes normally will be required once a week to remove the foulants that build up on the membrane surface. Three methods of cleaning are available. They are: 1) mechanical cleaning; 2) dispersing; and 3) solubilizing.

Mechanical cleaning, whether spongeball or brush cleaning, is only applicable in practice to large-diameter ultrafiltration membranes and is very effective in removing chemically precipitated species that adhere tenaciously to the membranes and are difficult to remove by any other method.

This method works best, though, when the adhesion between the fouling layer and the membrane is weak. The best use of spongeballs in oil/water systems is during a soap cleaning to increase the contact between the soap and the oil layer.

Dispersing methods of cleaning function by breaking up deposits in the membrane and dispersing them into colloidal sized particles. The most commonly used dispersants are detergents such as Ultraclean (a product of ABCOR, Inc.). Detergent cleaning is the most widely applicable cleaning method for oil/water ultrafiltration applications. Detergent cleaning methods are most effective when carried out at an elevated temperature (45-60°C) and as high a flow rate as possible to scour the membrane. It is also generally true that the effectiveness of these cleaning procedures is enhanced by raising the pH to as high a level as the membrane will tolerate (approximately 11-12). Cleaning usually takes from 1-2 hours.

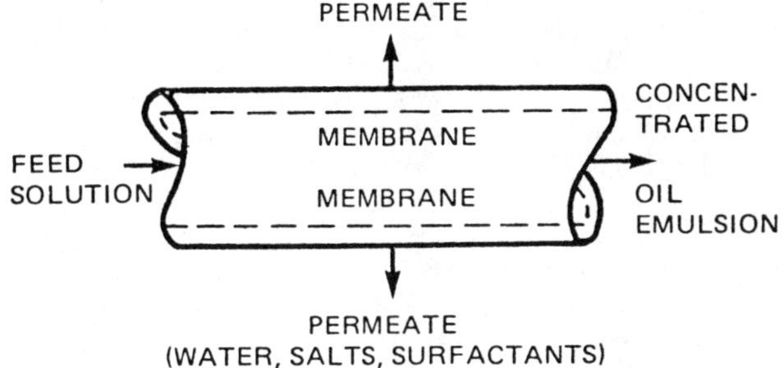

FIGURE 1
DIAGRAMMATIC PICTURE OF ULTRAFILTRATION

Cleaning by solubilizing consists of dissolving, by physical or chemical means, the fouling deposit. By dissolving we mean dispersing the deposit on a molecular rather than a colloidal level. Where applicable, this is the most effective of cleaning methods. It is most often used to clean the membranes of metal hydroxide or other chemical deposits. Solutions of acids and chelating agents are usually used for this purpose. Elevated temperatures (i.e., 40-60°C) are generally used in this cleaning method to shorten the cleaning time.

Application of Ultrafiltration to Treatment of Emulsified Oily Wastes

The majority of existing oil emulsion ultrafiltration plants process three types of emulsions: waste cutting and grinding oils, waste rolling oils, and alkaline degreasing baths.[4] Conditions are only moderately severe in ultrafiltration of waste cutting and grinding oil/water emulsions since the pH is generally less than 10, but the presence of various foulants requires a membrane that is tolerant of both acidic and caustic cleaning agents. The ABCOR HFD and HFM type membranes are examples of such membranes used in these services.

For ultrafiltration of steel rolling oils, the principal basis for choosing a membrane is its theoretical stability during processing at temperatures up to 70°C and its resistances to alkaline cleaners used to control fouling. In concentrating oil and dirt for removal from metal degreasing baths, while returning the permeate for reuse, the most important criterion for membrane selection is stability at high pH levels (9-13) and at temperatures around 70°C. The ABCOR HFM membrane is well suited for both of these applications.

Figure 2 shows a generalized flow schematic of an ultrafiltration system. The only pretreatment necessary for large diameter ultrafiltration membranes as produced by ABCOR is gravity separation of the waste emulsion to remove free floating oil and large solids. Large solids (i.e., rocks, glass, etc.) will have a tendency to puncture the membranes and also damage the pumping package. Free unemulsified oil, as mentioned previously, will tend to form an oil on the membrane and reduce the operating permeation rate and thereby increase the required membrane surface area. This pretreatment can be accomplished in a large holding tank equipped with an oil skimmer and/or by using an API-type oil separator.

Waste cutting oil emulsions, typically containing 1-5% oil and grease are concentrated by ultrafiltration to a concentrated solution containing 40-60% oil and solids. The normal method used for treatment of oil emulsions, as shown in Figure 2, is the modified feed and bleed semi-continuous batch method. A majority of the feed solution is recirculated in a loop in this method while five to twenty times the permeate rate of fresh feed is fed into the loop. The feed is used to make up for material that permeates through the membranes and that portion of the liquid in the circulation loop which is bled back to the feed tank (at approximately ten times the permeate flow rate). This method of operation saves on power consumption and piping size as compared to a normal batch method since the feed and bleed line sizes are reduced by using this recirculation technique. Fresh feed can be continuously added to the feed tank until some predetermined amount of feed has been added, and the concentration phase begins.

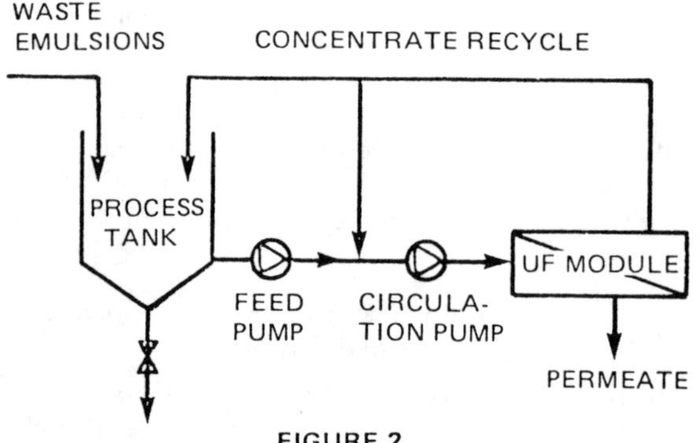

FIGURE 2
MODIFIED FEED AND BLEED SEMI-CONTINUOUS BATCH TREATMENT

In this manner 3-5 days of oily waste can be added to the feed tank before the feed line is closed off. After the feed addition is discontinued, the ultrafilter will continue to concentrate feed by removing water from the feed solution. The concentrated oil can usually be incinerated. In any event, the amount of waste that must be disposed of will be minimized. The waste volume can usually be reduced by 95-98%.

The permeate, or effluent, from the ultrafilter is normally of a quality to be reused (degreasing baths) or discharged directly to a sanitary sewer. In the ultrafiltration system, the only means of water removal is by passage through the ultrafiltration membranes. Since this provides for complete removal of oil particles and solids, the effluent water will always have a low oil and solids content. Operator error cannot cause poor quality of the discharge water as in processes in which the oil and water are in intimate contact. In ultrafiltration, the oil and water products are always separated by a mechanical barrier -- the membrane.

The permeate quality for this type of waste discharge is usually determined by measuring its "oil" content (freon extractable), solids content (suspended solids), and in some cases BOD level (5 day biochemical oxygen demand). The oil content is usually measured as oil and grease; that is, the level of both polar and non-polar organics, which are extractable in freon, that are contained in the waste.[5] The oil and grease content of ultrafiltered permeate is usually less than 100 mg/l. However, very high concentrations of polar surfactants in the wastestream may cause the level of these surfactants in the permeate to exceed 100 mg/l, since the permeate concentration is proportional to the feed concentration. The suspended solids level of the ultrafiltered permeate can always be expected to be below 10 mg/l because there is no way for solids to be "carried over" into the clean water. The BOD levels of the permeate will vary from one waste to another depending on the amount of small molecular weight organics in the feed stream, so no generalized level in the permeate is available.

If the ultrafiltered permeate is not of sufficient quality for discharge due to stringent local effluent guidelines, further polishing of the permeate can be done. Possible post treatment methods include biological degradation, carbon adsorption, or reverse osmosis.

Ultrafiltered permeate can contain high levels of biodegradable material which are measured as oils or BOD, but may not contain any other regulated substances. Thus, it may be possible to discharge ultrafiltered permeate to the plant sewer to be biodegraded by such methods as activated sludge treatment or trickling filters. The ultrafiltered permeate is also useful in plants that do not operate twenty-four hours per day and have an intermittent waste flow to the biological treatment system. The permeate could be used to stabilize the activated sludge by maintaining some feed to the bio-system 24 hours per day.

Activated carbon can be used to polish the ultrafiltered permeate to make it acceptable for discharge to streams. The use of activated carbon can produce a permeate containing less than 10 mg/l of oil and grease with no color bodies. Use of side-stream treatment through activated carbon can produce an effluent containing any desired level of quality. The use of activated carbon is usually limited to smaller flows (less than 5,000 gallons per day) so that the carbon can be disposed of when it is saturated.

Reverse osmosis (RO) is another membrane process that can be used to concentrate ultrafiltered permeate. As a rule, feed to a reverse osmosis unit must be free of suspended matter, thus ultrafiltration is usually a good pretreatment for reverse osmosis. RO is a process that removes approximately 90% of all dissolved matter from the ultrafiltered permeate. This process usually produces a permeate containing very little salts or oils and can, therefore, be recycled throughout the plant as make-up water for mixing the emulsions.

Table 1 illustrates the operating characteristics of oil emulsion ultrafiltration units based on an analysis of actual operating results.[4] The data are presented as physical quantities rather than as costs because unit costs vary widely among the countries where the units are located. The membrane replacement figures for the HFD and HFM membranes are still tentative until a statistically significant number of membranes have been replaced. Ultrafiltration is very often found to be the most cost effective method of treating oil emulsion wastestreams.

TABLE 1

OPERATING CHARACTERISTICS, UF OIL EMULSIONS*

Permeate Flux		45-90 L/m^{2}H 25-55 GFD
Energy Needs		10-15 KWH/m^3 Perm 38-56 KWH/10^3 ft^3 Perm
Cooling Water Requirements HFD/HFM		0
Detergent for Membrane Cleaning		25 g/m^3 of Permeate; 0.2 lb/1000 Gallons of Permeate
Manpower		7-10 H/Week
Membrane Replacement	HFD	> 12 Months
	HFM	> 12 Months

*Based on Commercial Installations

Summary

A need for a waste treatment system that could treat emulsified oily wastes and produce an effluent suitable for discharge to a sewer prompted the development of a new product. This product was the ultrafilter. An ultrafiltration system can produce an effluent water phase from a chemically emulsified oily waste that can be discharged to a sewer and a concentrated oily phase amounting to only 3-5% of the original feed volume, which can usually be incinerated. The major advantages of an ultrafiltration system are:

1. An ultrafiltration system requires little or no chemicals for operation.
2. An ultrafilter is a physical barrier between the emulsified oily waste and the clean water, and thereby prevents any of the oily waste from contaminating the effluent.
3. An ultrafiltration system is a very easy system to operate.
4. An ultrafiltration system is usually less costly than a chemical treatment system.

References

1. Hockenberry, H. R., and Lieser, J. E.: "Practical Applications of Membrane Techniques of Waste Oil Treatment," *Journal of the American Society of Lubrication Engineers* Volume 33, 5. pp. 247-251.
2. Goldsmith, R. L., et al; "Soluble Oil Waste Treatment Using Ultrafiltration,"46th Annual Water Pollution Control Federation, Cleveland, Ohio 10/4/1973.
3. deFilippi, R. P. and Goldsmith, R. L., "Application and Theory of Membrane Processes for Biological and Other Macromolecular Solutions," *Membrane Science and Technology,* Plenum Press, (1970).
4. Mir, L., Eykamp, W., Goldsmith, R. L., "Current and Developing Applications for Ultrafiltration," *Industrial Water Engineering,* May/June, 1977.
5. *Standard Methods for the Examination of Water and Wastewater,* APHA, 13th Edition, Procedure No. 137, pp. 254-256, (1971).

INDEX

A

B

C

D

E

F

G

H

I

L

M

N

O

P

Q

R

S

T

U

V

W

Y

Z